Highlights in Theoretical Chemistry

Vol. 4
Series Editors: Ch.J. Cramer • D.G. Truhlar

For further volumes:
http://www.springer.com/series/11166

Fernando R. Ornellas • Maria João Ramos
Volume Editors

Marco Antonio Chaer Nascimento

A Festschrift from Theoretical Chemistry Accounts

With contributions from

Adelia J. A. Aquino • Xavier Assfeld • Mario Barbatti
Patricia Barragán • María M. Branda • Benedito J. Costa Cabral
Sylvio Canuto • Nuno M. F. S. A. Cerqueira • Kaline Coutinho
Rachel Crespo-Otero • Marcus V. A. Damasceno
Gerardo Delgado-Barrio • Mostafa A. El-Sayed
Pedro A. Fernandes • Tertius L. Fonseca • Silvia Fuente
Herbert C. Georg • Rodrigo M. Gester • Francesc Illas
Kenneth Irving • P. Lazzeretti • Hans Lischka • Antonio Monari
Irina S. Moreira • Vudhichai Parasuk • Rita Prosmiti
Patricio F. Provasi • Maria João Ramos • João V. Ribeiro
Jean-Louis Rivail • Cristina Sanz-Sanz • Marc E. Segovia
Kanjarat Sukrat • Paul Szymanski • Daniel Tunega
Alvaro Valdés • Oscar N. Ventura • Thibaut Very
Pablo Villarreal

Springer

Volume Editors
Fernando R. Ornellas
Departamento de Química Fundamental
Instituto de Química
University of São Paulo
São Paulo, Brazil

Maria João Ramos
REQUIMTE - Depart. de Química e Bioquímica
Faculty of Science
University of Porto
Porto, Portugal

Originally Published in Theor Chem Acc, Volume 131 (2012)

ISSN 2194-8666 ISSN 2194-8674 (electronic)
ISBN 978-3-642-41162-5 ISBN 978-3-642-41163-2 (eBook)
DOI 10.1007/978-3-642-41163-2
Springer Heidelberg New York Dordrecht London

Printed on acid-free paper

Springer is part of Springer Science+Business Media (www.springer.com)

Contents

Theor Chem Acc (2013) 132:1319
DOI 10.1007/s00214-012-1319-3

PREFACE

Preface

Fernando R. Ornellas · Maria João Ramos

Published online: 12 January 2013

This issue of Theoretical Chemistry Accounts is dedicated to Professor Marco Antonio Chaer Nascimento on the occasion of his 65th birthday. Professor Chaer played a pioneering and active role in the early stages and latter developments of research activities in theoretical chemistry in Brazil. As part of this commemoration, an international scientific meeting also took place in Rio de Janeiro, Brazil, in the week of June 11–13, 2012. This special volume contains a selected sample of contributions from his former students, colleagues, and collaborators.

After successfully completing his doctorate at Caltech under the supervision of Professor William A. Goddard in 1977, Professor Chaer was faced with the decision of pursuing an academic career in United States or to return home and work on the establishment of a graduate research program in the Physical Chemistry Department of the Federal University of Rio de Janeiro (UFRJ, in Portuguese). Fortunately, for us, he chose the latter one. This decision had a profound impact on the academic activities of the department which, although excelling in undergraduate teaching, had no research activity whatsoever. Notwithstanding the fact that it was not the first physical

chemistry graduate research program in Brazil, it had definitely a character of its own, being strongly focused on theoretical chemistry and spectroscopy. This initiative sets a high level standard in human resources formation that served as model for similar research programs established later in the country. It is a motive of pride for us to see former students of his group working all over Brazil and even abroad.

One of his first initiative immediately after his return to Brazil, together with Professor D. Guenzburger, was to organize a national meeting putting together Brazilian theoretical chemists along with some distinguished international speakers to know each other research interests and to discuss and implement actions to farther the quality of the work in the field. This first meeting in Rio de Janeiro in 1981 was the nucleus of the biannual Brazilian Symposium of Theoretical Chemistry (SBQT, in Portuguese), now in its 17th edition and organized by national and local committees. Professor Chaer was also the Coordinator of the 10th edition of SBQT in 1999, and also of two other international Molecular Modeling Conferences (1992 and 1994), held in Rio de Janeiro, that were instrumental in showing

Published as part of the special collection of articles celebrating the 65th birthday of Professor Marco Antonio Chaer Nascimento.

F. R. Ornellas (✉)
Departamento de Química Fundamental, Instituto de Química, Universidade de São Paulo, Av. Prof. Lineu Prestes, 748, São Paulo 05508-000, Brazil
e-mail: frornell@usp.br

M. J. Ramos
Requimte, Departamento de Química e Bioquímica, Faculdade de Ciências, Universidade do Porto, Rua do Campo Alegre s/n, 4169-007 Porto, Portugal
e-mail: mjramos@fc.up.pt

the growing impact that theoretical chemistry techniques may have in helping solving large-scale chemical and chemical engineering problems.

All these actions, together with his intense participation in the advisory board of funding agencies in Brazil, certainly helped to recognize theoretical chemistry as an independent sub-area of research in physical chemistry by the Brazilian most important federal funding agencies (CNPq and CAPES) in the last decades.

In summary, Professor Nascimento helped to shape Brazilian theoretical chemistry research by participating directly or indirectly in the formation of numerous students, some of which are now independent professionals with their own groups. While he is still working with absolutely no sign of upcoming retirement, some of his former students and collaborators decided to pay this tribute to him on the occasion of his 65th birthday. Certainly, the best honors are those received in life, during the peak of activity. Within that spirit, in the name of all people that worked with him and were positively influenced by him, we take this opportunity to make this just homage.

André G. H. Barbosa, Clarissa O. da Silva,
Márcio Soares Pereira
TheoChem in Rio Committee

Maria João Ramos
Fernando R. Ornellas
Guest Editors

List of Publications of Professor Marco Antonio Chaer Nascimento

1. Fernandez-Lima FA, Nascimento MAC, da Silveira EF (2012) Nuclear Instr & Meth Phys Res B 273: 102–104.
2. Barbatti M, Nascimento MAC (2012) Int J Quantum Chem 112:3169–3173.
3. Freitas GN, Garrido JD, Ballester MY, Nascimento MAC (2012) J Phys Chem A 112:7677–7685.
4. Fernandez-Lima FA, Henkes AV, da Silveira EF, Nascimento MAC (2012) J Phys Chem C 116: 4965–4969.
5. Fantuzzi F, Messias Cardozo T, Nascimento MAC (2012) Phys Chem Chem Phys 14:5479–5488.
6. Fernandez-Lima FA, Henkes AV, da Silveira EF, Nascimento MAC (2012) J Phys Chem C 116: 4965–4969.
7. Pereira MS, da Silva AM, Nascimento MAC (2011) J Phys Chem 115:10104–10113.
8. Arrate JDG, Nascimento MAC, Ballester MY (2010) Int J Quantum Chem 110:549–557.
9. Cardozo TM, Nascimento-Freitas G, Nascimento MAC (2010) J Phys Chem A 114:8798–8805.
10. Cardozo TM, Nascimento MAC (2009) J Chem Phys 130:104102-1-104102-8.
11. Fernandez-Lima FA, Ponciano CR, Nascimento MAC, Silveira EF (2009) J Phys Chem A 113:1813–1821.
12. Liberti L, Lavor CC, Maculan N, Nascimento MAC (2009) Discrete Appl Math 157:1309–1318.
13. Fernandez-Lima FA, Cardozo TM, Silveira EF, Nascimento MAC (2009) Chem Phys Lett 474:185–189.
14. Barros PR, Stassen H, Freitas MS, Carlini CR, Nascimento MAC, Follmer C (2009) Biochim Biophys Acta: Proteins and Proteomics 1794:1848–1854.
15. Cardozo TM, Nascimento MAC (2009) J Phys Chem A 113:12541.
16. Fernandez-Lima FA, Vilela-Neto OP, Pimentel AS, Pacheco MAC, Ponciano CR, Nascimento, MAC, Silveira EF (2009) J Phys Chem A 113:15031–15040.
17. Milas I, Silva AM, Nascimento MAC (2008) Appl Catalysis A 333:17–22, 2008.
18. Nascimento MAC (2008) J Brazilian Chem Soc 19:245–256.
19. Sobrinho AMC, Nascimento MAC (2008) Int J Quantum Chem 108:2595–2602.
20. Andrade MD, Nascimento MAC (2008) Int J Quantum Chem 108:2486–2498.
21. Silva AM, Nascimento MAC (2008) J Phys Chem A 112:8916–8919.
22. Oliveira HCB, Nascimento MAC (2008) Int J Quantum Chem 108:2540–2549.
23. Fernandez-Lima FA, Becker C, Gilli K, Russell WK, Nascimento MAC, Russell DH (2008) J Phys Chem A 112:11061–11066.
24. Floriano WB, Domont G, Nascimento MAC (2007) J Phys Chem B 111:1893–1899.
25. Cardozo TM, Nascimento MAC (2007) J Mol Struct - Theochem 811:337–343.
26. Lavor CC, Liberti L, Maculan N, Nascimento MAC (2007) Europhys Lett 77:50006-1-50006-5.
27. Fernandez-Lima FA, Ponciano CR, Nascimento MAC (2007) Chem Phys Lett 445:147–151.
28. Fernandez-Lima FA, Cardozo TM, Ponciano CR, Nascimento MAC (2007) J Phys Chem A 111:8302–8307.
29. Fernandez-Lima FA, Ponciano CR, Silveira EF, Nascimento MAC (2007) Chem Phys 340:127–133.
30. Bitzer R, Pereira R, Rocco AM, Santos OS, Nascimento MAC, Filgueiras CA J Organomet Chem (2006) 691:2005–2013.
31. Pereira MS, Nascimento MAC (2006) J Phys Chem B 110:3231–3238.
32. Henriques E, Nascimento MAC, Ramos MJ (2006) Int J Quantum Chem 106:2107.
33. Fernandez-Lima FA, Ponciano CR, Silveira EF, Nascimento MAC (2006) J Phys Chem B 110: 10018–10024.
34. Fernandez-Lima FA, Ponciano CR, Silveira EF, Nascimento MAC (2006) Chem Phys Lett 426:351–356.
35. Andrade MD, Mundin K, Nascimento MAC, Malbouisson L (2006) Int J Quantum Chem 106:2700–2705.
36. Pereira MS, Nascimento MAC (2005) Chem Phys Lett 406:446–451.
37. Collado V, Fernandez-Lima FA, Ponciano CR, Nascimento MAC, Velazquez L, Silveira EF (2005) Phys Chem Chem Phys 7:1971–1976.
38. Lavor CC, Cardozo TM, Nascimento MAC (2005) Int J Quantum Chem 103:500–504.
39. Cardozo TM, Nascimento MAC (2005) J Mat Sci Lett 40:3549–3551.
40. Bitzer R, Barbosa AGH, Silva CO, Nascimento MAC (2005) Carbohydrate Res 340:2171–2184.
41. Milas I, Nascimento MAC (2005) Chem Phys Lett 418:364–368.
42. Silva CO, Barbosa AGH, Silva EL, Nascimento MAC (2004) Theor Chem Acc 111:231–236.
43. Barbosa AGH, Nascimento MAC (2004) Int J Quantum Chem 99:317–324.
44. Silva CO, Nascimento MAC (2004) Carbohydrate Res 339:113–122.

45. Floriano WB, Nascimento MAC (2004) Brazilian J Phys 34:38–41.
46. Silva CO, Nascimento MAC (2004) Theor Chem Acc 112:342–348.
47. Silva AM, Nascimento MAC (2004) Chem Phys Lett 393: 173–178.
48. Lins JOMA, Nascimento MAC (2004) Chem Phys Lett 391: 9–15.
49. Barbatti M, Nascimento MAC (2003) J Chem Phys 119:5444–544803.
50. Milas I, Nascimento MAC (2003) Chem Phys Lett 373:379–384.
51. Silva CO, Silva EC, Nascimento MAC (2003) Chem Phys Lett 381:244–246.
52. Nascimento MAC, Barbosa AGH Progr Theor Chem & Phys (2003) 12:247–267.
53. Barbatti M, Nascimento MAC (2003) Brazilian J Phys 33:792–797.
54. Pereira MS, Nascimento MAC (2003) Theor Chem Acc 110:441–445.
55. Barbatti M, Jalbert G, Nascimento MAC (2002) J Phys Chem A 106:551–555.
56. Silva CO, Nascimento MAC (2002) Adv Chem Phys 123:423–468.
57. Barbosa AGH, Nascimento MAC (2002) Mol Phys 100:1677–1680.
58. Barbosa AGH, Nascimento MAC (2002) Theor Comput Chem 10:117–142.
59. Nascimento MAC (2001) Chem Phys Lett 343:15–20.
60. Nascimento MAC (2001) Chem Phys Lett. 338:67–73.
61. Nascimento MAC (2001) J Chem Phys 114:7066–7072.
62. Nascimento MAC (2001) J Chem Phys 114:2213–2218.
63. Nascimento MAC (2001) J Computer-Aided Mol Design 15:309–322.
64. Nascimento MAC (2001) Phys Status Solidi A 187:1–14.
65. Nascimento MAC (2001) Progr Theor Chem & Phys 7:39–76.
66. Nascimento MAC, Silva EC, Silva CO (2000) J Phys Chem B 104:2402–2409.
67. Nascimento MAC (2000) J Chem Phys 113:4230–4237.
68. Nascimento MAC (1999) J Mol Struct - Theochem 464:239–247.
69. Nascimento MAC, Silva EC, Silva CO (1999) J Phys Chem A 103:11194–11199.
70. Nascimento MAC, Esteves PM, Mota CJA (1999) J Phys Chem B 103:10417–10420.
71. Nascimento MAC, Ramos MJ, Floriano WB (1999) Int J Quantum Chem 74:299–314.
72. Nascimento MAC, Silva EC, Silva CO (1999) Int J Quantum Chem 74:417–422.
73. Melo A, Ramos MJ, Floriano WB, Gomes JANF, Leão JFR, Magalhães AL, Maigret B, Nascimento MAC, Reuter N. (1998) J Mol Struct 463:81–90.
74. Nascimento MAC, Floriano WB, Domont G, Goddard WA (1998) Protein Sci 7:2301–2313.
75. Nascimento MAC, Lins JOMA (1997) Mol Eng 7:309–316.
76. Nascimento MAC (1997) Mol Eng 7:87–108.
77. Nascimento MAC, Barbosa AGH (1997) Chem Phys Lett 279:119–121.
78. Nascimento MAC, Silva CO, Silva EC, Azevedo JA (1996) Int J Quantum Chem 60:433–438.
79. Nascimento MAC, Blaszkowsli SR, Santen RV (1996) J Phys Chem 100:3463–3472.
80. Nascimento MAC, Lins JOMA (1996) J Mol Struct 371:237–243.
81. Nascimento MAC, Mota-Neto JD (1996) J Phys Chem 100:15105–15110.
82. Nascimento MAC, Silva EC, Silva CO (1995) Astrophys J 439:1044–1045.
83. Nascimento MAC, Miranda MP, Bielschowsky CE (1995) J Phys B 28:L15–18.
84. Nascimento MAC, Blazskowski SR, Floriano WB (1995) J Mol Struct 335:51–57.
85. Nascimento MAC, Silva EC, Silva CO (1995) Int J Quantum Chem S29:639–646.
86. Nascimento MAC, Blaskowski SR, Santen RV (1994) J Phys Chem 98:12938–12944.
87. Nascimento MAC, Hollauer E (1993) J Chem Phys 99:1207–1214.
88. Nascimento MAC, Craw, JS, Pavão AC (1993) Int J Quantum Chem 48:219–224.
89. Nascimento MAC, Hollauer E (1993) Chem Phys 174:79–83
90. Nascimento MAC, Blaskowski SR (1993) J Mol Struct 287:67–75.
91. Nascimento MAC, Silva SC (1992) J Mol Struct 282:51–57.
92. Nascimento MAC, Hollauer E, Bielchowsky CE (1992) Phys Rev A 45:7942–7947.
93. Nascimento MAC, Hollauer E (1991) Chem Phys Lett 184:470–478.
94. Nascimento MAC, Hollauer E (1991) Chem Phys Lett 181:463–466.
95. Craw JS, Nascimento MAC, Ramos MN (1991) J Chem Soc Faraday Trans 87:1293–1296.
96. Craw JS, Nascimento MAC, Neves MR (1991) Spectrochim Acta A 47:69–73.
97. Nascimento MAC, Craw JS (1990) Chem Phys Lett 172:265–269.

98. Nascimento MAC, Hollauer E (1990) Phys Rev A 42:6608–6614.
99. Nascimento MAC, Craw JS (1990) Chem Phys Lett 168:423–427.
100. Nascimento MAC, Hollauer E, Bielschowsky CE (1990) Phys Rev A 42:5223–5227.
101. Nascimento MAC, Hollauer E, Bielschoswsky CE (1990) J Phys B At Mol Opt Phys 23:L783–789.
102. Nascimento MAC, Hollauer E, Bielschowsky CE (1990) Química Nova 12:225–229.
103. Nascimento MAC, Mota J (1990) Química Nova12: 384–385.
104. Nascimento MAC, Hollauer E (1988) J Chem Phys 88:7245–7246.
105. Nascimento MAC (1998) Canadian J Chem 66: 2884–2887.
106. Nascimento MAC, Silva EC (1987) Chem Phys Lett 138:509–511.
107. Nascimento MAC (1985) Química Nova 9:143–145.
108. Nascimento MAC (1985) Química Nova 9:135–137.
109. Nascimento MAC (1985) J Mol Struct 120: 227–240.
110. Nascimento MAC(1985) J Phys Chem 89: 2713–2714.
111. Nascimento MAC, Richer G, Sandorfy C (1984) J Electron Spectry Rel Phen 34:327 335.
112. Nascimento MAC (1983) Chem Phys 74:51–66.
113. Nascimento MAC (1983) Int J Quantum Chem 23:1011–1016.
114. Nascimento MAC, Goddard WA (1981) Chem Phys 53:251–263.
115. Nascimento MAC, Goddard WA (1980) Chem Phys 53:265–277.
116. Nascimento MAC, Goddard W (1979) Chem Phys Lett 60:197–200.
117. Nascimento MAC, Goddard WA (1979) Chem Phys 36:147–160.
118. Nascimento MAC, Goddard WA (1977) Phys Rev A 16:1559–1567.
119. Nascimento MAC, Yeager D, McKoy V (1975) Phys Rev A 11:1168–1174.
120. Nascimento MAC (1969) Eai J Res Develop 2:1–6.
121. Nascimento MAC (1969) Eai J Res Develop 2:48–54.
122. Nascimento MAC, Costa-Neto A, Costa-Neto C (1968) Anais Acad Brasil Ciências 40:404–408.

Theor Chem Acc (2012) 131:1202
DOI 10.1007/s00214-012-1202-2

FEATURE ARTICLE

Some recent developments in photoelectrochemical water splitting using nanostructured TiO_2: a short review

Paul Szymanski · Mostafa A. El-Sayed

Received: 17 January 2012 / Accepted: 3 March 2012 / Published online: 30 May 2012

Abstract Photoelectrochemical cells with TiO_2 electrodes to convert sunlight and water into gaseous hydrogen and oxygen are a source of clean and renewable fuel. Despite their great potential, far-from-ideal performance and poor utilization of the solar spectrum have prevented them from becoming a widespread and practical technology. We review recent experimental work that uses dynamics measurements to study limitations of photoelectrochemical cells from a fundamental level and the use of TiO_2 nanotube arrays as a superior alternative to TiO_2 nanoparticles. Through a combination of nanoscale size control, doping, composite materials, and the incorporation of noble-metal nanoparticles, improved performance and light harvesting are demonstrated.

Keywords Titanium dioxide · Nanotubes · Transient absorption · Plasmonics · Solar energy · Hydrogen · Photoelectrochemistry

Abbreviations

AM 1.5 G	Air mass 1.5 global
FDTD	Finite-difference time-domain
NP	Nanoparticle
NT	Nanotube
PEC	Photoelectrochemical
SPR	Surface plasmon resonance
TA	Transient absorption
XPS	X-ray photoelectron spectroscopy

Dedicated to Professor Marco Antonio Chaer Nascimento and published as part of the special collection of articles celebrating his 65th birthday.

P. Szymanski · M. A. El-Sayed (✉)
Laser Dynamics Laboratory, Georgia Institute of Technology, School of Chemistry and Biochemistry, Atlanta, GA 303332, USA
e-mail: melsayed@gatech.edu

1 Introduction

It gives me a special and great pleasure to contribute a review type manuscript honoring the great contributions of an excellent theorist and a long-time close friend, Professor Nascimento. Soon after his studies at Cal Tech, Marco made excellent and well-recognized theoretical contributions to the field of nonlinear spectroscopy, in which my group and I were working on from the experimental side. We learned a great deal from his papers and from interacting with him on a personal level. He was always willing to help in an effective way in explaining the theoretical basis of his important work. We became very good friends. I very much enjoyed my visit to Rio a number of years ago, mostly because of the great hospitality of Marco and his family and my scientific interaction with Marco and his group. I do wish him and his family a healthy and enjoyable future.

Green and efficient energy technologies are a key area where nanoscience could accelerate the transition from fossil fuels to renewable energy sources. One attractive possibility is conversion of solar energy to electricity or chemical fuel. Sunlight is an abundant and free energy source, which can provide the earth's surface with an amount of energy equivalent to that consumed by the whole world's population in an entire year in just 1.5 h [1]. Semiconductor nanomaterials may play a prominent role as they can function as photocatalysts promoting various oxidation and reduction reactions under sunlight [2].

One of the primary research and technology areas, the use of metal oxide nanomaterials in dye-sensitized solar cells to generate electricity, is not covered here. We instead focus on the related problem of "water splitting"—harnessing solar energy to obtain hydrogen and oxygen from water—through a combination of photocatalysis and electrochemistry. We emphasize recent developments in optimizing the growth of TiO_2 nanotubes to improve their catalytic properties, new strategies toward combining TiO_2 with other semiconductors to form higher-performance composite materials, the use of plasmonics to increase device efficiency and the ongoing work to gain a detailed understanding of the underlying mechanisms and dynamics in TiO_2 photocatalysis. Although this review is dedicated to experimental work, we hope that it will prove inspiring and informative to the theoretical community.

Hydrogen is an attractive potential fuel source for future vehicles and other applications. Unlike fossil fuels, combustion of H_2 liberates energy without releasing carbon dioxide into the atmosphere. In order for hydrogen to meet the demand for a fuel that has minimal impact on the environment, significant amounts of highly pure H_2 are required. As such sources are not found in nature, they must be produced artificially. In the ideal case, "water splitting" could provide a supply of pure H_2. As combustion of pure H_2 produces only water vapor, it would qualify as renewable energy if the fuel production could be powered by an abundant and non-polluting energy source.

One particularly promising approach is photoelectrochemical (PEC) water splitting as it was first demonstrated by Fujishima and Honda [3]. This method spatially separates water oxidation (producing O_2) and hydrogen reduction (producing H_2) by having each process occur at a separate electrode (Fig. 1). Light absorption by a semiconductor electrode provides energy for the reaction,

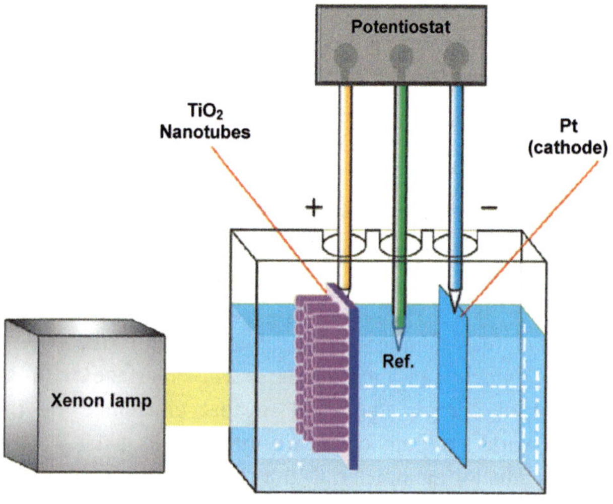

Fig. 1 Schematic of a three-electrode (anode, cathode, and reference electrode) photoelectrochemical (PEC) cell, employing TiO_2 nanomaterials in the anode. Reprinted with permission from Hamedani et al. [36]. Copyright 2011 American Chemical Society

assisted by an applied (electrical or chemical) bias to compensate for insufficient PEC cell voltage and overcome slow kinetics (in electrochemical nomenclature, the additional voltage required to drive a reaction at a desired rate or current density is known as *overpotential*). In an efficient system, the energy stored as fuel (H_2) will exceed the energy consumed by applying the bias.

Fujishima and Honda used an anode of bulk TiO_2 in their PEC cell. Modern designs typically make use of inexpensive TiO_2 nanoparticles (NPs), spread onto a conductive substrate and sintered to form a thin mesoporous film [4], similar to what is used in dye-sensitized solar cells [5]. Although the band gap of TiO_2 only allows absorption of UV light, comprising $\sim$4 % of the solar spectrum, the combination of low cost and chemical stability has proved difficult to surpass in other materials [6]. The efficiency of converting the energy from incident light to H_2 fuel, however, is less than 0.4 % [7] taking into account the entire air mass 1.5 global (AM 1.5 G, $\sim$100 mW/cm^2) solar spectrum [8]. Even considering only the spectrum from 320 to 400 nm around the TiO_2 absorption edge, a PEC made using TiO_2 NPs is at best $\sim$12 % efficient [7]. This is clearly far from ideal, and given the numerous advantages of TiO_2, multiple strategies have been employed to improve upon the original design.

As we will demonstrate, engineering the size and shape of nanoscale TiO_2 can have a significant impact on the conversion efficiency. To better understand the origin of these effects, we first discuss in more detail the mechanism of water splitting to uncover what fundamentally limits the efficiency. As another means of improving performance, incorporating other elements into the TiO_2 crystal lattice can enable the PEC to respond to visible light and we give examples of this in Sect. 4. Integrating advances made in another field of nanoscience, plasmonics has great promise to improve efficiency even further, which is discussed in Sect. 5. We conclude with some thoughts on future research directions in the field of PEC hydrogen production from water.

2 Mechanistic studies

Despite numerous studies of the catalytic efficiency of TiO_2 and other materials and the electron dynamics within photoexcited TiO_2, the mechanism of the water splitting itself has not been widely investigated. Because of the possibly long lifetimes of the photoelectrons and photoholes, the competing process of electron–hole recombination, and surface modifications that may occur when the catalyst is exposed to light for an extended duration, these measurements are extremely challenging to perform [4]. All of the results highlighted in this section used anodes containing TiO_2 NPs; however, it is expected that the findings will provide insight into the behavior of other nanostructures such as nanotubes (NTs).

Transient absorption (TA) spectroscopy has greatly improved our understanding of solar-powered devices, particularly in the areas of photocatalysis and photovoltaics using TiO_2. In the technique, a "pump" pulse, typically from a laser, places the sample in an excited state with new electronic and/or chemical properties. A time-delayed "probe" pulse measures the absorption spectrum of the sample after it is altered by the first pulse. Measuring the change in absorption compared to the initial value as a function of the delay between the pulses gives the dynamics of the system (Fig. 2).

When TiO_2 is photoexcited with energies greater than its band gap, an electron–hole pair, or exciton, is created. The electron and hole each become trapped at the surface within ~200 fs, giving rise to broad absorption features in the visible and near-IR regions of the spectrum [9]. Electrons gradually diffuse from these "shallow trap" states into lower-energy states within the bulk, causing absorption that is weak in the near-IR but grows increasingly more intense as the wavelength increases [9]. The induced absorption from the photoelectrons and photoholes is observed readily in TA experiments.

In all dynamics measurements meant to mimic the effects of sunlight, it is important to keep the incident laser fluence low. For example, Murai et al. [10] detect TA signals for electrons and holes in TiO_2 that decay at similar rates at fluences up to 0.28 mJ cm^{-2}. These weak excitation densities correspond to an average of one electron–hole pair or less per nanoparticle. At 3.4 mJ cm^{-2}, far above fluences produced by sunlight, the NPs relax much more rapidly. The results suggest bulk recombination of electron and holes begins to dominate once a sufficient density of electron–hole pairs is generated.

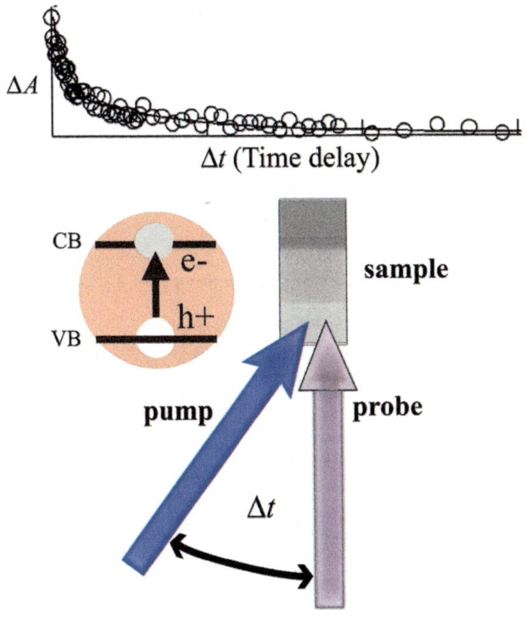

Fig. 2 Pump-probe transient absorption (TA) spectroscopy

Under open-circuit conditions, there is little or no oxygen evolution [4] or change in the photoelectron and photohole dynamics [10] when TiO_2 NPs in water are irradiated by UV light. This is strong indirect evidence that water splitting is much slower than electron–hole recombination.

To obtain direct evidence, recombination must be suppressed in a functioning PEC cell. One strategy is to introduce a scavenger species to rapidly remove either electrons or holes from the vicinity of the TiO_2. In work by Tang et al. [11], the film of TiO_2 NPs is immersed in $AgNO_3$ solution rather then pure water, and Ag^+ ions are reduced efficiently by photoelectrons in under 10 ns. Without the photoelectrons present, the photoholes decay with a half-life of 0.27 s mainly due to their reaction with water. Therefore, formation of O_2 requires on the order of 1 s under simulated sunlight (AM 1.5 G).

However, additives that would not be found in an actual device can affect the dynamics, as shown dramatically in studies of the effects of electrolyte composition on dye-sensitized solar cells [12]. This is true also in water splitting, as replacing Ag^+ with Pt metal, deposited onto the TiO_2 surface, in the experiments of Tang et al. [11] decreased the hole half-life to 0.5 ms. Recent work by Cowan et al. applied simultaneous TA and photoelectrochemistry to a complete PEC cell made from TiO_2 NPs [4, 13]. This allows both reaction products and dynamics to be studied under an applied bias voltage, reproducing the conditions within an actual working device. No scavenger species are introduced, to avoid competing side-reactions or unwanted changes of the TiO_2 surface properties.

Without an applied bias, no photocurrent or gaseous product is produced and electrons and holes have roughly equal lifetimes [4]. This shows that electron–hole recombination is a primary factor limiting PEC efficiency. Under a positive bias, where the cell produces O_2 and H_2, greatly increased hole lifetimes are observed (Fig. 3) [4, 13]. Temperature has no observable effect on the hole lifetime, meaning that the activation energy for water oxidation is too small to be of consequence [13]. The main problem, therefore, is extremely slow kinetics due to the four-hole process required to form O_2. The applied bias affects the energies of the charge carriers and likely aids the diffusion of electrons away from holes. Improving spatial separation between charge carriers to maintain a larger fraction of long-lived holes can be achieved by several strategies, most critically controlling the size and shape of TiO_2 structures on nanometer length scales.

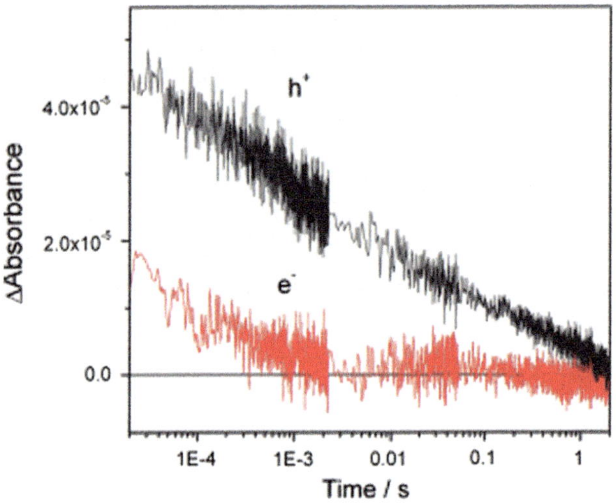

Fig. 3 Transient absorption signal decay for hole and electron probed at 460 and 900 nm, respectively, measured at a TiO_2 within a PEC cell containing 0.05 M NaOH/0.5 M $NaClO_4$ [13]. Under working conditions of positive bias (245 mV vs Ag/AgCl) and low excitation (50 μJ/cm^2, 355 nm), the hole greatly outlives the electron. Reprinted with permission from Cowan et al. [13]. Copyright 2011 American Chemical Society

3 Titania nanotubes

3.1 Properties

Given the problem of water splitting being much slower than electron–hole recombination, improving charge separation is vital to increasing yields from photocatalytic and PEC reactions. Can controlling the size and shape of TiO_2 on the nanoscale provide a path toward a solution? The electrical conductivity of a film made from a connected network of TiO_2 nanomaterials is limited by electron trapping to a far greater extent than single-crystalline bulk semiconductors [14]. Extensive theoretical modeling combined with careful experimentation has determined that the electron diffusion length in a film of TiO_2 NPs is ~10 μm [15]. For UV light, absorbed strongly by TiO_2, electron–hole pairs will be created at distances from the semiconductor particle surface that are within the diffusion length. If the electron can be transferred to another particle, the probability of which is influenced by the applied electric field (see previous section), and then, recombination can be avoided and the hole will remain available for chemistry. As we will see later, incorporating other materials into the TiO_2 particle network can extend the absorption range out to visible wavelengths, although the absorptivity in the visible often remains much weaker than in the UV. Weaker absorptivity requires that light be harvested over length scales, which may exceed the diffusion length, such that a greater number of electrons will recombine with holes before water oxidation can occur.

The efficiency of TiO_2 mesoporous films for water splitting also depends greatly on the degree of electrical connectivity within the film. Films made through sol–gel methods show better connectivity and an order of magnitude higher efficiency than films of sintered NPs [7]. Taken with the above results, this study suggests that a more conductive particle network can improve charge separation and potentially allow for more effective power conversion efficiency even when complete light harvesting requires distances longer than 10 μm. At the same time, holes have a diffusion length of only ~20 nm and thus must have ready access to the particle surface [16]. One type of nanostructure that satisfies all these constraints is the nanotube (NT), and as a result, interest in TiO_2 NTs has exploded in recent years [17–20].

3.2 Fabrication

A number of methods have been developed to form TiO_2 NTs, but the resulting materials are often of poor structural quality or have a tendency to form aggregates [21]. The most widely used procedure is currently potentiostatic anodization of Ti metal, first reported by Zwilling et al. [22]. Anodization is performed typically on thin (0.25 mm) high-purity Ti foils in a two-electrode electrochemical cell containing fluoride ions (Fig. 4). Titanium is used as the working electrode with Pt foil, the most stable choice for the counter electrode. An applied potential causes the formation of an array of hollow TiO_2 NTs that grows outward from the surface of the Ti foil (Fig. 5). The fact that the tubes are in a regular array oriented largely normal to the surface has attracted a lot of attention to this method. The electrolyte composition, anodization potential, and temperature during growth affect significantly the growth rate and the final dimensions of the NTs [17, 18, 20, 23, 24]. The resulting NTs are generally amorphous and require annealing after growth to form crystalline domains.

Mor et al. [25] reported the formation of 120-nm-long NTs with a wall thickness of 9 nm during anodization at

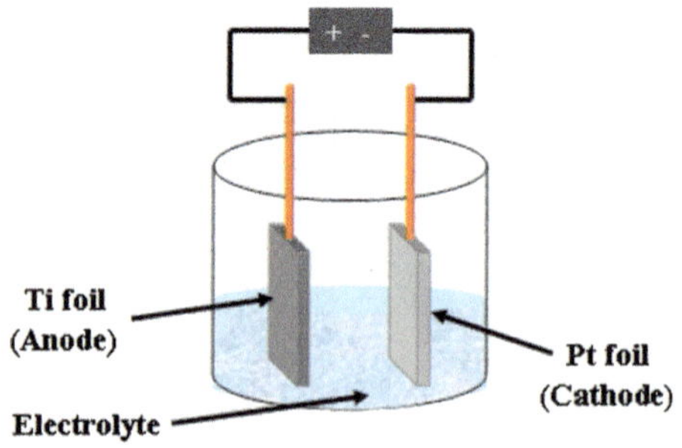

Fig. 4 Schematic diagram of the anodization set-up used to fabricate TiO_2 NT arrays. Reprinted with permission from Allam and Grimes [70]. Copyright 2009 American Chemical Society

Fig. 5 Field emission scanning electron microscope images of TiO_2 NTs formed after anodizing Ti foil at 20 V for 20 h in formamide containing 0.2 M NH_4F, 1 M H_3PO_4, and 3 vol% H_2O: **a** top view, **b** cross-sectional view. Reprinted with permission from Allam and El-Sayed [29]. Copyright 2010 American Chemical Society

10 V in 50 °C electrolyte made from a mixture of HF and acetic acid. The tube length increased to 224 nm if the anodization was performed at 5 °C, but the wall thickness increased to 34 nm. Longer tubes were formed with less acidic electrolytes, in which TiO_2 dissolves more slowly [20]. Using a neutral electrolyte of 1 M Na_2SO_4 with 5 wt% NH_4F, NTs 3 μm in length were formed in 30 min of anodization at 20 V [26]. The wall thickness was ~10 nm, but the diameter of the tubes was larger at the bottom (100 nm) than at the top (~50 nm) due to the pH gradient that forms during anodization.

Producing longer NTs requires the use of organic solvent mixtures with low water content. As the solvent viscosity increases, ion mobility decreases, which slows the chemical dissolution of TiO_2 by fluoride [24]. Decreasing the water fraction also decreases the concentration of fluoride, so fewer reactive ions will be present at the top of the nanotube. The reduced dissolution rate allows higher potentials to be applied, so that Ti oxidation can occur more quickly. As a result, growth rates increase and longer tube structures can be formed. A more extensive discussion of the formation mechanism can be found in papers by Macak et al. [18] and Su and Zhou [20].

The exact NT dimensions depend on the kinetic balance that is established for a particular electrolyte composition and anodization potential. Yin et al. [24] observed growth rates as high as 100 μm/h in 0.09 M NH_4F in ethylene glycol containing 1 vol% H_2O, anodizing at 150 V and 20 °C. The outer diameter of the NTs was 300 nm. The growth rate dropped to <5 μm/h when the water content was increased to 10 vol%, but an outer diameter of 600 nm was obtained. Shankar et al. [23] reported tubes with length 223 μm, wall thickness 25 nm, and outer diameter 160 nm using 0.25 wt% NH_4F in ethylene glycol with 2 vol% H_2O. The NT growth, however, required 17 h at 60 V, with anodization at 80 V not leading to formation of NT structures. If more viscous solvents, such as polyethylene glycol, are used, the rate of dissolution is slowed to such an extent that partially crystalline NTs can form during anodization [21]. Although most studies have used NH_4F, there are recent reports of the successful use of ionic liquids as electrolytes which provides further possibilities to tune NT growth [27, 28].

Crystallization of the amorphous NTs into the photoactive anatase phase of TiO_2 begins at annealing temperatures of ~280 °C in a variety of atmospheres [30], with conversion to the rutile phase beginning at annealing temperatures above 500 °C [26, 29–31]. As the rutile phase is more dense than anatase, this phase transition is accompanied by significant degradation and cracking of the NT array [29, 32]. Incorporation of dopants/impurities from the electrolyte, such as P ions, into the TiO_2 can retard the anatase-to-rutile phase transition and allow higher annealing temperatures to be used [29].

3.3 Performance in water splitting

For both TiO_2 NPs and NTs, the size and phase of the crystalline domains as well as the connectivity between domains has a profound effect on the power conversion efficiency to produce hydrogen fuel from solar energy. Unannealed, amorphous samples have little to no ability to be used for water splitting [21]. Park et al. [33] compared directly 3-μm-long NTs with a 15-μm-thick film of NPs (P-25, a mixture of anatase and rutile TiO_2 phases), both samples annealed at 450 °C. In a PEC cell with 1 M KOH electrolyte, the photocurrent under an applied bias voltage is ~10 times greater for the NT array compared to the NP film. Decreasing the NT length lowered the photocurrent, as both the light absorption and surface area in contact with electrolyte decreased.

For a water splitting PEC cell, the efficiency (η) may be calculated from

$$\eta = \frac{(1.229\mathrm{V} - V_{\mathrm{bias}})I_{\mathrm{p}}}{P} \tag{1}$$

where 1.229 V is the standard potential required to form H_2 and O_2 from H_2O, V_{bias} is the applied bias voltage, I_p is the measured photocurrent density in A/m^2, and P is the intensity of the incident light expressed in W/m^2 [19]. For broadband illumination under simulated AM 1.5 G sunlight, P is $\sim$1,000 W/m^2. Since pure TiO_2 has negligible absorptivity above 400 nm, efficiency is often calculated from the UV portion of the solar spectrum alone, typically 320–400 nm [34]. Care must be taken when comparing results that the measurements were taken under similar conditions. There is also a dependence of I_p on V_{bias}, although a plateau in I_p is typically reached resulting in a distinct optimal value for V_{bias} that maximizes η [7, 19].

There is a general trend toward increased cell performance as the temperature at which the NTs are annealed increases. This is expected due to greater crystallinity, which improves charge transport. Arrays of 7-μm-long NTs, whose structural properties are shown in Fig. 6, exhibit conversion efficiencies (Fig. 7) as high as 10 % under UV light (320–400 nm, filtered Xe arc lamp, 1,000 W/m^2) [29]. Interestingly, the optimum annealing temperature of 580 °C causes a reduction in the average grain size of the anatase phase and the appearance of the rutile phase. This may indicate that the anatase-to-rutile phase transition occurs more readily at the largest anatase grains. Control experiments on bare Ti foil show complete oxidation to rutile at the same temperature. As the substrate of Ti metal remains after anodization, this suggests that a significant portion of the rutile signal (Fig. 6) is originating from the oxidized substrate and not the NTs. Above 580 °C, the NT array is significantly damaged and η decreases.

Conversion efficiencies as high as 16.25 % have been reported [23]. This result was achieved with 30-μm-long

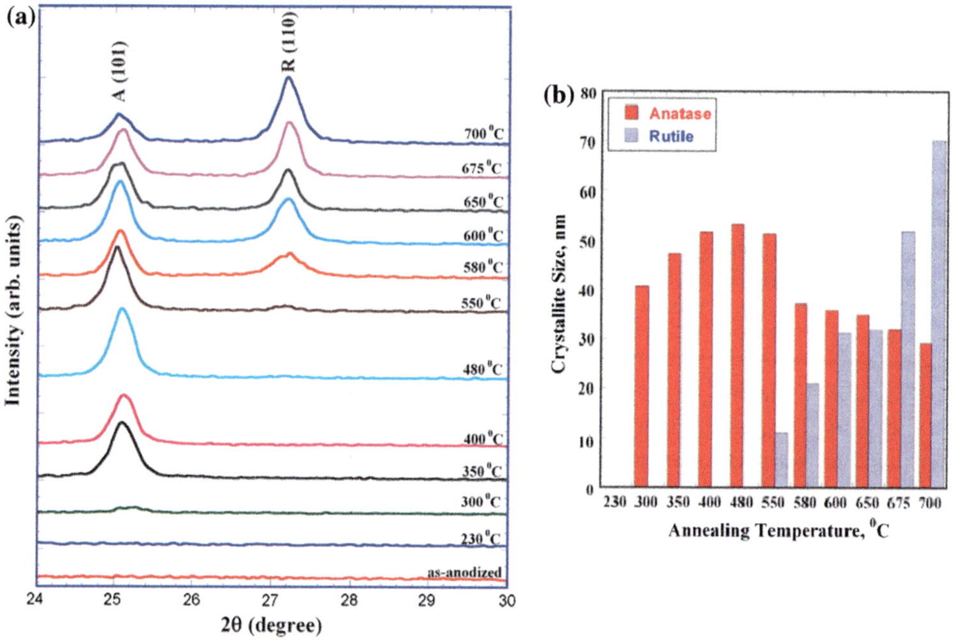

Fig. 6 Effects of annealing on crystallinity of 7-μm-long TiO_2 NTs formed after anodizing Ti foil at 20 V for 20 h in formamide containing 0.2 M NH_4F, 1 M H_3PO_4, and 3 vol% H_2O: **a** Glancing angle X-ray diffraction patterns and **b** corresponding crystallite sizes. The films are initially amorphous, with anatase (101) (*A*) and rutile (110) (*R*) phases forming after annealing [29]. Reprinted with permission from Allam and El-Sayed [29]. Copyright 2010 American Chemical Society

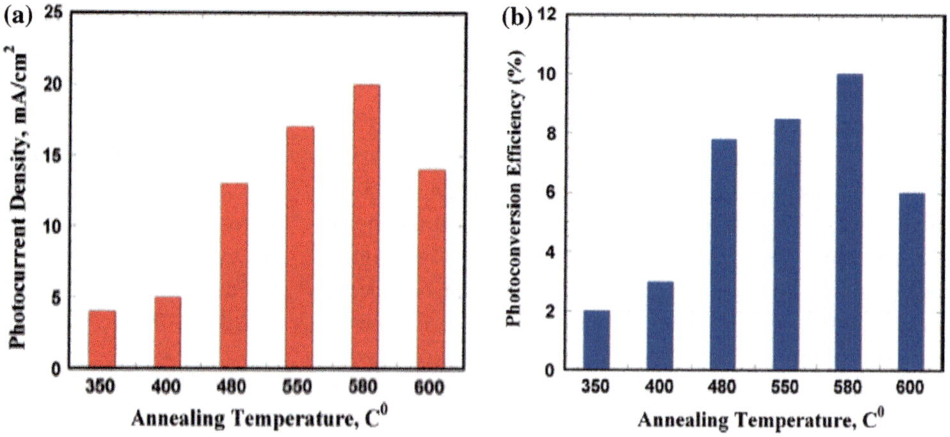

Fig. 7 **a** The steady-state photocurrent generated from 7-μm-long TiO_2 nanotube arrays under an applied voltage of 0.5 V (vs saturated Ag/AgCl) in 1 M KOH with respect to annealing temperature and **b** the corresponding photoconversion efficiencies under 320–400 nm light [29]. Reprinted with permission from Allam and El-Sayed [29]. Copyright 2010 American Chemical Society

NTs in 1 M KOH under light from a filtered Hg arc lamp (320–400 nm, 980 W/m^2). The optimal NT dimensions for water splitting, and the anodization and annealing conditions required to achieve them, are by no means established. With continuing improvements in NT fabrication and crystallization providing higher material quality and better size control, the balance between light harvesting and charge diffusion (Sect. 3.1) that makes the most efficient use of UV light will hopefully be achieved.

4 Composite nanotubes

Despite the advantages of TiO_2 as a material for solar energy conversion, the high (>3.0 eV) band gap still presents a significant drawback. Incorporating components with lower band gaps or lower-energy defect states into photoelectrodes can potentially improve conversion efficiency by extending the absorption edge into the visible region of the spectrum. Even in the case where UV photons are absorbed by TiO_2, placing TiO_2 in contact with a material with a lower-energy conduction-band edge may improve cell performance [35]. Electrons can transfer to lower-energy sites within the electrode, and the spatial separation of electron and hole reduces the rate of recombination.

Additional elements may be incorporated into NTs by anodizing Ti foil with an appropriately modified electrolyte. For example, we have recently fabricated TiO_2 NTs doped with Sr by using an aqueous electrolyte containing NH_4F, H_3PO_4, and $Sr(OH)_2$ [36]. Pure TiO_2 NTs, 1.2 μm long, displayed a maximum η of 0.2 % (in 1 M KOH under AM 1.5 G simulated sunlight, 1,000 W/m^2). By contrast, NTs of similar length (1.4 μm) with just 0.41 % Sr content were found to have a maximum η of 0.69 % under the same conditions. Compared with approaches that add $SrTiO_3$ to the surface of the NTs after anodization, incorporating Sr directly into the tube growth preserves the advantageous structural features of NTs.

Phosphorus-doped TiO_2 NTs have been formed by anodizing Ti foil in formamide containing NH_4F and H_3PO_4 [29]. This has the benefit of stabilizing the anatase phase allowing higher annealing temperatures to be used (see Sect. 3.3). X-ray photoelectron spectroscopy (XPS) revealed incorporated or doped P atoms, which was accompanied by a slightly redshifted absorption edge. The redshift was previously observed in P-doped sol–gel TiO_2 made by Xu et al. [37] with samples containing 16.7 % P having an absorption edge at 447 nm consistent with a band gap reduction of 0.43 eV. Density functional theory predicted a mixing of the O 2p states in the valence band of TiO_2 with P 3p states, causing the narrower band gap consistent with experimental results.

A similar approach has been used to form N-doped TiO_2 NTs. Shankar et al. performed anodization in an aqueous electrolyte of HF and NH_4NO_3 [38]. The N atoms were distributed inhomogeneously with a higher concentration near the tops of the tubes. The doped NTs showed a second absorption edge near 510 nm and produced roughly 5 times more photocurrent under broadband illumination (AM 1.5 G simulated sunlight). Similar absorption features were observed in N-doped TiO_2 thin films produced by DC magnetron sputtering [39, 40], where it was determined that the dopant gave rise to surface states above the TiO_2 valence band [39]. Although the surface states produce a PEC response to visible light, they can accelerate electron–hole recombination and decrease performance under UV light [40].

Li and Shang formed NTs using an electrolyte solution comprised on NH_4F, NH_4Cl, glycerin, and water [41]. The resulting NT array was ~500 μm long after 3 h of anodization. X-ray photoelectron spectra showed a sample enriched with N and F atoms, with some belonging to adsorbed species, while others were bonded to Ti in the crystal lattice. Annealing in an oxygen-free N_2 atmosphere preserved some of the F-containing and most of the N-containing species, yielding a sample with broad, weak absorption from 400 to 780 nm. Similar results were obtained by Liu et al. [42].

A very promising strategy for synthesizing composite NTs containing TiO_2 is anodizing metal alloys to form mixed metal oxides. Anodization has been reported on alloys of Ti with Fe [43], Cu [44], Pd [45], W [46], Al [47, 48], Mn [49], Nb [50, 51], and Zr [52] to successfully form NTs. Mor et al. [43] tested an array of Ti–Fe–O NTs as the photoanode for PEC water splitting, motivated by the lower band gap of Fe_2O_3. The most efficient material was obtained starting from a Ti foil containing 6.6 % Fe. The ~1.5-μm-long NTs were active at wavelengths out to ~600 nm, but no more than 7 % of absorbed photons at ~450 nm, where maximum visible light conversion efficiency was obtained, produced current. Photocurrent was lower if NTs with greater Fe content were used, possibly because of the low electron mobility in Fe_2O_3.

We have fabricated vertically oriented Ti–Nb–Zr–O mixed oxide NT arrays with wall thicknesses of 10 ± 2 nm [53]. This material showed enhanced power conversion efficiency for water splitting as compared to pure TiO_2 nanotubes (Fig. 8). Under UV light (filtered Xe arc lamp, 320–400 nm, 1,000 W/m^2) in 1 M KOH, the mixed metal oxides are ~17.5 % more efficient and require less applied bias. Oxides of Zr and Nb, formed on the surface of the NTs, may help to isolate electrons in the TiO_2 conduction band from species in solution, reducing recombination losses and consequently improving PEC efficiency.

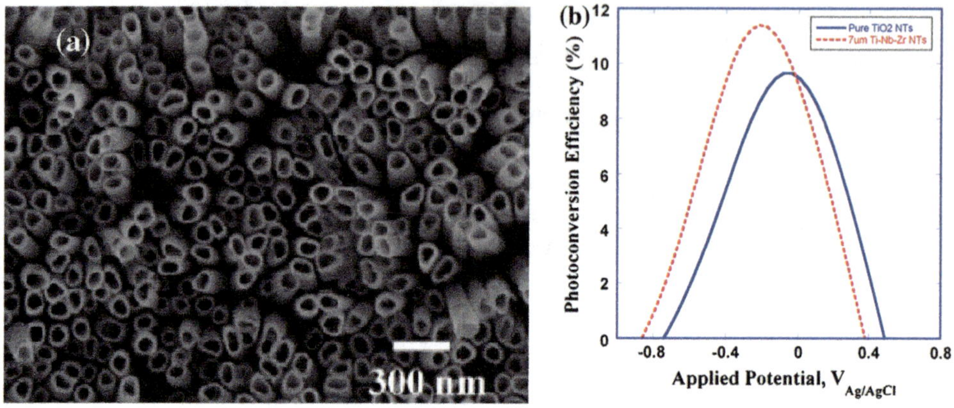

Fig. 8 **a** Field emission scanning electron microscope image of Ti–Nb–Zr mixed oxide NTs and **b** the UV power conversion efficiency of the mixed oxide NTs as compared to TiO_2 NTs of the same length (7 μm) [53]. Reprinted with permission from Allam et al. [53]. Copyright 2010 American Chemical Society

Recently, we have combined the use of metal alloys with doping to great effect [45]. We anodized Ti–Pd alloy, as Pd is an excellent catalyst that absorbs visible light, to form NTs 6 μm in length. The alloy NTs were then annealed in NH_3 to dope them with nitrogen. Increasing the annealing temperature to 550 °C caused a profound change in the X-ray photoelectron spectrum confirming the presence of Ti–O–N structures rather than unbonded nitrogen dopants. The Ti–Pd oxynitride NTs exhibited an absorption band edge redshifted to 577 nm and over 4 times greater power conversion efficiency compared to pure TiO_2 NTs.

Finally, it is worth noting that absorption properties can be modified greatly by the creation of large amounts of defect states without introducing dopants into the bulk. This was recently demonstrated by annealing TiO_2 NPs [54], nanowires [55], and NTs [55] in a hydrogen atmosphere. The H_2-treated TiO_2 absorbed visible light, turning gray or black depending on the annealing temperature. The mechanism is controversial. In the NP samples, the color change was attributed to a disorder-induced shift in the valence band position [54]; no such shift was measured for the nanowires and NTs [55]. The color change in the nanowires and NTs imparted only extremely weak visible light PEC activity but increased the UV response by a factor of 2 or more [55]. The increased UV activity was explained by the measured increase in carrier density following H_2 treatment, due to the resulting oxygen vacancies acting as electron donors. The vacancy states themselves, located within the band gap, are believed to be too high in energy and too electronically localized to be significantly active in water splitting.

5 Plasmonic effects in solar water splitting

As we have seen, methods that extend the absorption range of TiO_2 into the visible and produce PEC activity to visible light do so only weakly. In addition to increasing the absorptivity of the material itself, one can also increase the electric field strength to increase the number of photons absorbed. This can be done on the nanoscale by the use of noble-metal NPs that exhibit plasmonic effects [56, 57].

A surface plasmon resonance (SPR) is a collective oscillation of electrons confined to the surface of a metal NP. The lowest order effect gives rise to a local electric field due to a light-induced dipole (Fig. 9). At the surface of the NP, the dipole enhances the electric field to several orders of magnitude greater than the field of the light itself. This near-field enhancement can benefit a variety of optical processes. The energy of the SPR depends on multiple factors, such as the electronic properties of the metal, the dielectric properties of the surrounding medium, the size and shape of the NP, and the position of neighboring NPs. Only Ag, Au, and Cu possess SPRs that are always located in the visible or redder regions of the spectrum, but only Ag and Au are widely utilized due to their greater chemical stability.

Liu et al. [42] evaporated a Au film onto a F- and N-doped TiO_2 NT array, forming islands of metal with plasmonic properties of NPs. A significant enhancement in photocurrent compared to an array not containing Au was

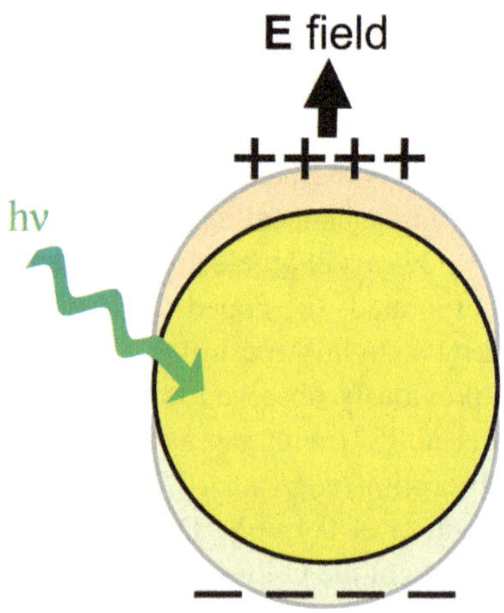

Fig. 9 Surface plasmon resonance (SPR) in a metal nanoparticle

reported when a visible light source was used. The authors attributed this to a local field enhancement at wavelengths that excited the SPR of Au and were also absorbed by TiO_2. The absorption spectrum of the TiO_2 NTs contained a broad tail in the visible region of the spectrum due to the impurity doping. Photocurrent caused by visible light, 1–2 orders of magnitude higher than that produced by the NTs without Au, was still lower than photocurrent produced by UV light. In fact, photocurrent caused by UV absorption was significantly reduced by Au NPs, explained by the authors as the effect of reduced absorption by TiO_2 and less oxide surface area in contact with the electrolyte.

It is known that Au [58, 59] and Ag [60] NPs can act as sensitizers, injecting electrons into TiO_2 after absorbing visible light. Liu et al. [42] did not observe photocurrent under visible light, however, if Au NPs were adsorbed on undoped TiO_2 that lacked visible light absorption. This result shows that the combination of doping and the plasmonic effect are likely responsible for the enhanced current in PEC cells. Finite-difference time-domain (FDTD) simulations of the Au/TiO_2 film show regions of enhanced electric field strength in the gaps between closely spaced Au NPs, consistent with the observed plasmonic effect [42]. Due to the regions of enhanced field, electron–hole pairs are created near the TiO_2 surface thus facilitating water oxidation by the photoholes.

Similar effects were seen for Ag nanocubes, prepared through colloidal synthesis and deposited on N-doped TiO_2 NTs [61]. Ingram and Linic observed current enhancements of an order of magnitude in a Ag/TiO_2 composite water splitting cell when light from 400 to 500 nm was used. Light in this range of wavelengths is resonant with both the absorption from the Ag SPR and the N dopants in TiO_2. A similar PEC cell, constructed using Au nanospheres instead of Ag nanocubes, showed little or no enhancement because the SPR did not overlap with the absorption of their NTs. At 370 nm, on the edge of the SPR absorption but still resonant with TiO_2 absorption, an enhancement factor between 3 and 4 was still observed. Whether the Ag nanocubes caused a reduction in current at lower wavelengths, as occurred with the Au film used by Liu et al. [42], was not reported.

In addition to FDTD simulations, which showed significantly enhanced electric fields at the sharp corners of the nanocubes, power dependence studies helped to confirm the plasmonic nature of the photocurrent enhancement [61]. In studies of the $TiO_2(110)$ surface using a beam of electrons, which have shallow penetration depths, to excite electron–hole pairs near the surface, the concentration of holes depends linearly on the excitation intensity [62]. The current generated through water splitting using the plasmonic photoelectrode also depended linearly on the excitation intensity (of visible light, in this case), suggesting

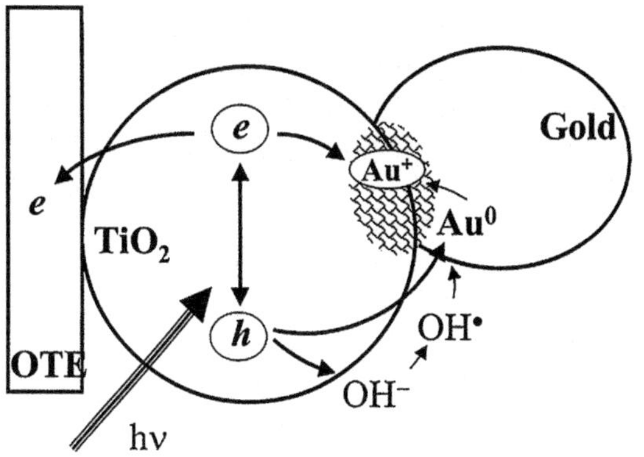

Fig. 10 Proposed reaction scheme for the decomposition of Au NPs on TiO_2 surfaces in the presence of OH^-. Reprinted with permission from Subramanian et al. [63]. Copyright 2001 American Chemical Society

that holes were being produced near the surface in this system as well [61]. Near-surface excitation with visible light is due to the locally enhanced electric field caused by the plasmonic effect. For the PEC without Ag nanocubes, the power dependence is nonlinear and consistent with excitation deeper in the bulk of TiO_2 [61, 62].

For plasmonic effects from noble-metal NPs to be fully utilized in water splitting PEC cells, the NPs must be chemically stable during extended use. Although Subramanian et al. [63] reported enhanced photocurrent after depositing Au NPs onto TiO_2 films, the current dropped rapidly during 60 min of continuous illumination. The working conditions were in 0.05 M NaOH solution under UV light. Changes in the UV–visible absorption spectrum of the electrode were consistent with the gradual incorporation of ions, possibly Au^+, in the TiO_2 matrix [63, 64]. X-ray absorption fine structure measurements confirmed the accumulation of Au atoms or ions [64]. The proposed mechanism (Fig. 10) was oxidation of hydroxide ions by photoholes to form highly reactive OH radicals, which could react with Au metal to form Au ions. One possible remedy is the presence of an electron donor, such as a metal chloride, in contact with the metal particle to neutralize metal ions and stabilize the NPs [60]. Another promising approach is metal core/oxide-shell NPs [65–68] to prevent oxidation of the metal NPs by radicals in solution.

6 Conclusions

In this review, we have presented some of the highlights in the area of research dedicated to developing a practical PEC system to turn solar energy and water into hydrogen and oxygen using the highly abundant and chemically stable material TiO_2. Performing TA spectroscopy on PEC cells under operating conditions reveals dynamics on multiple timescales, providing further mechanistic understanding. Controlled anodization of metal foils to produce metal oxide NTs has the potential to develop highly efficient photoelectrodes for "water splitting", incorporating dopants or phases of other materials to improve charge separation and light harvesting. Finally, plasmonic NPs, incorporated into the photoelectrodes, can further improve light harvesting provided that issues of chemical stability can be addressed and performance under UV light is not adversely affected.

A considerable challenge will be to measure and interpret the dynamics of photoexcited charge carriers in 1D nanostructures such as TiO_2 NTs. The main experimental issue is that anodization of metal foils produces an NT array atop an opaque metal foil, preventing the measurement of light transmission (and absorption). Considerable effort has been invested in forming ordered NT arrays on transparent conductive substrates for photovoltaic systems [19], producing transmissive electrodes that would enable many spectroscopic measurements. Alternatively, TA could be measured through diffuse reflectance from an opaque sample as demonstrated on dye-sensitized solar cells [69]. Ultimately, materials synthesis, spectroscopy, and modeling and computational studies should inform each other and influence their research directions, pointing the way toward increasingly more efficient and practical systems for solar water splitting.

Acknowledgments The authors would like to thank the financial support of the Office of Basic Energy Sciences of the US Department of Energy under contract number DE-FG02-97ER14799.

References

1. Lewis NS, Crabtree G, Nozik AJ, Wasielewski MR, Alivisatos AP (2006) Basic research needs for solar energy utilization. US Department of Energy, Washington
2. Fukushima A, Hasimoto K, Watanabe T (1999) TiO_2 photocatalysis: fundamentals and applications, 1st edn. BKC, Tokyo
3. Fujishima A, Honda K (1972) Electrochemical photolysis of water at a semiconductor electrode. Nature 238(5358):37–38
4. Cowan AJ, Tang J, Leng W, Durrant JR, Klug DR (2010) Water splitting by nanocrystalline TiO_2 in a complete photoelectrochemical cell exhibits efficiencies limited by charge recombination. J Phys Chem C 114(9):4208–4214
5. Grätzel M (2005) Solar energy conversion by dye-sensitized photovoltaic cells. Inorg Chem 44(20):6841–6851
6. Chen X, Shen S, Guo L, Mao SS (2010) Semiconductor-based photocatalytic hydrogen generation. Chem Rev 110(11):6503–6570
7. Hartmann P, Lee D-K, Smarsly BM, Janek J (2010) Mesoporous TiO_2: comparison of classical sol–gel and nanoparticle based photoelectrodes for the water splitting reaction. ACS Nano 4(6):3147–3154
8. Gueymard CA, Myers D, Emery K (2002) Proposed reference irradiance spectra for solar energy systems testing. Sol Energy 73(6):443–467
9. Tamaki Y, Furube A, Murai M, Hara K, Katoh R, Tachiya M (2007) Dynamics of efficient electron-hole separation in TiO_2 nanoparticles revealed by femtosecond transient absorption spectroscopy under the weak-excitation condition. Phys Chem Chem Phys 9(12):1453–1460
10. Murai M, Tamaki Y, Furube A, Hara K, Katoh R (2007) Reaction of holes in nanocrystalline TiO_2 films evaluated by highly sensitive transient absorption spectroscopy. Catal Today 120(2): 214–219
11. Tang J, Durrant JR, Klug DR (2008) Mechanism of photocatalytic water splitting in TiO_2. reaction of water with photoholes, importance of charge carrier dynamics, and evidence for four-hole chemistry. J Am Chem Soc 130(42):13885–13891
12. Listorti A, O'Regan B, Durrant JR (2011) Electron transfer dynamics in dye-sensitized solar cells. Chem Mater 23(15):3381–3399
13. Cowan AJ, Barnett CJ, Pendlebury SR, Barroso M, Sivula K, Grätzel M, Durrant JR, Klug DR (2011) Activation energies for the rate-limiting step in water photooxidation by nanostructured α-Fe_2O_3 and TiO_2. J Am Chem Soc 133(26):10134–10140
14. de Jongh PE, Vanmaekelbergh D (1996) Trap-limited electronic transport in assemblies of nanometer-size TiO_2 particles. Phys Rev Lett 77(16):3427
15. Leng WH, Barnes PRF, Juozapavicius M, O'Regan BC, Durrant JR (2010) Electron diffusion length in mesoporous nanocrystalline TiO_2 photoelectrodes during water oxidation. J Phys Chem Lett 1(6):967–972
16. Grimes CA, Varghese OK, Ranjan S (2008) Light, water, hydrogen: the solar generation of hydrogen by water photoelectrolysis. Springer US, Boston
17. Grimes CA (2007) Synthesis and application of highly ordered arrays of TiO_2 nanotubes. J Mater Chem 17(15):1451–1457
18. Macak JM, Tsuchiya H, Ghicov A, Yasuda K, Hahn R, Bauer S, Schmuki P (2007) TiO_2 nanotubes: self-organized electrochemical formation, properties and applications. Curr Opin Solid State Mater Sci 11(1–2):3–18
19. Shankar K, Basham JI, Allam NK, Varghese OK, Mor GK, Feng X, Paulose M, Seabold JA, Choi K-S, Grimes CA (2009) Recent advances in the use of TiO_2 nanotube and nanowire arrays for oxidative photoelectrochemistry. J Phys Chem C 113(16):6327–6359
20. Su Z, Zhou W (2011) Formation, morphology control and applications of anodic TiO_2 nanotube arrays. J Mater Chem 21 (25):8955–8970
21. Allam NK, Grimes CA (2009) Room temperature one-step polyol synthesis of anatase TiO_2 nanotube arrays: photoelectrochemical properties. Langmuir 25(13):7234–7240
22. Zwilling V, Darque-Ceretti E, Boutry-Forveille A, David D, Perrin MY, Aucouturier M (1999) Structure and physicochemistry of anodic oxide films on titanium and TA6V alloy. Surf Interface Anal 27(7):629–637
23. Shankar K, Mor GK, Prakasam HE, Yoriya S, Paulose M, Varghese OK, Grimes CA (2007) Highly-ordered TiO_2 nanotube arrays up to 220 μm in length: use in water photoelectrolysis and dye-sensitized solar cells. Nanotechnology 18(6):065707
24. Yin H et al (2010) The large diameter and fast growth of self-organized TiO_2 nanotube arrays achieved via electrochemical anodization. Nanotechnology 21(3):035601
25. Mor GK, Shankar K, Paulose M, Varghese OK, Grimes CA (2004) Enhanced photocleavage of water using Titania nanotube arrays. Nano Lett 5(1):191–195
26. Lockman Z, Ismail S, Sreekantan S, Schmidt-Mende L, MacManus-Driscoll JL (2010) The rapid growth of 3 μm long Titania nanotubes by anodization of titanium in a neutral electrochemical bath. Nanotechnology 21(5):055601

27. Li H, Qu J, Cui Q, Xu H, Luo H, Chi M, Meisner RA, Wang W, Dai S (2011) TiO_2 nanotube arrays grown in ionic liquids: high-efficiency in photocatalysis and pore-widening. J Mater Chem 21(26):9487–9490
28. Wender H, Feil AF, Diaz LB, Ribeiro CS, Machado GJ, Migowski P, Weibel DE, Dupont J, Teixeira SrR (2011) Self-organized TiO_2 nanotube arrays: synthesis by anodization in an ionic liquid and assessment of photocatalytic properties. ACS Appl Mater Interfaces 3(4):1359–1365
29. Allam NK, El-Sayed MA (2010) Photoelectrochemical water oxidation characteristics of anodically fabricated TiO_2 nanotube arrays: structural and optical properties. J Phys Chem C 114(27): 12024–12029
30. Varghese OK, Gong D, Paulose M, Grimes CA, Dickey EC (2003) Crystallization and high-temperature structural stability of titanium oxide nanotube arrays. J Mater Res 18(01):156–165
31. Sun Y, Yan K, Wang G, Guo W, Ma T (2011) Effect of annealing temperature on the hydrogen production of TiO_2 nanotube arrays in a two-compartment photoelectrochemical cell. J Phys Chem C 115(26):12844–12849
32. Hardcastle FD, Ishihara H, Sharma R, Biris AS (2011) Photoelectroactivity and Raman spectroscopy of anodized titania (TiO_2) photoactive water-splitting catalysts as a function of oxygen-annealing temperature. J Mater Chem 21(17):6337–6345
33. Park JH, Kim S, Bard AJ (2005) Novel carbon-doped TiO_2 nanotube arrays with high aspect ratios for efficient solar water splitting. Nano Lett 6(1):24–28
34. Varghese OK, Grimes CA (2008) Appropriate strategies for determining the photoconversion efficiency of water photoelectrolysis cells: a review with examples using titania nanotube array photoanodes. Sol Energy Mater Sol Cells 92(4):374–384
35. Im JS, Lee SK, Lee Y-S (2011) Cocktail effect of Fe_2O_3 and TiO_2 semiconductors for a high performance dye-sensitized solar cell. Appl Surf Sci 257(6):2164–2169
36. Hamedani HA, Allam NK, Garmestani H, El-Sayed MA (2011) Electrochemical fabrication of strontium-doped TiO_2 nanotube array electrodes and investigation of their photoelectrochemical properties. J Phys Chem C 115(27):13480–13486
37. Xu L, Tang C-Q, Qian J, Huang Z-B (2010) Theoretical and experimental study on the electronic structure and optical absorption properties of P-doped TiO_2. Appl Surf Sci 256(9):2668–2671
38. Shankar K, Tep KC, Mor GK, Grimes CA (2006) An electrochemical strategy to incorporate nitrogen in nanostructured TiO_2 thin films: modification of bandgap and photoelectrochemical properties. J Phys D Appl Phys 39(11):2361–2366
39. Sakthivel S, Janczarek M, Kisch H (2004) Visible light activity and photoelectrochemical properties of nitrogen-doped TiO_2. J Phys Chem B 108(50):19384–19387
40. Lindgren T, Mwabora JM, Avendaño E, Jonsson J, Hoel A, Granqvist C-G, Lindquist S-E (2003) Photoelectrochemical and optical properties of nitrogen doped titanium dioxide films prepared by reactive DC magnetron sputtering. J Phys Chem B 107(24):5709–5716
41. Li Q, Shang JK (2009) Self-organized nitrogen and fluorine co-doped titanium oxide nanotube arrays with enhanced visible light photocatalytic performance. Environ Sci Technol 43(23): 8923–8929
42. Liu Z, Hou W, Pavaskar P, Aykol M, Cronin SB (2011) Plasmon resonant enhancement of photocatalytic water splitting under visible illumination. Nano Lett 11(3):1111–1116
43. Mor GK, Prakasam HE, Varghese OK, Shankar K, Grimes CA (2007) Vertically oriented Ti–Fe–O nanotube array films: toward a useful material architecture for solar spectrum water photoelectrolysis. Nano Lett 7(8):2356–2364
44. Mor GK, Varghese OK, Wilke RHT, Sharma S, Shankar K, Latempa TJ, Choi K-S, Grimes CA (2008) p-Type Cu–Ti–O nanotube arrays and their use in self-biased heterojunction photoelectrochemical diodes for hydrogen generation. Nano Lett 8(7):1906–1911
45. Allam NK, Poncheri AJ, El-Sayed MA (2011) Vertically oriented Ti–Pd mixed oxynitride nanotube arrays for enhanced photoelectrochemical water splitting. ACS Nano 5(6):5056–5066
46. Nah Y-C, Ghicov A, Kim D, Berger S, Schmuki P (2008) TiO_2–WO_3 composite nanotubes by alloy anodization: growth and enhanced electrochromic properties. J Am Chem Soc 130(48): 16154–16155
47. Bayoumi FM, Ateya BG (2006) Formation of self-organized titania nano-tubes by dealloying and anodic oxidation. Electrochem Commun 8(1):38–44
48. Berger S, Tsuchiya H, Schmuki P (2008) Transition from nanopores to nanotubes: self-ordered anodic oxide structures on titanium–aluminides. Chem Mater 20(10):3245–3247
49. Mohapatra SK, Raja KS, Misra M, Mahajan VK, Ahmadian M (2007) Synthesis of self-organized mixed oxide nanotubes by sonoelectrochemical anodization of Ti-8Mn alloy. Electrochim Acta 53(2):590–597
50. Ghicov A, Aldabergenova S, Tsuchyia H, Schmuki P (2006) TiO_2–Nb_2O_5 nanotubes with electrochemically tunable morphologies. Angew Chem Int Ed 45(42):6993–6996
51. Dongyan D et al (2009) Anodic fabrication and bioactivity of Nb-doped TiO_2 nanotubes. Nanotechnology 20(30):305103
52. Yasuda K, Schmuki P (2007) Electrochemical formation of self-organized zirconium titanate nanotube multilayers. Electrochem Commun 9(4):615–619
53. Allam NK, Alamgir F, El-Sayed MA (2010) Enhanced photoassisted water electrolysis using vertically oriented anodically fabricated Ti–Nb–Zr–O mixed oxide nanotube arrays. ACS Nano 4(10):5819–5826
54. Chen X, Liu L, Yu PY, Mao SS (2011) Increasing solar absorption for photocatalysis with black hydrogenated titanium dioxide nanocrystals. Science 331(6018):746–750
55. Wang G, Wang H, Ling Y, Tang Y, Yang X, Fitzmorris RC, Wang C, Zhang JZ, Li Y (2011) Hydrogen-treated TiO_2 nanowire arrays for photoelectrochemical water splitting. Nano Lett 11(7): 3026–3033
56. Jain PK, Huang X, El-Sayed IH, El-Sayed MA (2008) Noble metals on the nanoscale: optical and photothermal properties and some applications in imaging, sensing, biology, and medicine. Acc Chem Res 41(12):1578–1586
57. Jain PK, El-Sayed MA (2010) Plasmonic coupling in noble metal nanostructures. Chem Phys Lett 487(4–6):153–164
58. Du L, Furube A, Yamamoto K, Hara K, Katoh R, Tachiya M (2009) Plasmon-induced charge separation and recombination dynamics in gold–TiO_2 nanoparticle systems: dependence on TiO_2 particle size. J Phys Chem C 113(16):6454–6462
59. Furube A, Du L, Hara K, Katoh R, Tachiya M (2007) Ultrafast plasmon-induced electron transfer from gold nanodots into TiO_2 nanoparticles. J Am Chem Soc 129(48):14852–14853
60. Yu J, Dai G, Huang B (2009) Fabrication and characterization of visible-light-driven plasmonic photocatalyst Ag/AgCl/TiO_2 nanotube arrays. J Phys Chem C 113(37):16394–16401
61. Ingram DB, Linic S (2011) Water splitting on composite plasmonic-metal/semiconductor photoelectrodes: evidence for selective plasmon-induced formation of charge carriers near the semiconductor surface. J Am Chem Soc 133(14):5202–5205
62. Zhang Z, Yates JT (2010) Direct observation of surface-mediated electron–hole pair recombination in TiO_2(110). J Phys Chem C 114(7):3098–3101
63. Subramanian V, Wolf E, Kamat PV (2001) Semiconductor–metal composite nanostructures. To what extent do metal nanoparticles improve the photocatalytic activity of TiO_2 films? J Phys Chem B 105(46):11439–11446

64. Lahiri D, Subramanian V, Shibata T, Wolf EE, Bunker BA, Kamat PV (2003) Photoinduced transformations at semiconductor/metal interfaces: X-ray absorption studies of titania/gold films. J Appl Phys 93(5):2575–2582
65. Hirakawa T, Kamat PV (2005) Charge separation and catalytic activity of Ag@TiO_2 core−shell composite clusters under uv−irradiation. J Am Chem Soc 127(11):3928–3934
66. Sakai H, Kanda T, Shibata H, Ohkubo T, Abe M (2006) Preparation of highly dispersed core/shell-type titania nanocapsules containing a single Ag nanoparticle. J Am Chem Soc 128(15): 4944–4945
67. Awazu K, Fujimaki M, Rockstuhl C, Tominaga J, Murakami H, Ohki Y, Yoshida N, Watanabe T (2008) A plasmonic photocatalyst consisting of silver nanoparticles embedded in titanium dioxide. J Am Chem Soc 130(5):1676–1680
68. Chuang H-Y, Chen D-H (2009) Fabrication and photocatalytic activities in visible and UV light regions of Ag@TiO_2 and NiAg@TiO_2 nanoparticles. Nanotechnology 20(10):105704
69. Furube A, Wang Z-S, Sunahara K, Hara K, Katoh R, Tachiya M (2010) Femtosecond diffuse reflectance transient absorption for dye-sensitized solar cells under operational conditions: effect of electrolyte on electron injection. J Am Chem Soc 132(19):6614–6615
70. Allam NK, Grimes CA (2009) Effect of rapid infrared annealing on the photoelectrochemical properties of anodically fabricated TiO_2 nanotube arrays. J Phys Chem C 113(19):7996–7999

Theor Chem Acc (2012) 131:1190
DOI 10.1007/s00214-012-1190-2

REGULAR ARTICLE

Role of step sites on water dissociation on stoichiometric ceria surfaces

Silvia Fuente · María M. Branda · Francesc Illas

Received: 22 December 2011 / Accepted: 9 February 2012 / Published online: 10 March 2012

Abstract The adsorption and dissociation of water on $CeO_2(111)$, $CeO_2(221)$, $CeO_2(331)$, and $CeO_2(110)$ has been studied by means of periodic density functional theory using slab models. The presence of step sites moderately affects the adsorption energy of the water molecule but in some cases as in $CeO_2(331)$ is able to change the sign of the energy reaction from endo- to exothermic which has important consequences for the catalytic activity of this surface. Finally, no stable molecular state has been found for water on $CeO_2(110)$ where the reaction products lead to a very stable hydroxylated surface which will rapidly become inactive.

Keywords Water gas shift · Ceria · CeO2 · DFT · GGA + *U*

1 Introduction

Since the early forties, the water–gas shift reaction ($CO + H_2O \rightarrow CO_2 + H_2$) constitutes an important step in the industrial production of CO-free hydrogen [1] to be subsequently used in hydrodesulfuration processes in oil refineries, in ammonia synthesis through the Bosch–Haber process or in fuel cells [2]. The water–gas shift (WGS) reaction is also involved in other important industrial processes such as the methanol synthesis [3] or in the methanol steam reforming process [4]. In the chemical industry, the WGS reaction is carried out in two steps at high (623–673 K) and low (463–503 K) temperature [5]. The low temperature step uses Cu [6]- or Au [7–9]-based catalysts which often constitute the catalyst active phase [10–12]. Nevertheless, this apparently simple reaction is more complex that imagined and other factors must be considered such as the nature of the support [7, 13–16], the existence of point defect such as oxygen vacancies [17, 18], or the catalyst preparation process [19]. Likewise, subtle modifications of the catalyst by doping with traces of other metals [20, 21] or by formation of alloys [22, 23] have been found to considerably improve the catalytic performance. Nevertheless, the reaction mechanism, at least for the metallic phase and the low temperature step, is rather well understood, especially after a series of recent papers reporting microkinetic studies based mainly on the rate constants derived from density functional calculations [24–26] and the work of Fajin et al. [27] highlighting the important role of step sites.

Rather recently, inverse catalysts where an inactive noble metal surface such as Au(111) acts as support for CeO_2 or TiO_2 nanoparticles have proven to be active for the WGS reaction and almost as good catalysts as Cu extended surfaces [23, 28]. X-ray photoelectron spectroscopy (XPS) and scanning tunneling microscopy (STM) experiments [23] suggest that the catalytic activity of these systems toward the WGS reaction is strongly related to the direct participation of the oxide–metal interface in the catalytic process. Moreover, these experiments have shown that water can easily dissociate on either TiO_{2-x}/Au(111)

Dedicated to Professor Marco Antonio Chaer Nascimento and published as part of the special collection of articles celebrating his 65th birthday.

S. Fuente · F. Illas (✉)
Departament de Química Física and Institut de Química Teòrica i Computacional (IQTCUB), Universitat de Barcelona, C/Martí i Franquès 1, 08028 Barcelona, Spain
e-mail: francesc.illas@ub.edu

S. Fuente · M. M. Branda
Departamento de Física, Universidad Nacional del Sur, Bahía Blanca, Argentina

or CeO_{2-x}/Au(111) but that no water dissociation is seen when there are no O vacancies in the supported oxide nanoparticles. These findings are in agreement with surface science experiments showing strong adhesion of molecularly adsorbed water to stoichiometric CeO_2(111) [29] and further surface reduction when reduced CeO_{2-x}(111) was exposed to water with a concomitant presence of hydroxyl groups [30, 31] and are also in agreement with theoretical studies based on density functional calculations for the stoichiometric CeO_2(111) and reduced CeO_{2-x}(111) surfaces indicating that water does not dissociate on the clean surface, whereas the process becomes thermodynamically favorable on the oxygen vacancies containing surface [32–34].

From the discussion above, one can readily see that comparison between the experiments for the inversed catalyst models and the surface science systems coincides in evidencing the important role of the oxygen vacancies on the catalyzed dissociation of water. However, one must also realize that the ceria nanoparticles supported on Au(111) in the inverse catalysts necessarily posses a large number of edge-like sites which are not present in either the CeO_2(111) or the CeO_{2-x}(111) surfaces. It is reasonable to argue that the presence of low-coordinated sites will somehow influence the reactivity of these systems toward water dissociation. This is especially the case since it has been suggested that the presence of step edges can lead to the appearance of Ce^{3+} centers even without the presence of oxygen vacancies [35].

The interaction of water with ceria surfaces has been the object of several theoretical studies although all consider the CeO_2(111) surface only. Thus, Fronzi et al. [32] considered water adsorption on stoichiometric and reduced CeO_2(111) surfaces using the standard PBE, pure GGA functional, which is adequate for the stoichiometric surface but questionable for the reduced one. They found that the most stable configuration for water is when the O atoms is bonded directly to a Ce surface cation and involving two H-bonds between the hydrogen atoms and the surface oxygen atoms. The adsorption energy reported by these authors for the stoichiometric surface is −0.49 eV. Clearly, the adsorption energy of water appears to be stronger when oxygen vacancies are present although this is not considered in the present work. These authors also find that water does not spontaneously dissociate on the clean stoichiometric surface, while on the surface with oxygen vacancies, this process becomes thermodynamically favorable. A smaller value of the adsorption energy (−0.35 eV) for water on the perfect CeO_2(111) stoichiometric surface was reported by Watkins et al. [36] but, as noticed by Fronzi et al. [32], this is because the equilibrium geometry configuration obtained by these authors is not the most stable one, the reason being the existence of only one hydrogen bond between the adsorbed molecule and the ceria surface. Hence, the reason behind this discrepancy can be attributed to the difficulty to locate the most stable adsorption. Note also that different choices of surface unit cell induce different lateral interactions among the adsorbates and to a strong dependence of the binding energy with respect to the coverage. More recently, Yang et al. [37, 38] studied the interaction of a water molecule with the (111) surfaces of stoichiometric and reduced ceria using DFT + *U*. For the stoichiometric surface, their results are similar to those of Fronzi et al. [32] and also coincide with the results of Kumar et al. [15] and also of Chen et al. [39] using PW91 and PW91 + *U*, respectively.

In order to investigate the role of low coordinated step sites in the dissociation of water catalyzed by ceria without interfering with possible effects derived from the size of the nanoparticles, such as their size dependence facility to promote oxygen vacancy formation [40], a series of density functional calculations have been carried out to establish the energy profile of water dissociation on CeO_2(111), CeO_2(221), CeO_2(331), and CeO_2(110) which have been found to be, in this order, the most stable surfaces [41]. We will show that the presence of low-coordinated sites has a moderate influence on the energy barrier for water dissociation except for the later which is found to be especially reactive.

2 Computational details

Self-consistent density functional theory (DFT) calculations using slab periodic models with large enough supercells have been carried out to study the adsorption and dissociation of H_2O on the regular CeO_2(111), (221), (331), and (110) surfaces. The calculations have been carried out using the PW91 [42, 43] form of the Generalized Gradient Approximation (GGA) corrected with the so called Hubbard parameter (*U*) [44]. The one-electron wave functions are expanded in a basis of periodic plane waves with a cut-off of 415 eV for the kinetic energy. The PAW method [45] in the implementation of Kresse and Joubert [46] was used to represent the effect of the inner cores on the valence density. The integration in the Brillouin zone was performed on a proper Monkhorst–Pack grid [47] of $5 \times 5 \times 1$ special k-points. The total energy tolerance defining self-consistency of the electron density was 10^{-4} eV. The structures of the system under study were optimized until the maximum forces acting on each atom became less than 0.01 eV/Å. All density functional calculations were carried out with the *Vienna* Ab Initio *Simulation Package* (VASP) [48–51].

The introduction and choice of the numerical value for the *U* parameter deserves a further comment. It penalizes the double occupation of *4f* orbital and thus allows for a proper description of reduced ceria [52–54]. In the present work, a value of $U = 4$ eV has been chosen on the basis of previous experience and several considerations. Note that this value is slightly larger than the value of 3 eV proposed [54] to achieve a sufficiently balanced description of both CeO_2 and Ce_2O_3 but closer to the value of $U = 5$ eV suggested by other authors [52, 53, 55]. In addition, $U = 4$ eV has been found to be necessary to properly describe localized solutions for Ce^{3+} cations in ceria nanoparticles [40, 56–59] and, also, largely facilitates convergence toward self-consistency of the Kohn–Sham equations. In the following, we will use the notation G4 to indicate DFT calculations carried out with the PW91 implementation of GGA and using an effective U value of 4 eV.

The perfect CeO_2(111), (221), (331), and (110) surfaces were represented by slabs models and proper unit cells. However, the size of the unit cell depends on the type of surface and also on the size and shape of the adsorbed molecule and the dissociation fragments. Thus, to model the stepped CeO_2(221) and (331) surfaces, it is necessary to employ slab models with larger number of layers. Therefore, 2 × 2, 4 × 2, 3 × 2, and 4 × 2 units cells with 3, 8, 6, 4 layers of CeO_2 units (three atomic layers each) have been used to represent the CeO_2(111), (221), (331), and (110) surfaces, respectively. Using this slab models, the minor distance between adsorbed water molecules is 7.7 Å. Therefore, we can consider that the initial adsorbed molecules and the final dissociated fragments on the surface do not interact. A vacuum width larger than 10 Å was used to avoid interaction between the periodically repeated slabs. A water molecule has been added to each surface mode and the resulting structure obtained from total energy minimization, a similar relaxation procedure has been used to obtain the structure of dissociated water molecule fragments (H and OH). In all cases, the outermost external layers of the CeO_2 were fully relaxed, i.e., 1, 2, 2, and 1 layers of CeO_2 units for CeO_2(111), (221), (331), and (110) surfaces, respectively. The transition state structures have been located using the climbing image nudged elastic band (CI-NEB) method of Henkelman et al. [60, 61]. The stable configuration of the adsorbed molecules and the TS structures was characterized by appropriate vibrational analysis.

Once the optimum geometries for the reactants and products have been obtained and the transition state structure characterized, calculations for a representative set, these structures have been also carried out allowing spin polarization. In all cases, the calculations converged to the non-spin polarized solution. Consequently, the analysis of results in the next section is based always on the results of non-spin polarized calculations.

3 Results and discussion

3.1 Water molecular adsorption on ceria surfaces

Here, we discuss the results for molecular water adsorption on the different surfaces studied in the present work. The adsorption energy is calculated as

$$E_{ads} = E(H_2O/CeO_2) - E(H_2O) - E(CeO_2) \tag{1}$$

where negative energies indicate an exothermic process. For the stoichiometric CeO_2(111) surface, the present results coincide almost quantitatively with those of Fronzi et al. [32]. Thus, the present estimate for E_{ads} from G4 calculations differs only 0.03 eV (Table 1) from those reported Fronzi et al. [32] using PBE, which is a pure GGA functional and is within 0.01 eV of those reported by Chen et al. [39] using a similar (PW91 + U; $U = 6.3$) approach. The agreement between the different sets of calculations suggests that, in fact, the introduction of the U term is not necessary as far as no oxygen vacancies (i.e., Ce^{3+} cations) are present.

From the preceding discussion, it appears that the description of the interaction of water with the stoichiometric CeO_2(111) surface provided by different density functional approaches and slightly different surface models is almost the same excepting perhaps the results of Watkins et al. [36] although the reason for the discrepancy is also understood (see above). In addition, the calculated value of the adsorption energy compares well with experimental estimate of 0.53 eV for 0.2 ML coverage arising from thermal programmed desorption data for CeO_2 thin films grown on ytria stabilized zirconia [62], and it is not far from the 0.61 eV value obtained from CeO_2 powders [63]. Therefore, one may expect a similarly accurate prediction

Table 1 Relevant structural data and adsorption energy (E_{ads}) of H_2O on the ceria (111), (221), and (331) surfaces and note that no stable molecular adsorption state has been found for water on the (110)

Surface	d_{O-Ce} (Å)	d_{O-H} (Å)	$d_{O(sup)-H}$ (Å)	∠HOH (°)	E_{ads} (eV)
(111)	2.62	0.99/0.98	2.03/2.10	107.5	−0.52
(221)	2.68	0.99/0.98	2.06/2.18	107.8	−0.56
(331)	2.64	0.99/0.98	1.98/2.33	106.8	−0.72

for the rest of ceria surfaces considered in the present work. Results for the (221) and (331) in Table 1 indicate that the adsorption energy of water does only moderately depend on the crystal face. For the (221) surface, it is only slightly larger than for the most stable (111) surface, and going to the (331) surface does only increase E_{ads} up to 0.72 eV which, interestingly enough, is close to the thermodynamic heat of adsorption measured for CeO_2 powders. The closeness between the calculated values for the adsorption energy of water on these three surfaces strongly suggests that the bonding mode is very similar. This is indeed the case as one can see from the structural data reported in Table 1 and the corresponding structures in Fig. 1. Note that the structure of adsorbed water is almost the same in the three surfaces and that the only difference is the slight difference in the distance from the O atom to the Ce surface atom.

The case of the $CeO_2(110)$ surface deserves a separate discussion since all attempts to locate a stable molecularly adsorbed state have been unsuccessful. In fact, all geometry optimization calculations converged to a situation where the water molecule is spontaneously dissociated into OH and H (see next section). This result seems to be rather surprising since there is no experimental evidence of water dissociation on ceria unless oxygen vacancies are present. However, one may argue that this is because ceria samples used in the experiments have predominantly (111) facets which are rather unreactive. A different situation may occur if samples with other more reactive crystal planes can be synthesized. This has been precisely the approach followed by Zhou et al. [64]. These authors have used a solution-based hydrothermal method to obtain single crystalline CeO_2 nanorods exhibiting dominantly (100) and (110) surfaces. These authors show that these nanorods are more reactive for CO oxidation that ceria nanoparticles with more stable terminations. In the view of the present results, one may speculate that these types of crystals are active toward water dissociation. However, it is also likely that the resulting systems will not lead to good catalysts because the surface, even initially active toward water dissociation, will become rapidly deactivated because of the extremely strong binding of the reaction products as it will be further commented in the next section. The resulting surface will become fully hydroxylated and chemically inactive. Note that the catalytic activity of these special ceria nanorods obtained from wet chemistry requires the precipitate to be filtrated, washed with deionized water, dried at 60 °C for 24 h, and then calcined at 350 °C for 4 h [64]. It is also worth mentioning that very recently Yang et al. succeeded in synthesizing ceria nanoparticles exposing either (111) or (100) faces [65].

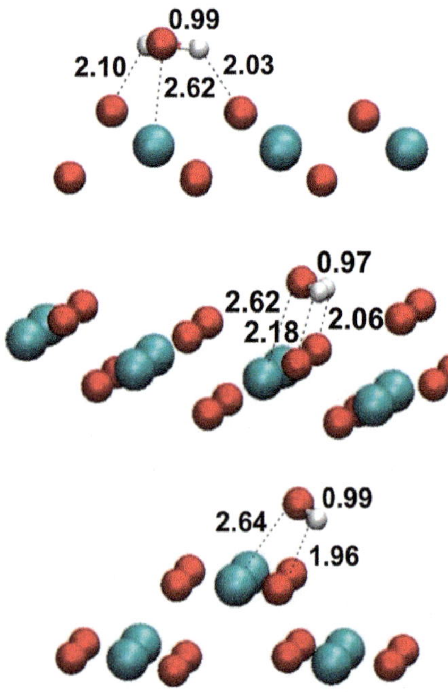

Fig. 1 Schematic representation of the structure of the most stable configuration of water adsorbed on the (111), (221), and (331) ceria surfaces (*top*, *middle* and *bottom panels*, respectively). The interatomic distances are expressed on Å

3.2 Water dissociation on ceria surfaces

In order to discuss the molecular mechanism of water dissociation on the different well-defined surfaces of stoichiometric ceria considered in the present work, we have first determined the most stable sites for the coadsorption of H and OH (see Fig. 2) and used these and the structure of adsorbed water to run the CI-NEB calculations and thus locate the corresponding transition state (TS) structures. From these calculations, it has been possible to construct the energy profiles. The relevant energy data were reported in Table 2. In particular, we focus on the activation (E_{act}) and reaction (E_{reac}) energy which are defined as in Eqs. 2 and 3

$$E_{act} = E(H_2O/CeO_2)_{TS} - E(H_2O/CeO_2) \quad (2)$$

$$E_{reac} = E(H + OH/CeO_2) - E(H_2O/CeO_2) \quad (3)$$

with this definition positive activation energy values indicate an energy barrier and negative reaction energy values indicate that the process is exothermic. Finally, we consider the adsorption energy of the dissociation products (E'_{ads}) as

$$E'_{ads} = E(H + OH/CeO_2) - E(H_2O) - E(CeO_2). \quad (4)$$

Let us start by considering the most stable, and also more studied, $CeO_2(111)$ surface. The reaction is predicted to be slightly endothermic (Table 2) and, hence, the reaction products, H and OH adsorbed on top of a surface

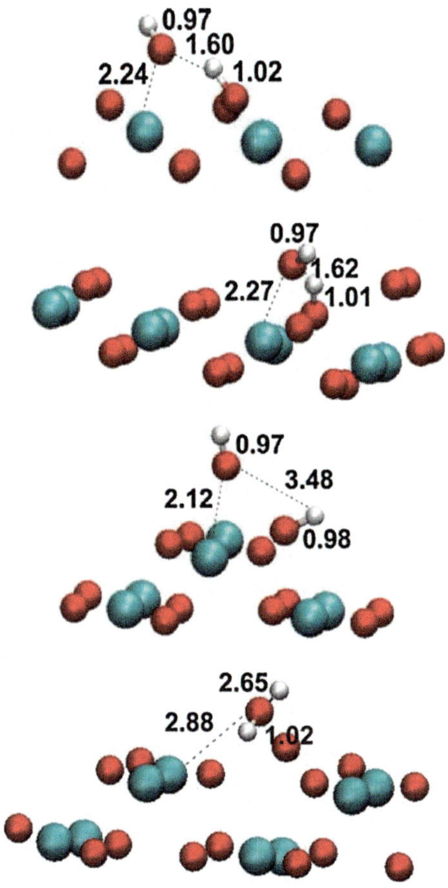

Fig. 2 Schematic representation of the structure of the dissociated water adsorbed on the (111), (221), (331), and (110) ceria surfaces from *top* to *bottom*, respectively. The interatomic distances are expressed on Å

Table 2 Activation (E_{act}), reaction (E_{reac}), and H + OH coadsorption (E'_{ads}) energy (in eV) for water dissociation on several ceria surfaces

Surface	E_{act}	E_{reac}	E'_{ads}
(111)	+0.19	+0.14	−0.38
(221)	+0.31	+0.27	−0.29
(331)	+0.21	−0.14	−0.86
(110)	–	–	−4.57

oxygen and of a surface ceria, respectively, are to some extent less stable than adsorbed water. On this surface, water dissociation faces a moderate energy barrier of 0.19 eV which is significantly lower than the desorption energy indicating that the process could take place. However, the presence of the energy barrier requires an extra energy cost. Consequently, water dissociation will not be facilitated, in agreement with the experimental observation that water does not spontaneously dissociate on this surface unless oxygen vacancies are present [23, 28]. From the data on Table 2, one can also see that even if the energy barrier to dissociate water on this surface can be overcome at a moderately high temperature, the inverse reaction will be faster orders of magnitude since the corresponding energy barrier is of 0.05 eV only. Note that the energy barrier for water dissociation on this stoichiometric surface is noticeably smaller than the one reported by Fronzi et al. [32] for the reduced $CeO_{2-x}(111)$ surface. These authors report a value larger than 2 eV for reduced which is against the experimental evidence that dissociation of water on this surface. Note also that for the stoichiometric surface, the present values for E_{ads} and E_{reac} are almost identical to those reported by Fronzi et al. which leads us to conclude that the algorithm used by these authors to locate the TS converged to a wrong structure. Finally, one must realize that the activation energy reported by these authors is significantly larger than the desorption energy (1.28 eV) reported in the same work.

Let us now focus on the stepped surfaces which allow us to investigate in detail the role of the low-coordinated edge site. For the $CeO_2(221)$ surface, the situation is similar to the (111) surface and it even appears to be less reactive. Thus, the energy barrier is slightly larger but, again smaller than water adsorption energy on this surface, and the reaction is also slightly more endothermic. In this case, the presence of step sites does not largely changes the reactivity, probably because the presence of step edges does only destabilize the surface to a rather small extent, and the atomic structure of adsorbed water, coadsorbed H and OH and transition state for water dissociation are also very similar indicating also that the chemical bond between these two ceria surfaces and these adsorbates has a strong local character. Next, we consider the $CeO_2(331)$ surface which is only moderately less stable than the previous one. However, the rather subtle change in atomic structure and surface stability has a more pronounced effect on the molecular mechanism for water dissociation. In fact, the reaction is now predicted to be fairly exothermic which is consistent with the larger adsorption of molecular water. In any case, the additional thermodynamic driving force for the dissociation reaction does not appear to be enough to facilitate it. In fact, the calculated activation energy is only 0.02 eV larger than the value predicted for the most stable $CeO_2(111)$ surface. However, the fact that the reaction is now exothermic has implications for the reverse step. Now, assuming that the system receives (e.g., by heating) the energy necessary to overcome the barrier for water dissociation, the reaction will proceed because the reverse step has now to surmount an energy barrier of 0.35 eV. Therefore, one can conclude that this type of stepped stoichiometric surface has the potential to act as a rather good catalyst for water dissociation, for instance in the WGS reaction. There is, of course, the difficulty to prepare it but the recent advances in synthetic methods already allow to grow ceria nanoparticles with preferential (111) or (100) facets [65]. It is then likely than nanoparticles with (331) could be also obtained.

Finally, we consider the case of the $CeO_2(110)$ surface where all attempts to locate a molecularly adsorbed state for water have failed and lead to the situation where the molecule spontaneously splits. The resulting products, H and OH, are strongly bound to the surface with an adsorption energy of 4.57 eV (Table 2) which indicates that once the water molecule is broken, a very stable hydroxylated surface will be formed which from now on will be chemically inactive. This high surface activity could be attributed to the specific position that O and Ce atoms occupy on this surface. In fact, in this case, both Ce and O surface atoms are located at the same outermost atomic layer (see Fig. 1 of [41]). This allows the simultaneous formation of strong Ce_{sup}–OH and O_{sup}–H bonds and thus facilitates water dissociation. This interpretation is consistent with the fact that the topmost atomic layer of the least active (111) surface exposes only oxygen atoms favoring the formation of O_{sup}–H bonds but difficulting the interaction of OH species with the surface. On the other hand, (221) and (331) surfaces show also oxygen atoms on the top layer but here atoms at the step edges provide the necessary Ce and O surface sites.

4 Conclusions

The adsorption and dissociation of water on the four most stable surfaces of stoichiometric ceria has been studied by means of periodic density functional theory using slab models. The analysis of the energy profile for the corresponding molecular mechanism allows us to extract important conclusions about the role of step sites in this important chemical reaction. In particular, present values for the stoichiometric surfaces provide a valuable reference for further modeling of reduced surfaces where experiment indicate that the process occurs spontaneously and, hence, necessarily with energy barriers smaller than those corresponding to the stoichiometric surfaces studied in the present work.

The presence of step sites does only moderately affect the adsorption energy of the water molecule which varies from −0.52 eV for $CeO_2(111)$ to −0.72 eV for $CeO_2(331)$. However, molecular water does not seem to be stable on the CeO2(110) surface where spontaneous dissociation is predicted. In agreement with the results for water adsorption energy, the energy barrier for water dissociation exhibits little variation with the crystal face. The calculated values for the energy barrier are roughly of 0.2 eV, and hence, one can argue that it may be easily surmounted by moderate heating of the system. However, for the (111) and (221) surfaces, the reaction is predicted to be endothermic, and consequently, the energy barrier for the recombination of adsorbed H and OH is smaller than that corresponding to water dissociation, and consequently, even if energy is provided to the system so that water molecules could dissociate, recombination will be much faster making the process completely inefficient from a catalytic point of view. A different situation occurs on the $CeO_2(331)$ surface. Here, the reaction is moderately exothermic; the energy barrier for water dissociation is similar to the one calculated for the other two surfaces but the recombination step faces now a higher energy barrier. Consequently, the stoichiometric $CeO_2(331)$ surface could act as a reasonable good catalyst for water dissociation. Finally, no stable molecular state has been found for water on $CeO_2(110)$. Nevertheless, this surface cannot be considered as catalytically active since the reaction products lead to a very stable hydroxylated surface which will rapidly become inactive.

The prediction that $CeO_2(331)$ can result in a catalytically active surface is appealing although at present the interest is mainly academic. Recent advances in the synthesis of ceria nanoparticles or nanorods with defined crystal faces [63, 64] will perhaps allow to experimentally verify this prediction and may open the way toward a new family of designed catalysts.

Acknowledgments Financial support by the Spanish MICINN (grants FIS2008-02238, CTQ2007-30547-E/BQU, CTQ2009-07647/BQU), *Generalitat de Catalunya* (grants 2009SGR1041 and XRQTC) is gratefully acknowledged. S.F. and M.M.B acknowledge support from Argentinean CONICET and F.I. acknowledges additional support through the ICREA Academia award for excellence in research. Computational time has been generously provide by the Barcelona Supercomputing Center.

References

1. Ladebeck JR, Wagner JP (2003) Handbook of fuel cells—fundamentals, technology and applications (ISBN: 0-471-49926-9). Vielstich W, Lamm A, Gasteiger HA (eds), vol 3, Part 2: 190–201. Wiley, Chichester
2. Lee SHD, Applegate DV, Ahmed S, Calderone SG, Harvey TL (2005) Int J Hydrogen Energy 30:829–842
3. Rozovskii AY, Lin GI (2003) Top Catal 22:137–150
4. LarrubiaVargas MA, Busca G, Costantino U, Marmottini F, Montanari T, Patrono P, Pinzari F, Ramis V (2007) J Mol Catal A Chem 266:188–197
5. Newsome DS (1980) Catal Rev Sci Eng 21:275–315
6. Schumacher N, Boisen A, Dahl S, Gokhale AA, Kandoi S, Grabow LC, Dumesic JA, Mavrikakis M, Chorkendorff I (2005) J Catal 229:265–275
7. Burch R (2006) Phys Chem Chem Phys 8:5483–5500
8. Mendes D, Garcia H, Silva VB, Mendes A, Madeira LM (2009) Ind Eng Chem Res 48:430–439
9. Rodríguez JÁ, Evans J, Graciani J-B, Park J, Liu P, Hrbek J, Sanz JJF (2009) Phys Chem C 113:7364–7370

10. Li L, Zhan Y, Zheng Q, Zheng Y, Lin X, Li D, Zhu J (2007) Catal Lett 118:91–97
11. Boccuzzi F, Chiorino A, Manzoli M, Andreeva D, Tabakova T (1999) J Catal 188:176–185
12. Liu Z-P, Jenkins SJ, King DA (2005) Phys Rev Lett 94:196102 (1–7)
13. Yahiro H, Murawaki K, Saiki K, Yamamoto T, Yamaura H (2007) Catal Today 126:436–440
14. Rodríguez JA, Liu P, Hrbek J, Pérez M, Evans J (2008) J Mol Catal A: Chem 281:59–65
15. Kumar P, Idem R (2007) Energy Fuels 21:522–529
16. Rodriguez JA, Liu P, Wang X, Wen W, Hanson J, Hrbek J, Pérez M, Evans J (2009) Catal Today 143:45–50
17. Wang X, Rodríguez JA, Hanson JC, Gamarra D, Martínez-Arias A, Fernández-García MJ (2006) Phys Chem B 110:428–434
18. Chen Y, Cheng J, Hua P, Wang H (2008) Surf Sci 602:2828–2834
19. Zhang L, Wang X, Millet J-MM, Matter PH, Ozkan US (2008) App Catal A Gen 351:1–8
20. Du X, Yuana Z, Cao L, Zhanga C, Wanga S (2008) Fuel Processing Technol 89:131–141
21. Nishida K, Atake I, Li D, Shishido T, Oumi Y, Sano T, Takehira K (2008) App Catal A Gen 337:48–57
22. Knudsen J, Nilekar AU, Vang RT, Schnadt J, Kunkes EL, Dumesic JA, Mavrikakis M, Besenbacher FJ (2007) Am Chem Soc 129:6485–6490
23. Zhao X, Ma S, Hrbek J, Rodriguez JA (2007) Surf Sci 601:2445–2452
24. Liu P, Rodriguez JA (2007) J Chem Phys 126:164705–164712
25. Gokhale AA, Dumesic JA, Mavrikakis M (2008) J Am Chem Soc 130:1402–1414
26. Madon RJ, Braden D, Kandoi S, Nagel P, Mavrikakis M, Dumesic JA (2011) J Catal 281:1–11
27. Fajin JLC, Cordeiro MNDS, Illas F, Gomes JRB (2009) J Catal 268:131–141
28. Rodriguez JA, Ma S, Liu P, Hrbek J, Evans J, Pérez M (2007) Science 318:1757–1760
29. Gritschneder S, Iwasawa Y, Reichling M (2007) Nanotechnology 18:044025–044030
30. Henderson MA, Perkins CL, Engelhard MH, Thevuthasan S, Peden CHF (2003) Surf Sci 526:1–18
31. Berner U, Schierbaum K, Jones G, Wincott P, Haq S, Thornton G (2000) Surf Sci 467:201–213
32. Fronzi M, Piccinin S, Delley B, Traversa E, Stampfl C (2009) Phys Chem Chem Phys 11:9188–9199
33. Yang Z, Wang Q, Wei S, Ma D, Sun Q (2010) J Phys Chem C 114:14891–14899
34. Yang Z, Xie L, Ma D, Wang G (2011) J Phys Chem C 115:6730–6740
35. Branda MM, Loschen C, Neyman KM, Illas F (2008) J Phys Chem C 112:17643–17651
36. Watkins MB, Foster AS, Shluger AL (2007) J Phys Chem C 111:15337–15341
37. Yang Z, Wang Q, Wei S, Ma D, Sun Q (2010) J Phys Chem C 114:14891–14899
38. Yang Z, Xie L, Ma D, Wang GJ (2011) Phys Chem C 115:6730–6740
39. Chen H-T, Choi YM, Liu M, Lin MC (2007) Chem Phys Chem 8:849–855
40. Migani A, Vayssilov GN, Bromley ST, Illas F, Neyman KM (2010) Chem Commun 46:5936–5938
41. Branda MM, Ferullo RM, Causà M, Illas F (2011) J Phys Chem C 115:3716–3721
42. Perdew JP, Chevary JA, Vosko SH, Jackson KA, Pederson MR, Singh DJ, Fiolhais C (1992) Phys Rev B 46:6671–6687
43. Perdew JP, Chevary JA, Vosko SH, Jackson KA, Pederson MR, Singh DJ, Fiolhais C (1993) Phys Rev B 48:4978
44. Dudarev SL, Botton GA, Savrasov SY, Humphreys CJ, Sutton AP (1998) Phys Rev B 57:1505–1509
45. Blöchl PE (1994) Phys Rev B 50:17953–17979
46. Kresse G, Joubert D (1999) Phys Rev B9(59):1758–1775
47. Monkhorst HJ, Pack JD (1976) Phys Rev B 13:5188–5192
48. Kresse G, Hafner J (1993) Phys Rev B 47:558–561
49. Kresse G, Hafner J (1993) Phys Rev B 48:13115–13118
50. Kresse G, Hafner J (1994) Phys Rev B 49:14251–14269
51. Kresse G, Furthmuller J (1996) Phys Rev B 54:11169–11186
52. Nolan M, Grigoleit S, Sayle DC, Parker SC, Watson GW (2005) Surf Sci 576:217–229
53. Fabris S, de Gironcoli S, Baroni S, Vicario G, Balducci G (2005) Phys Rev B 71:041102 (1–4)
54. Loschen C, Carrasco J, Neyman KM, Illas F (2007) Phys Rev B 75:035115 (1–8)
55. Lu Z, Yang Z (2010) J Phys: Condens Matter 22:475003 (1–10)
56. Loschen C, Migani A, Bromley ST, Illas F, Neyman KM (2008) Phys Chem Chem Phys 10:5730–5738
57. Migani A, Loschen C, Illas F, Neyman KM (2008) Chem Phys Lett 465:106–109
58. Migani A, Neyman KM, Illas F, Bromley ST (2009) J Chem Phys 131:64701 (1–7)
59. Migani A, Vayssilov GN, Bromley ST, Illas F, Neyman KM (2010) J Mater Chem 20:10535–10546
60. Henkelman G, Uberuaga BP, Jonsson H (2000) J Chem Phys 113:9901–9904
61. Henkelman G, Jonsson H (2000) J Chem Phys 113:9978–9985
62. Henderson MA, Perkins CL, Engelhard MH, Thevuthasan S, Peden CHF (2003) Surf Sci 526:1–18
63. Prin M, Pijolat M, Soustelle M, Touret O (1991) Thermochim Acta 186:273–283
64. Zhou K, Wang X, Sun X, Peng Q, Li Y (2005) J Catal 229:206–212
65. Yang F, Choi Y, Agnoli S, Liu P, Stacchiola DJ, Hrbek J (2011) Rodriguez J A J Phys Chem C 115:23062–23066

Theor Chem Acc (2012) 131:1232
DOI 10.1007/s00214-012-1232-9

REGULAR ARTICLE

Proton exchange reactions of C2–C4 alkanes sorbed in ZSM-5 zeolite

Kanjarat Sukrat · Daniel Tunega · Adelia J. A. Aquino · Hans Lischka · Vudhichai Parasuk

Received: 3 March 2012 / Accepted: 2 May 2012 / Published online: 30 May 2012

Abstract An extensive theoretical study has been carried out to determine barriers for the proton exchange reactions of C2–C4 alkanes in ZSM-5. It was found that cluster size and cavity structure are very important for predicting this barrier. A decrement of up to 20 kcal/mol was observed when employing the periodic model instead of using the small cluster model. Effects of basis set quality and electron correlation to the activation energy are positive and in combination could contribute up to 8 kcal/mol. An extrapolation scheme for estimating the reaction barrier that takes into account effects of cluster size, basis set quality, and electron correlation has been proposed. The regioselectivity and the chain length were discussed.

Keywords ZSM-5 zeolite · Proton exchange · Density functional theory · Activation barrier

Dedicated to Professor Marco Antonio Chaer Nascimento and published as part of the special collection of articles celebrating his 65th birthday.

K. Sukrat · V. Parasuk
Department of Chemistry, Faculty of Science, Chulalongkorn University, Phyathai Rd., Pathumwan, Bangkok 10330, Thailand

D. Tunega · H. Lischka (✉)
Institute of Theoretical Chemistry, University of Vienna, Währingerstraße 17, 1090 Vienna, Austria
e-mail: hans.lischka@univie.ac.at

D. Tunega · A. J. A. Aquino
Institute of Soil Research, University of Natural Resources and Life Sciences, Peter-Jordan-Straße 82, 1190 Vienna, Austria

A. J. A. Aquino · H. Lischka
Department of Chemistry and Biochemistry, Texas Tech University, Lubbock, TX 79409-1061, USA

1 Introduction

Currently, the fluid catalytic cracking (FCC) represents the largest volume of catalysts used in oil refineries. From 2008 to 2011, the catalytic cracking consumption was forecast to increase from 919 to 998 million US dollar [1–3]. The catalytic cracking process involves 3 basic reactions such as proton exchange ($Z^-H_B^+ + C_nH_{2n+2} \rightarrow Z^-H^+ + C_nH_{2n+1}H_B$), C–C bond cleavage ($Z^-H_B^+ + C_nH_{2n+2} \rightarrow Z^-C_{(n-x)}H^+_{2(n-x)+1} + C_xH_{2x+1}H_B$), and dehydrogenation ($Z^-H_B^+ + C_nH_{2n+2} \rightarrow Z^-C_nH^+_{2n+1} + HH_B$). The competition between these reactions determines the ratio of petrochemical products. Thus, the knowledge and predictability of these reactions are valuable. Many experimental studies have been carried out for the proton exchange reactions of small and large alkanes with the Brønsted acid site of zeolites [4–10]. However, the mechanism of the proton exchange reaction is far from conclusive. Two mechanisms, direct [11] and mediated [12] proton exchange, have been proposed. The direct mechanism is a single-step mechanism involving a direct transfer of the Brønsted acidic proton (H_B^+) from a zeolitic framework to the alkane molecule. The mediated mechanism involves two proton transfer steps. In first step, H_B^+ transfers to an alkene molecule (representing an impurity in the alkane sample). The second step involves the transfer of hydride ion from a nearby alkane to the alkyl cation. Generally, there are 3 carbon positions in alkanes for the H-exchange reaction with the Brønsted proton of zeolites. These positions will be referred to as primary, secondary, and tertiary, respectively. It was also found that there exists the regioselectivity in the proton exchange reaction of small alkanes [10, 13]. For example, the proton exchange rate at the primary position of propane was reported to be 1.5 times faster than at the secondary position [7]. Theoretical studies play an important role in the elucidation of the mechanism and the explanatory of the regioselectivity of the proton exchange reaction of alkanes. Zimmerman et al. [14] have suggested for the density functional theory (DFT) calculations that using large clusters of zeolite framework combining with a proper dispersion functional gives results in a good agreement with experimental results. They also found that in the larger clusters, the lower of energy barrier is a result from the effects of the electrostatic stabilization in cracking transition-state structures (TSs) and the long-range interactions, whereas both effects are missed in the small cluster models. A large number of theoretical works have been carried out in this light [15–19]. However, due to the size of zeolites, small-size cluster model (3T or 5T models where T represents alumina or silica tetrahedron) has been employed in most of those studies. The small-size cluster model could not appropriately describe the effect of the zeolites' porous structure or cavity toward the reactivity or selectivity of the catalyst. Recently, Bučko et al. [20] have carried out calculations on periodic models using PW91 with plane wave basis set and molecular dynamics simulations for the proton exchange reaction of small alkanes in chabazite. They concluded that the entropic effect is an important driving force for the regioselectivity of the proton exchange reaction of small alkanes and that *iso*-butane (*i*-butane) proceeds through the mediated mechanism, while for other alkanes, the direct mechanism is preferred. However, their calculations were performed in chabazite, zeolite with significantly smaller cavities than in the commercial and widely used catalyst ZSM-5.

In this work, the proton exchange reactions of four small alkanes, that is, ethane, propane, *i*-butane, and normal-butane (*n*-butane), in ZSM-5 were studied by means of quantum chemical methods. Effects of the size of the cluster model, the quality of basis set, and the importance of the level of electron correlation were systematically investigated. The proton exchange barriers and the regioselectivity of these alkanes sorbed in ZSM-5 were addressed. The summary of types of proton exchange reactions of studied alkanes is given in Table 1.

2 Computational details

2.1 Models of ZSM-5 and optimizations

Coordinates of ZSM-5 were taken from crystallographic data of silicalite-1 [21], and then, the Si atom at the intersection between the straight and sinusoidal (zigzag) channel (T7) was replaced by the Al atom. This single substitution provides a Si/Al ratio of 95. An acidic proton associated with the substitution was also added to one of the O atoms adjacent to the Al to balance the negative charge. Five cluster models, 5T, 20T, 28T, 38T, and 96T, were employed. The 96T model contains the whole ZSM-5 unit cell. In the 5T model, oxygen atoms at its edge (Si–O bonds) were saturated with hydrogen atoms. In case of the 20T, 28T, 38T, and 96T models, oxygen atoms at their edge were replaced by H atoms. Directly from the crystal lattice, the silicate components were locally fixed and there is no change even after substitution with other parts. In addition, the periodic model denoted as P was also considered. The periodic model was created from the unit cell of ZSM-5 obtained from crystallographic data [21]. Illustrations of five cluster models (5T, 20T, 28T, 38T, and 96T) and the periodic model are given in Fig. 1.

Table 1 Types of proton exchange reactions of C2–C4 alkanes, with notation for further reference given in parenthesis

Alkanes	Types of proton exchange reaction
Ethane	Primary (ethane)
Propane	Primary (propane/1); secondary (propane/2)
Iso-butane	Primary (*i*-butane/1); tertiary (*i*-butane/3)
Normal-butane	Primary (*n*-butane/1); secondary (*n*-butane/2)

For the 5T, 20T, 28T, and 38T clusters, geometry optimizations of alkanes, isolated clusters, reactant/product complexes (Rcx), and TS were performed using two approaches—RI-PBE/SVP (the Turburmole program's [22] implementation of the resolution of identity integral approximation (RI) [23] in DFT with the Perder–Burke–Ernzerhof functional [24, 25] and the SVP basis set [26–28]) and PBE/DNP [29–31] (Dmol3 program [21] implementation of PBE functional and the double numerical (DN) basis set plus polarization), respectively. Partial geometry optimization was performed by relaxing the acid site and O–H terminals for the 5T cluster model, while for the 20T, 28T, and 38T clusters, Si–H terminals were fixed. For both reactant complexes and transition-state structures, only positions of atoms involved in the reaction were optimized, while positions of atoms in the zeolite skeleton were fixed. For 96T and P models, optimized structures of Rcx with alumina tetrahedron were acquired in a similar manner to that of the smaller cluster but using PBE/DNP. While the structures of TS were obtained by performing single-point calculations on structures generated by embedding TS structures of 38T to 96T and P models. Frequency calculations were carried out on optimized structures of bare clusters, alkanes, and TS for 5T, 20T, 28T, and 38T using RI-PBE/SVP.

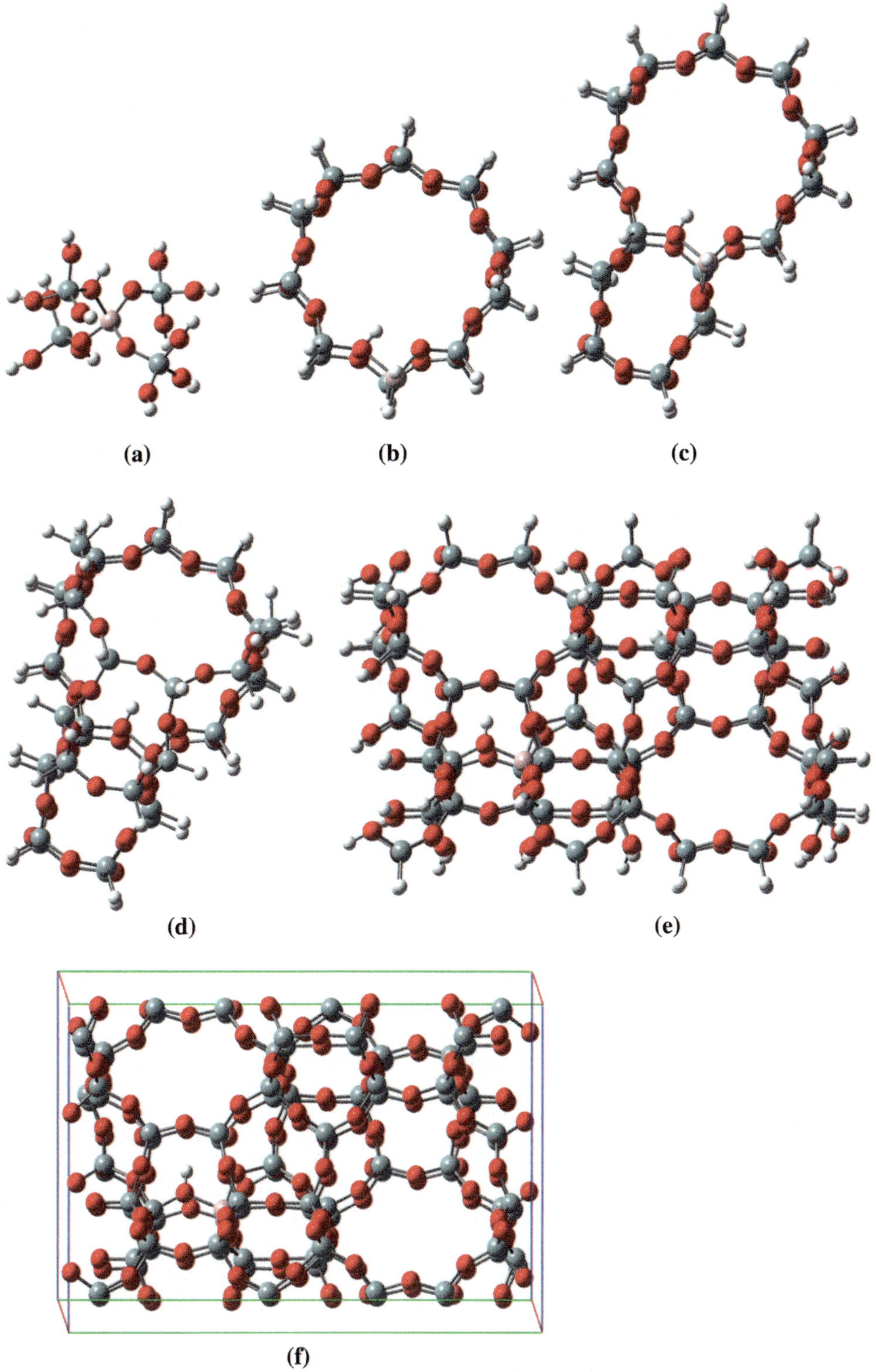

Fig. 1 Illustration of cluster models used in this study. **a** 5T, **b** 20T, **c** 28T, **d** 38T, **e** 96T; **f** periodic model

2.2 Adsorption energy and reaction barrier

Figure 2 described relations between adsorption energy (E_{ads}) and reaction barrier ($\Delta E^{\ddagger}$). The E_{ads} of C2–C4 alkanes in ZSM-5 were calculated according to

$$\text{ZSM-5} + C_nH_{2n+2} \rightarrow \text{ZSM-5} \cdots C_nH_{2n+2}.$$

Therefore,

$$E_{\mathrm{ads}} = E(\text{ZSM-5}) + E(\text{alkane}) - E(\text{Rcx}) \qquad (1)$$

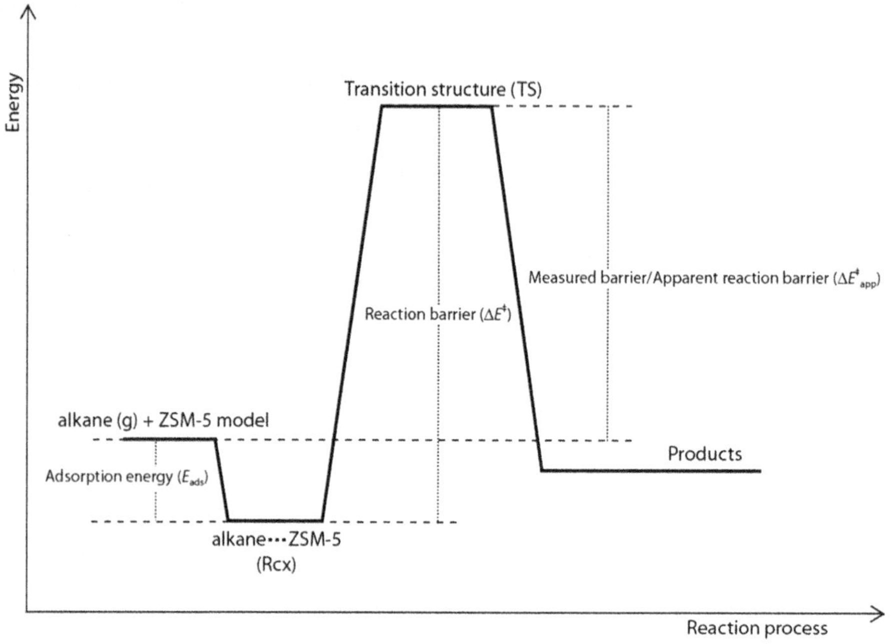

Fig. 2 Energy diagram for proton exchange reaction of alkanes C2–C4

where E(Rcx) is the energy of the reactant complex, E(ZSM-5) is the energy of the bare ZSM-5 structural model, and E(alkane) is the energy of the free alkane molecule.

The (intrinsic) reaction barrier ($\Delta E^{\ddagger}$) of the proton exchange reaction was calculated according to

$$\text{ZSM-5}\cdots C_nH_{2n+2} \rightarrow \text{ZSM-5}\cdots C_nH^{\ddagger}_{2n+2}$$

and thus

$$\Delta E^{\ddagger} = E(\text{TS}) - E(\text{Rcx}) \quad (2)$$

where E(TS) is the energy of the transition state.

The apparent reaction barrier ($\Delta E^{\ddagger}_{app}$), the value which is comparable to the measured activation energy, is therefore defined as

$$\Delta E^{\ddagger}_{app} = \Delta E^{\ddagger} - E_{ads} \quad (3)$$

Since E_{ads} according Eq. (1) is positive, $\Delta E^{\ddagger}_{app}$ is always smaller than $\Delta E^{\ddagger}$.

2.3 Complete basis set (CBS) extrapolation

To evaluate the effect of basis set, an extrapolation to the complete basis set (CBS) limit at MP2 and PBE levels of theory has been performed according to the scheme suggested by Helgaker et al. [32]

$$E_X = E_\infty + AX^{-\alpha} \quad (4)$$

where the cardinal number $X = 2$, 3, and 4 is corresponding to the SVP, TZVPPP, and QZVPP basis set, respectively. To obtain α and A, we carried out MP2 and PBE calculations for reactant complexes of ethane, propane, n-butane, and i-butane with the 5T model.

From our RI-PBE calculations, we found that various reactant complexes between alkanes and ZSM-5 have very similar value for α and the value of 6.56 gave the best fit. A similar observation could also be noticed for RI-MP2 calculations. In addition, there has been a report [23] that the α value depends mainly on method and basis set and it does not vary much for the molecular system with the same set of atoms. Thus, values of α from RI-PBE and RI-MP2 calculations were later used to determine CBS limit at 38T model through Eq. (5) proposed by Truhlar et al. [23].

$$E_\infty = \left(\frac{3^\alpha}{3^\alpha - 2^\alpha}E_3 - \frac{2^\alpha}{3^\alpha - 2^\alpha}E_2\right) \quad (5)$$

3 Results and discussions

3.1 Transition-state structures

The transition state (TS) of the proton exchange reaction of alkanes in ZSM-5 involves the formation of the pentacoordinated carbonium ion. Therefore, a single-step or direct mechanism is suggested. The example of the TS structure of the proton exchange reaction (5T model) is shown in Fig. 3.

Selected geometrical parameters for the TS structures of the 38T cluster model are provided in Table 2. Considering the TS structures for the reaction at the primary carbon (C_1) and C_1–H_B^+, C_1–H distances are 1.282 and 1.289 Å for ethane, 1.278 and 1.286 Å for propane, 1.285 and 1.282 Å for n-butane/1, and 1.283 and 1.271 Å for i-butane/1.

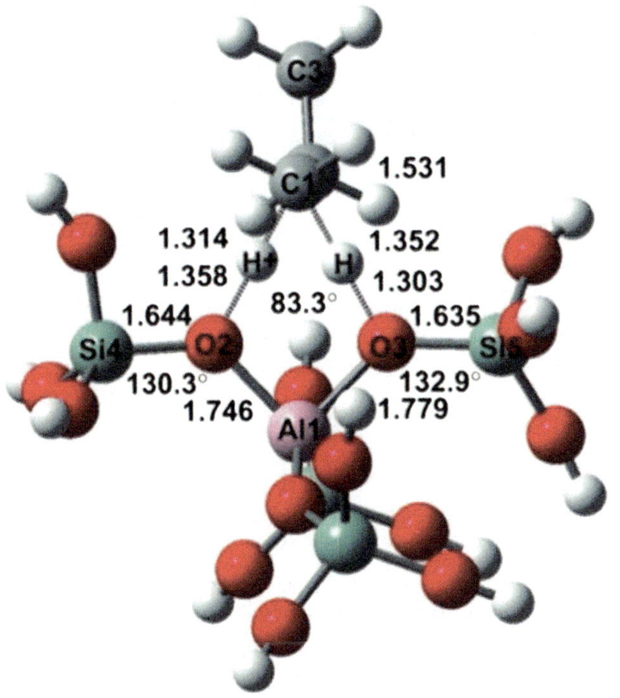

Fig. 3 The transition-state structure with selected geometrical parameters of the proton exchange reaction in propane/1

Apparently, the C1–H_B^+ and C1–H distances at the primary position are almost equal. For the proton exchange reaction on the secondary carbon (C_2), the C_2–H_B^+ distance in the transition state is shorter than C_2–H distance, that is, 1.251 and 1.338 Å for propane/2, and 1.249 and 1.331 Å for *n*-butane/2. For the reaction at the tertiary position (C_3) of *i*-butane/3, it is observed that C_3–H_B^+ is longer than the C_3–H distance, particularly 1.421 and 1.258 Å. The difference between C_x–H and C_x–H_B^+ distances is probably caused by the steric effect between the reacting center and neighboring methyl moieties. Also, Al_1–O_2 and Al_1–O_3 bond lengths and O_2–Al_1–O_3 bond angle were found to be dependent on the types of the reaction centers (primary, secondary, or tertiary). The transition states for reactions at primary (ethane/1, propane/1, *i*-butane/1, and *n*-butane/1) and secondary (propane/2 and *n*-butane/2) carbons have Al_1–O_2 shorter than Al_1–O_3 and O_2–Al_1–O_3 angle around 90°. However, the bond variance for the reaction at the secondary carbon is more evident. For the reaction with *i*-butane/3, Al_1–O_2 is longer than Al_1–O_3 bond length and O_2–Al_1–O_3 bond angle of 89.6°. All of these behaviors could be ascribed to the steric effect between the reacting center and the alkane molecule.

3.2 Adsorption energy

Adsorption of the alkane molecule to zeolites represents an important step of the proton exchange reaction and was extensively investigated. Using PBE/DNP, adsorption energies of the ethane, propane, *i*-butane, and *n*-butane with various cluster models were calculated and the results are displayed in Fig. 4.

A non-negligible dependency of adsorption energies on the cluster size is observed. As the size increases, computed adsorption energies increase for all alkanes studied. The adsorption energy as predicted using the periodic model was up to 3–7 kcal/mol larger than that using the 5T model. Even with the 38T model, adsorption energies are still underestimated. Only with the 96T cluster an acceptable difference of ~1 kcal/mol between the cluster and the periodic calculations is achieved. The dependency of calculated values versus basis sets (DNP, SVP, and CBS) and methods (PBE and MP2) are displayed in Table 3. The CBS values were obtained using formulas provided in Sect. 2.3. In addition, the dispersion correction of adsorption energies (E_{disp}) as proposed by Grimme et al. [33] was calculated for various cluster sizes. For the periodic model, the equation as suggested by Kerber et al. [34] was used. This correction was used to adjust PBE/DNP values referred to as PBE-D (PBE/DNP + E_{disp}), and their values are given in Table 3. MP2 adsorption energies are found to be consistently larger than the PBE values. Improvement of basis set decreases their values. Interestingly, PBE-D values show good agreement with values obtained at MP2 level for all cluster sizes. However, even with the E_{disp} adjustment to PBE/DNP at periodic model computed adsorption energies will be too large as compared to experimental heats of adsorption. With DFT-D, Zimmerman et al. [14] arrived with similar observation for the adsorption of propane on ZSM-5. Thus, the discrepancy

Table 2 Selected parameters for the transition states of the proton exchange reaction of all alkanes in clusters 38T

Alkanes	Al_1–O_2	Al_1–O_3	Si_4–O_2–Al_1	Si_5–O_3–Al_1	O_2–Al_1–O_3	O_2–H_B^+	C_{1-3}–H_B^+	O_3–H	C_{1-3}–H	Al_1–C_{1-3}
Ethane	1.744	1.759	130.4°	132.5°	89.9°	1.453	1.282	1.426	1.289	3.663
Propane/1	1.744	1.758	130.4°	133.4°	90.1°	1.456	1.278	1.419	1.286	3.651
Propane/2	1.734	1.766	131.2°	130.8°	90.0°	1.546	1.251	1.422	1.338	3.723
i-Butane/1	1.746	1.755	130.8°	134.3°	90.6°	1.421	1.283	1.442	1.271	3.624
i-Butane/3	1.755	1.739	127.2°	133.8°	89.6°	1.518	1.421	1.611	1.258	3.855
n-Butane/1	1.746	1.757	130.2°	133.1°	90.1°	1.447	1.285	1.444	1.282	3.644
n-Butane/2	1.734	1.765	131.2°	131.3°	90.1°	1.546	1.249	1.421	1.331	3.712

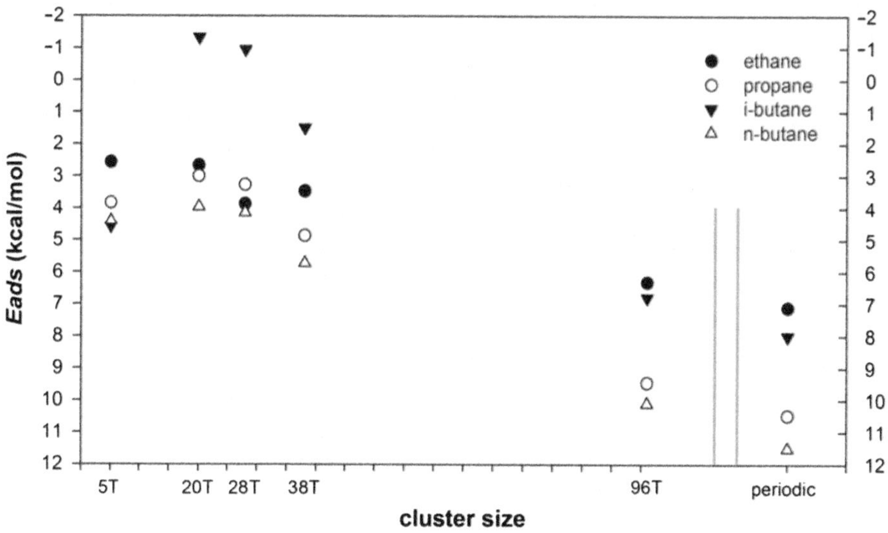

Fig. 4 Adsorption energy (E_{ads}) of C2–C4 alkanes in ZSM-5 as function of cluster size calculated using PBE/DNP

Table 3 Adsorption energies (E_{ads}) in kcal/mol of ethane, propane, *i*-butane, and *n*-butane on ZSM-5 computed using PBE/DNP, PBE/CBS, MP2/SVP, MP2/CBS, and PBE-D (PBE/DNP + E_{disp}) with 5T, 28T, 38T, and P in comparison with experimental enthalpies of adsorption

Alkanes	E_{ads}									Experiments
	PBE/DNP		PBE/CBS	MP2/SVP		MP2/CBS	PBE-D			
	38T	P	38T	5T	28T	5T	5T	28T	P	
Ethane	3.5	7.1	1.0	4.7	13.1	2.9	5.1	13.4	18.0	7.3 [36], 6.9 [37]
Propane	4.9	10.5	1.4	8.5	16.5	6.2	9.1	16.5	24.7	10.2 [4], 9.5 [38], 10.9 [39]
i-Butane	1.5	8.0	−3.2	8.8	18.6	6.6	10.0	18.4	26.7	11.6 [40], 12.4 [39]
n-Butane	5.7	11.5	1.1	9.7	19.2	7.4	10.5	19.1	29.8	14.7 [4], 11.9 [38], 14.3 [5]

should come from the overestimation of dispersion energy from the Grimme et al. [33] and Kerber et al. [34] formulae as well as from the MP2 calculations. De Moor et al. [35] using QM-Pot(MP2//B3LYP) have shown that adsorption enthalpies of *n*-alkanes (C2–C8) in H-FAU, H-BEA, H-MOR, and H-ZSM-5 at 341, 370, and 400 K are nearly independent of temperature and adsorption enthalpies at 373, 773, and 1139 K slightly depend on the temperature. In the subsequent discussion, we will use the experimental enthalpies of adsorption for calculations of apparent reaction barriers in the later section without worrying about their temperature dependency.

3.3 Proton exchange barrier

3.3.1 Cluster-size dependency

The effect of cluster size on reaction barriers of the proton exchange reactions discussed in this work is displayed in Fig. 5 where activation energies computed using PBE/DNP are plotted against the cluster size, that is, 5T, 20T, 28T, 38T, 96T, and P models. For all alkanes, proton exchange barriers are reduced by around 5 and 10 kcal/mol when expanding from 5T to 38T and from 38T to 96T, respectively. However, a reduction of less than 2 kcal/mol is observed when the model was further extended to periodic boundary conditions. It appears that performing calculations using the 38T cluster model is not sufficient to render the cluster-size effect, since the largest decrease of the predicted reaction barrier was observed when expanding from 38T to P model. The cluster-size effect becomes minimal when including all atoms in the unit cell (96T).

Table 4 displays theoretical (intrinsic) activation energies (in kcal/mol) for proton exchange reactions of C2–C4 alkanes in ZSM-5 compared with previous DFT calculations using the 3T cluster model [15–18]. Our PBE/DNP activation barriers for 5T clusters are in a relatively good agreement with the other theoretical results at 3T level. However, with the periodic model, activation energies for proton exchange reactions of C2–C4 alkanes in ZSM-5 are strongly reduced and range between 12.1 (ethane) and 28.4 (*i*-butane/3) kcal/mol. It is evident that predicted activation energies of all alkanes reduce drastically with the increment of the cluster size. The reduction is in the range of 10 (butane/1) to 20 (ethane) kcal/mol. This large reduction of activation energy has also been observed by Sauer et al. [41] who performed theoretical calculations for methylation reactions of alkanes in ZSM-5. Thus, the cluster-size

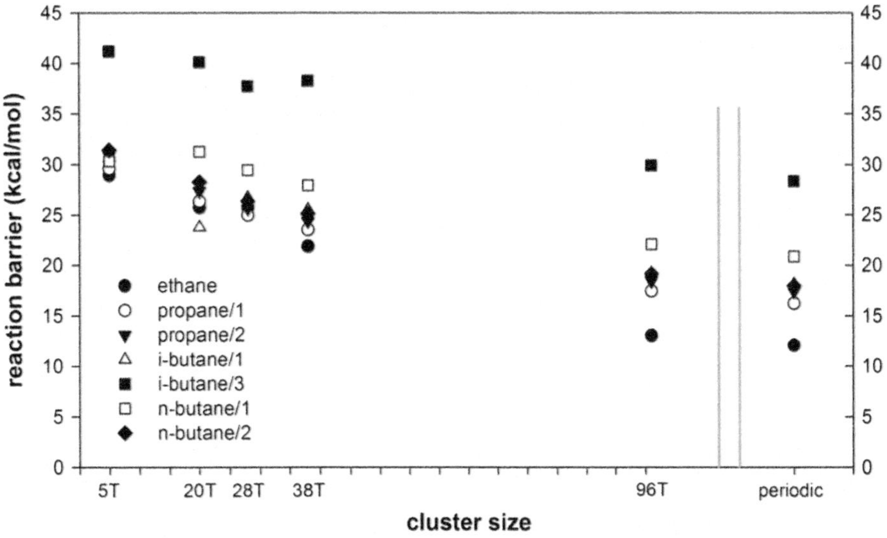

Fig. 5 Reaction barriers (intrinsic activation energies) of proton exchange reactions of C2–C4 alkanes in ZSM-5 calculated using PBE/DNP as a function of cluster size

Table 4 Intrinsic activation energies (in kcal/mol) of the proton exchange reactions for C2–C4 alkanes in ZSM-5 obtained using PBE/DNP for 5T, and P models in comparison with previous DFT calculations using the 3T model

Alkanes	5T	P	Previous DFT work (3T)
Ethane	28.9	12.1	28.2 [16], 32.3 [15], 31.0 [17]
Propane/1	29.6	16.3	32.2 [15], 30.4[17]
Propane/2	30.9	17.4	33.3 [15], 29.8 [17]
i-Butane/1	32.6	18.1	32.3 [15], 29.4[18]
i-Butane/3	41.2	28.4	36.2 [15], 29.9 [18]
n-Butane/1	30.3	20.9	29.9 [17]
n-Butane/2	31.4	18.0	28.3 [17]

effect is non-negligible and crucial. These reaction barriers at periodic model for the proton exchange reactions of C2–C4 alkanes in ZSM-5 are quite different from those in chabazite which were reported by Bučko et al. [20]. Unlike our predictions, proton exchange barriers in chabazite are larger and showed no linear dependency with the alkane chain length. There is the chain-length dependency with increment of ~4 kcal/mol. This is in agreement with experiments where the increments of 2–3 kcal/mol were reported [39, 42, 43]. The smaller cavity of chabazite as compared to ZSM-5 might be a reason for a difference in the chain-length dependency between 2 zeolites.

3.3.2 Basis set dependency

Table 5 lists reaction barriers for proton exchange reactions of C2–C4 alkanes for 5T and 38T cluster models calculated using RI-PBE with SVP and/or TZVPPP basis sets, respectively. The values at CBS limit were obtained using the procedure as described in Sect. 2.3. It appears that computed reaction barriers for all proton exchange reaction increase with the improvement of the basis set. For the 5T model, when changing from SVP to TZVPPP, the activation energies were raised between 1.9 and 4.6 kcal/mol. The increment from TZVPPP to CBS limit is smaller, only 0.1–0.4 kcal/mol. For the 38T model, computed reaction barriers increase between 2.5–3.6 (SVP to TZVPPP) and 0.2–0.3 kcal/mol (TZVPPP to CBS), respectively. It could be noticed that there is a smaller increase of the activation energy due to the improvement of the basis set for the larger 38T cluster. This is probably because of the more extended availability of functions in the larger cluster that makes the basis set more complete. Thus, for DFT calculations with PBE functional, the effect of basis set has a positive contribution to the obtained activation energies.

3.3.3 Effect of the level of electron correlation

Activation energies could be better estimated by incorporation of electron correlation. Generally, this could be achieved by performing MP2 calculations. However, the effect of the electron correlation is, as well, cluster-size dependent as pointed out by Sauer et al. [41]. Table 6 shows the reaction barriers of the proton exchange reaction of C2–C4 alkanes in ZSM-5 computed with the RI-MP2/SVP method for the 5T, 28T, and 38T models in comparison with previous MP2 calculations on 3T model. Our MP2 calculations gave activation energies in the same range as those reported by previous work. From Table 6, the cluster-size dependency as previously observed in PBE/DNP calculations (Fig. 5) is evident. The decrement of the activation energies from 2.9 (ethane) to 6.7 (*i*-butane/1) kcal/mol was obtained, and larger values of activation energies as compared to DFT calculations were reported.

Table 5 Reaction barriers (in kcal/mol) of proton exchange reactions of C2–C4 alkanes on 5T and 38T cluster model computed using RI-PBE with SVP and TZVPPP basis sets, and CBS limit

Alkanes	RI-PBE/SVP		RI-PBE/TZVPPP		RI-PBE/CBS limit	
	5T	38T	5T	38T	5T	38T
Ethane	24.4	20.2	28.4	22.8	28.7	23.0
Propane/1	26.9	22.1	29.2	24.6	29.4	24.8
Propane/2	27.4	22.6	30.5	25.3	30.8	25.5
i-Butane/1	26.6	23.0	30.2	26.6	30.4	26.9
i-Butane/3	36.1	35.5	40.7	39.1	41.1	39.3
n-Butane/1	28.0	26.0	29.9	29.1	30.0	29.3
n-Butane/2	28.3	23.5	31.1	26.3	31.3	26.5

To obtain accurate prediction of reaction barriers for the proton exchange reaction of C2–C4 in ZSM-5, it would be useful to perform MP2 calculations for the periodic model. However, such MP2 calculations are extremely costly and not feasible. From our calculations on adsorption energies, we found a good relation between dispersion energy computed using Grimme et al. [33] formula and the MP2–PBE energy difference. Therefore, dispersion interactions were adopted to adjust PBE activation energies. Table 7 shows the contribution of dispersion interactions to activation energy ($\Delta E^{\ddagger}_{\mathrm{disp}}$) for proton exchange reactions of C2–C4 alkanes in ZSM-5 as a function of cluster size in comparison with MP2–PBE activation energy differences for 5T, 28T, and 38T model. The MP2–PBE activation energy differences could be assigned to a different treatment of electron correlation.

As expected, $\Delta E^{\ddagger}_{\mathrm{disp}}$ are negative and cluster-size dependent for all alkanes. For all alkanes, $\Delta E^{\ddagger}_{\mathrm{disp}}$ for 38T and 96T clusters are appeared to be similar, which suggests the convergence of $\Delta E^{\ddagger}_{\mathrm{disp}}$ at 38T model. We found that $\Delta E^{\ddagger}_{\mathrm{disp}}$ at P remains the same as that at 96T. The dispersion correction seems to already converge at 96T. However, E_{disp} (of related species) at 96T and P are just around 3–4 kcal/mol different. In addition, the effect of dispersion interactions depends on the alkane molecule. The effect is larger (more negative) for larger alkane. The dispersion interactions are also governed by the shape of the TS complex. For example, the effect of dispersion interactions on the proton exchange barrier of *i*-butane is larger at the tertiary than the primary position.

Table 6 Reaction barriers (in kcal/mol) of proton exchange reactions of C2–C4 alkanes computed using RI-MP2/SVP on 5T, 28T, and 38T cluster models in comparison with previous MP2/6-31G**//HF/6-31G** calculations on 3T model

Alkanes	5T	28T	38T	Previous work (3T)
Ethane	32.5	32.4	29.6	31.4 [15]
Propane/1	35.8	30.5	29.2	30.6 [15]
Propane/2	34.8	32.3	29.1	30.6 [15]
i-Butane/1	34.8	32.5	28.1	29.8 [15]
i-Butane/3	42.9	42.3	39.2	31.5 [15]
n-Butane/1	37.0	29.2	31.2	–
n-Butane/2	35.8	29.2	30.0	–

MP2–PBE activation energy differences are positive and, similarly to $\Delta E^{\ddagger}_{\mathrm{disp}}$, also cluster-size dependent. However, the difference between MP2–PBE activation energy differences and $\Delta E^{\ddagger}_{\mathrm{disp}}$ does not depend on the cluster size. For example, the differences in kcal/mol are −7.8 at 5T, −7.4 at 28T, and −8.1 at 38T for ethane, and −7.4 at 5T, −7.3 at 28T, and −7.3 at 38T for *i*-butane/1. Knowing this, such constant difference or "E_{adj}" can be used to adjust $\Delta E^{\ddagger}_{\mathrm{disp}}$ for MP2–PBE activation energy differences at a larger cluster size in a following way

$$\left[\Delta E^{\ddagger}_{\mathrm{disp}}\right]_c = \left[\Delta E^{\ddagger}_{\mathrm{MP2}}\right]_c - \left[\Delta E^{\ddagger}_{\mathrm{DFT}}\right]_c + E_{\mathrm{adj}} \quad (6)$$

This is somehow in accordance with the suggestion by Sauer et al. [34, 41].

3.3.4 Zero-point-energy correction

The zero-point-energy (ZPE) corrections were performed using RI-PBE/SVP for the 5T, 20T, 28T, and 38T cluster models. Their values as a function of cluster size are illustrated in Fig. 6. According to Fig. 6, the ZPE correction has a negative contribution to the computed reaction barrier and the values vary with the cluster size. However, the size dependency seems to converge very fast. For the proton exchange reactions of C2–C4 alkanes in ZSM-5, the ZPE correction already converges at 28T model. ZPEs for 38T model are ranging between −1.6 and −2.5 kcal/mol.

3.3.5 Apparent reaction barriers

In previous sections, it has been shown that apart from the cluster size, other effects such as basis set and level of electron correlation are non-negligible. The augmentation of larger basis sets and electron correlation raises values of predicted proton exchange barriers. However, it is hard to

Table 7 RI-MP2/SVP and PBE/DNP activation energy differences in kcal/mol for proton exchange reaction of C2–C4 alkanes in ZSM-5 for 5T, 28T, and 38T cluster model and dispersion contributions ($\Delta E^{\ddagger}_{disp}$) in kcal/mol for 5T, 20T, 28T, 38T, 96T, and periodic models

Alkanes	$\Delta E^{\ddagger}$(RI-MP2/SVP) – $\Delta E^{\ddagger}$(PBE/DNP) 5T	$\Delta E^{\ddagger}$(RI-MP2/SVP) – $\Delta E^{\ddagger}$(PBE/DNP) 28T	$\Delta E^{\ddagger}$(RI-MP2/SVP) – $\Delta E^{\ddagger}$(PBE/DNP) 38T	$\Delta E^{\ddagger}_{disp}$ 5T	20T	28T	38T	96T	P
Ethane	3.6	6.7	7.7	−4.2	0.0	−0.7	−0.4	−0.4	−0.4
Propane/1	6.2	5.5	5.6	−1.7	−2.1	−1.9	−2.4	−2.1	−2.1
Propane/2	3.9	6.0	4.8	−3.1	−1.0	−0.3	−2.2	−2.1	−2.1
i-Butane/1	4.2	5.8	2.6	−3.2	2.9	−1.5	−4.7	−4.1	−4.1
i-Butane/3	1.7	4.6	0.9	−5.9	−2.8	−2.7	−6.0	−5.7	−5.7
n-Butane/1	6.7	−0.2	3.3	−1.6	−7.1	−6.8	−5.0	−4.6	−4.6
n-Butane/2	4.4	2.8	4.8	−3.0	−3.9	−3.6	−2.8	−2.6	−2.6

accommodate all of these effects without resourcing to enormous amounts of computation time. Here, an extrapolated scheme where effects of electron correlation and basis set could be addressed with modest computing efforts was proposed. For periodic model, the extrapolated reaction barrier, $\Delta E^{\ddagger}_{ex}$, is

$$\left[\Delta E^{\ddagger}_{ex}\right]_p = \left[\Delta E^{\ddagger}_{DFT}\right]_P + \left[\Delta E^{\ddagger}_{disp}\right]_{96T} + \left[\Delta E^{\ddagger}_{basis}\right]_{38T} - E_{adj}. \quad (7)$$

The $[\Delta E^{\neq}_{basis}]_{38T}$ values were obtained from $\Delta E^{\neq}$(38T, PBE/CBS) – $\Delta E^{\neq}$(38T, PBE/DNP). The proposed extrapolation scheme given by Eq. (7) is in the same light as it has been previously suggested by Sauer et al. [41] Extrapolated proton exchange barriers (intrinsic) are ranging from 20.9 (ethane) to 30.5 (*i*-butane/3) kcal/mol, and their values as well as the decomposition energies are shown in Table 8.

The apparent reaction barrier ($\Delta E^{\ddagger}_{app}$) was calculated from $\Delta E^{\ddagger}_{ex}$ plus the adsorption enthalpies from experiments given in Table 3. The apparent reaction barriers and the apparent reaction barrier plus ZPE correction for proton exchange reaction of C2–C4 alkanes in ZSM-5 are also given in Table 8. The ZPE correction was computed using RI-PBE/SVP for the 38T cluster model. $\Delta E^{\ddagger}_{app}$ + ZPE values range around 9–12 kcal/mol for all proton exchange reactions with exception of *i*-butane/3, which has the barrier of 17.4 kcal/mol. This different behavior of *i*-butane/3 is possibly due to its shape. While other alkanes possess a linear or quasi-linear shape, *i*-butane/3 has a globular-like structure. Apparent enthalpies, entropies, and Gibbs free energies of activation for the proton exchange reaction of C2–C4 alkanes in ZSM-5 at 500 K were calculated and are listed in Table 9.

Statistical mechanics models were applied for computations of translation, rotation, vibration (with vibration frequency from 38T model), and electronic contributions to enthalpy and entropy [44]. For the enthalpy, the electronic contribution ($\Delta H^{\ddagger}_{elec}$) was determined from apparent reaction barrier with ZPE correction. Entropies of activation are negative and have values between 40.6 (ethane) and 47.6 (*n*-butane/1) cal $K^{-1}mol^{-1}$ for the proton exchange reaction in ZSM-5. From the Arrhenius equation, the entropy term contributes to the prefactor [35] and, thus, it slows down the proton exchange rate. Apparent Gibbs energies of activation at 500 K are in the range of 30.3 (ethane) to 40.6 (*i*-butane/3) kcal/mol.

Computed $\Delta H^{\ddagger}$ values are compared in Table 9 with experimental results obtained from Arrhenius plots. In most cases, our best values, including extrapolation techniques are significantly too small. Interestingly, the computed barriers obtained from small cluster sizes agree quite well with experiment. However, as our investigations show, this good agreement is fortuitous. It should be noticed that the works of Hansen et al. [46] on alkylation of benzene and Tuma et al. [47] on deprotonation of a tert-butyl cation using similar computational scheme to ours showed good results compared with experiments. More investigations will be needed to explain this discrepancy.

3.4 Regioselectivity

Performing calculations in chabazite, Bučko et al. [20] suggested the entropy contribution to be crucial for describing the regioselectivity of the proton exchange reaction [13]. This is because in chabazite, the potential energy for proton exchange reactions is the same for all reactions. Unlike the report of Bučko et al. [20], we could observe the regioselectivity of the proton exchange reaction when consider either the entropy contribution or enthalpy. However, the direction is different. Considering only the entropy contribution, the proton exchange reaction for all alkanes proceeds faster at the primary position than at the secondary and tertiary positions. However, the enthalpies of activation predict the proton exchange reaction to proceed faster at the primary position for *i*-butane but at the secondary position for propane and *n*-butane. At

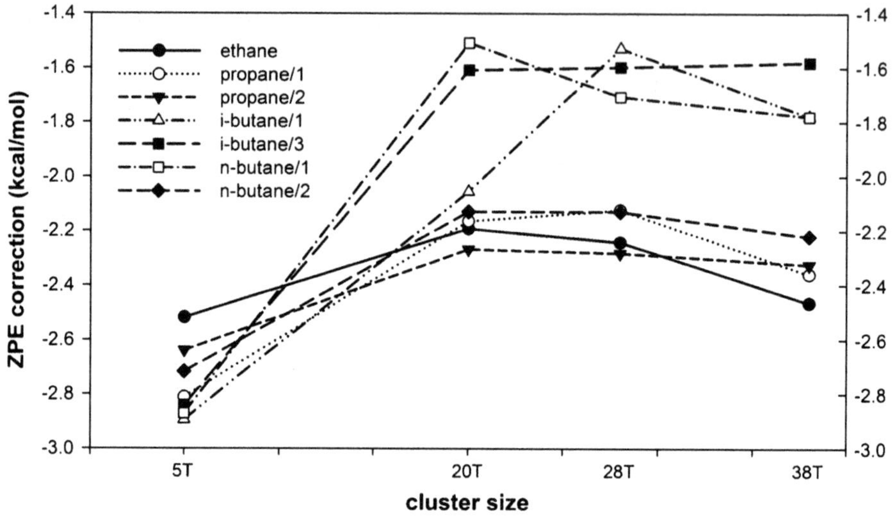

Fig. 6 ZPE corrections calculated using RI-PBE/SVP for reaction barriers of proton exchange reactions of C2–C4 alkanes in ZSM-5 as a function of cluster size

Table 8 Extrapolated intrinsic ($\Delta E^{\ddagger}_{\text{ex}}$) and apparent ($\Delta E^{\ddagger}_{\text{app}}$) reaction barriers of proton exchange reactions of C2–C4 alkanes in ZSM-5 as well as $\Delta E^{\ddagger}_{\text{disp}} - E_{\text{adj}}$, basis set ($\Delta E^{\ddagger}_{\text{basis}}$), and ZPE in comparison with experiments (all in kcal/mol)

	$\Delta E^{\ddagger}_{P}$	$\Delta E^{\ddagger}_{\text{disp}} - E_{\text{adj}}$	$\Delta E^{\ddagger}_{\text{basis}}$	$\Delta E^{\ddagger}_{\text{ex}}$	$\Delta E^{\ddagger}_{\text{app}}$	ZPE	$\Delta E^{\ddagger}_{\text{app}}$ + ZPE
Ethane	12.1	7.7	1.1	**20.9**	13.6	−2.5	**11.1**
Propane/1	16.3	5.8	1.3	**23.3**	13.1	−2.4	**10.8**
Propane/2	17.4	4.8	1.1	**23.3**	13.1	−2.3	**10.8**
i-Butane/1	18.1	3.2	1.4	**22.7**	11.1	−1.8	**9.3**
i-Butane/3	28.4	1.1	1.0	**30.5**	18.9	−1.6	**17.4**
n-Butane/1	20.9	3.6	1.4	**25.8**	13.9	−1.8	**12.1**
n-Butane/2	18.0	5.0	1.4	**24.4**	12.5	−2.2	**10.8**

Bold values indicate that they result from summation of the previous columns

Table 9 Apparent reaction enthalpies ($\Delta H^{\ddagger}$), entropy contributions ($T\Delta S^{\ddagger}$), and Gibbs free energies of activation ($\Delta G^{\ddagger}$) (in kcal/mol) for proton exchange reactions of C2–C4 alkanes in ZSM-5 at 500 K

	$\Delta H^{\ddagger}$	$-T\Delta S^{\ddagger}$	$\Delta G^{\ddagger}$	Experiments (Arrhenius barriers)
Ethane	12.3	18.0	30.3	
Propane/1	12.2	18.8	31.0	25.7 ± 1.6 [6], 25.5 ± 2.4 [7]
Propane/2	12.5	19.1	31.5	27.8 ± 1.6 [6], 29.7 ± 1.4 [7]
i-Butane/1	11.1	20.9	32.0	12.0 ± 0.5 [12], 13.6 [45]
i-Butane/3	19.8	20.8	40.6	
n-Butane/1	13.9	21.7	35.6	19.0 [4, 8], 20.2 [5], 27.5 ± 7.2 [9]
n-Butane/2	12.1	19.6	31.7	28.9 ± 8.4 [9]

higher temperature, the entropy term will become dominant and govern the direction of the Gibbs energy of activation. This does not mean that the entropy dictates the direction of regioselectivity. The activation Gibbs free energy for *i*-butane/3 proton exchange is almost 8 kcal/mol larger than *i*-butane/1, which suggests the proton exchange reaction at the tertiary position as impossible for *i*-butane. However, the entropy contribution for *i*-butatne/3 is the same as for *i*-butane/1. Therefore, Gibbs free energy of activation should be rather considered when addressing the regioselectivity. The data in Table 9 are the values at 500 K. To compare with experiments, we determined Gibbs free energies of activation for the proton exchange of propane in ZSM-5 at 553 K. For propane/1 and propane/2 reactions, Gibbs free energies of activation are 35.9 and 36.3 kcal/mol, respectively. Using transition-state theory [44], $k = \frac{k_B T}{h} e^{\Delta G^{\ddagger}/RT}$ where k is the rate constant, the proton transfer rate at the primary position is found to be 1.4 times faster than at the tertiary position in good agreement with experiments (1.5 times) [7]. Thus, the regioselectivity of the proton exchange reaction of alkanes in ZSM-5 is well predicted.

4 Conclusion

The TS structures for the reactions on the primary and secondary carbon have shorter or almost equivalent C_x–H_B^+ and C_1–H distances (~ 1.3 Å), while for the tertiary carbon, the C_x–H_B^+ distance is longer (1.4 Å). Steric effects are used to describe this behavior. The TS structures reflect the stability of the TS complex and hence the activation energies for the proton transfer of alkanes in ZSM-5. The effect of cluster size reduces the activation energy of proton exchange reactions of C2–C4 alkanes in ZSM-5 computed when using small cluster model by 20 kcal/mol. For ZSM-5, the cluster-size effect is converged at 96T model where the ZSM-5 cavity is well-represented. The chain-length dependency of the proton exchange barrier of alkanes in ZSM-5 could be predicted. Therefore, it is important to choose a model that could describe the cavity structure of zeolites. Contributions representing effects of level of basis set and electron correlation on computed activation energies could be as high as +8 kcal/mol. An extrapolated scheme that includes effects of cluster size, basis set, and electron correlation for the estimation of proton exchange barrier in ZSM-5 has been proposed. Computed energy barriers based on our most extended extrapolations are found to be significantly too low as compared to experiment data. The origin of this discrepancy is still unclear. The Gibbs free energies of activation are used to predict the regioselectivity for the proton exchange reaction of propane, *n*-butane, and *i*-butane. The preference of the primary position to other position is observed. From Gibbs free energies of activation at 553 K, the rate of proton transfer for propane in ZSM-5 is 1.4 times faster at the primary position than at the secondary position, in agreement with experiment.

Acknowledgments Thanks to the Office of the Higher Education Commission, Ministry of Education, Thailand for supporting Ms. Kanjarat Sukrat under the program "Strategic Scholarships for Frontier Research Network for the Joint Ph.D. Program Thai Doctoral degree." Special thanks to the Radchadapisek Sompot Endowment grants of Chulalongkorn University, The National Research University Project of CHE, and the Ratchadapisek Somphot Endowment Fund (AM1078I). The authors are also grateful to the Computational Chemistry Unit Cell, Faculty of Science, Chulalongkorn University, the Institute for Theoretical Chemistry, University of Vienna, the Vienna Scientific Cluster (VSC-1, project No. 70055), and to National Electronics and Computer Technology Center (NECTEC), Thailand, for providing research facilities, software packages, and computing times. The authors are finally grateful for financial support from the Austrian Sciences Fund (project P20893-N19), and the German Research Foundation (DFG), priority program SPP 1315, projects No. GE 1676/1-1 and SCHA849/8-1. Support was also provided by the Robert A. Welch Foundation under Grant No. D-0005.

References

1. Vermeiren W, Gilson JP (2009) Impact of zeolites on the petroleum and petrochemical industry. Top Catal 52(9):1131–1161
2. The intelligence report: business shift in the global catalytic process industries 2005–2011. The Catalyst Group Resources, Inc, May 2006
3. World catalysts. The Freedonia Group, Inc, Jan 2007
4. Narbeshuber TF, Vinek H, Lercher JA (1995) Monomolecular conversion of light alkanes over H-ZSM-5. J Catal 157(2):388–395
5. Lercher JA, Santen RA, Vinek H (1994) Carbonium ion formation in zeolite catalysis. Catal Lett 27(1):91–96
6. Stepanov A, Ernst H, Freude D (1998) In situ ^{1}H MAS NMR studies of the H/D exchange of deuterated propane adsorbed on zeolite H-ZSM-5. Catal Lett 54(1):1–4
7. Arzumanov SS, Reshetnikov SI, Stepanov AG, Parmon VN, Freude D (2005) In Situ ^{1}H and ^{13}C MAS NMR kinetic study of the mechanism of H/D exchange for propane on zeolite H-ZSM-5. J Phys Chem B 109(42):19748–19757. doi:10.1021/jp054037n
8. Narbeshuber TF, Stockenhuber M, Brait A, Seshan K, Lercher JA (1996) Hydrogen/deuterium exchange during *n*-butane conversion on H-ZSM-5. J Catal 160(2):183–189
9. Arzumanov SS, Stepanov AG, Freude D (2008) Kinetics of H/D exchange for *n*-butane on zeolite H-ZSM-5 studied with ^{1}H MAS NMR in situ. J Phys Chem C 112(31):11869–11874. doi:10.1021/jp802162h
10. Zheng A, Deng F, Liu S-B (2011) Regioselectivity of carbonium ion transition states in zeolites. Catal Today 164(1):40–45
11. Kramer GJ, van Santen RA, Emeis CA, Nowak AK (1993) Understanding the acid behaviour of zeolites from theory and experiment. Nature 363(6429):529–531
12. Sommer J, Habermacher D, Jost R, Sassi A, Stepanov AG, Luzgin MV, Freude D, Ernst H, Martens J (1999) Activation of small alkanes on solid acids. An H/D exchange study by liquid and solid-state NMR: the activation energy and the inhibiting effect of carbon monoxide. J Catal 181(2):265–270
13. Stepanov AG, Arzumanov SS, Luzgin MV, Ernst H, Freude D, Parmon VN (2005) In situ ^{1}H and ^{13}C MAS NMR study of the mechanism of H/D exchange for deuterated propane adsorbed on H-ZSM-5. J Catal 235(1):221–228
14. Zimmerman PM, Head-Gordon M, Bell AT (2011) Selection and validation of charge and Lennard-Jones parameters for QM/MM simulations of hydrocarbon interactions with zeolites. J Chem Theory Comput 7(6):1695–1703. doi:10.1021/ct2001655
15. Esteves PM, Nascimento MAC, Mota CJA (1999) Reactivity of alkanes on zeolites: a theoretical ab initio study of the H/H exchange. J Phys Chem B 103(47):10417–10420. doi:10.1021/jp990555k
16. Blaszkowski SR, Nascimento MAC, van Santen RA (1996) Activation of C–H and C–C bonds by an acidic zeolite: a density functional study. J Phys Chem 100(9):3463–3472. doi:10.1021/jp9523231
17. Zheng X, Blowers P (2005) A computational study of alkane hydrogen-exchange reactions on zeolites. J Mol Catal A: Chem 242(1–2):18–25
18. Zheng X, Blowers P (2006) Reactivity of isobutane on zeolites: a first principles study. J Phys Chem A 110(7):2455–2460. doi:10.1021/jp056707v
19. Tuma C, Sauer J (2006) Treating dispersion effects in extended systems by hybrid MP2: DFT calculations–protonation of isobutene in zeolite ferrierite. Phys Chem Chem Phys 8(34):3955–3965
20. Bŭcko T, Benco L, Hafner J, Ángyán JG (2007) Proton exchange of small hydrocarbons over acidic chabazite: Ab initio study of entropic effects. J Catal 250(1):171–183
21. Delley B (2000) From molecules to solids with the DMol3 approach. J Chem Phys 113(18):7756–7764
22. Ahlrichs R, Bär M, Häser M, Horn H, Kölmel C (1989) Electronic structure calculations on workstation computers: the program system turbomole. Chem Phys Lett 162(3):165–169
23. Truhlar DG (1998) Basis-set extrapolation. Chem Phys Lett 294 (1–3):45–48

24. Hohenberg P, Kohn W (1964) Inhomogeneous electron gas. Phys Rev 136(3B):B864–B871
25. Perdew JP, Burke K, Ernzerhof M (1996) Generalized gradient approximation made simple. Phys Rev Lett 77(18):3865–3868
26. Schafer A, Horn H, Ahlrichs R (1992) Fully optimized contracted Gaussian basis sets for atoms Li to Kr. J Chem Phys 97(4):2571–2577
27. Schafer A, Huber C, Ahlrichs R (1994) Fully optimized contracted Gaussian basis sets of triple zeta valence quality for atoms Li to Kr. J Chem Phys 100(8):5829–5835
28. Eichkorn K, Weigend F, Treutler O, Ahlrichs R (1997) Auxiliary basis sets for main row atoms and transition metals and their use to approximate Coulomb potentials. Theor Chem Acc 97(1): 119–124
29. Delley B (1990) An all-electron numerical method for solving the local density functional for polyatomic molecules. J Chem Phys 92(1):508–517
30. Delley B (1991) Analytic energy derivatives in the numerical local-density-functional approach. J Chem Phys 94(11):7245–7250
31. Delley B (1996) Fast calculation of electrostatics in crystals and large molecules. J Phys Chem 100(15):6107–6110. doi:10.1021/jp952713n
32. Helgaker T, Klopper W, Koch H, Noga J (1997) Basis-set convergence of correlated calculations on water. J Chem Phys 106(23):9639–9646
33. Grimme S (2006) Semiempirical GGA-type density functional constructed with a long-range dispersion correction. J Comput Chem 27(15):1787–1799
34. Kerber T, Sierka M, Sauer J (2008) Application of semiempirical long-range dispersion corrections to periodic systems in density functional theory. J Comput Chem 29(13):2088–2097. doi:10.1002/jcc.21069
35. De Moor BA, M-F Reyniers, Gobin OC, Lercher JA, Marin GB (2010) Adsorption of C2–C8 *n*-alkanes in zeolites. J Phys Chem C 115(4):1204–1219. doi:10.1021/jp106536m
36. Hampson JA, Rees LVC (1993) Adsorption of ethane and propane in silicalite-1 and zeolite NaY: determination of single components, mixture and partial adsorption data using an isosteric system. J Chem Soc, Faraday Trans 89:3169–3176
37. Stach H, Fiedler K, Janchen J (1993) Correlation between initial heats of adsorption and structural parameters of molecular sieves with different chemical composition—a calorimetric study. Pure Appl Chem 65(10):2193–2200
38. Anderson BG, Schumacher RR, van Duren R, Singh AP, van Santen RA (2002) An attempt to predict the optimum zeolite-based catalyst for selective cracking of naphtha-range hydrocarbons to light olefins. J Mol Catal A: Chem 181(1–2):291–301
39. Eder F, Stockenhuber M, Lercher JA (1997) Bronsted acid site and pore controlled siting of alkane sorption in acidic molecular sieves. J Phys Chem B 101(27):5414–5419. doi:10.1021/jp9706487
40. Yanping S, Brown TC (2000) Kinetics of isobutane dehydrogenation and cracking over HZSM-5 at low pressures. J Catal 194(2):301–308
41. Svelle S, Tuma C, Rozanska X, Kerber T, Sauer J (2009) Quantum chemical modeling of zeolite-catalyzed methylation reactions: toward chemical accuracy for barriers. J Am Chem Soc 131(2):816–825. doi:10.1021/ja807695p
42. Eder F, Lercher JA (1997) Alkane sorption in molecular sieves: the contribution of ordering, intermolecular interactions, and sorption on Brønsted acid sites. Zeolites 18(1):75–81
43. Eder F, Lercher JA (1997) On the role of the pore size and tortuosity for sorption of alkanes in molecular sieves. J Phys Chem B 101(8):1273–1278. doi:10.1021/jp961816i
44. Jensen F (2006) Introduction to computational chemistry, 2nd edn. Wiley, Chichester
45. Truitt MJ, Toporek SS, Rovira-Truitt R, White JL (2006) Alkane C–H bond activation in zeolites: evidence for direct protium exchange. J Am Chem Soc 128(6):1847–1852. doi:10.1021/ja0558802
46. Hansen N, Kerber T, Sauer J, Bell AT, Keil FJ (2010) Quantum chemical modeling of benzene ethylation over H-ZSM-5 approaching chemical accuracy: a hybrid MP2: DFT study. J Am Chem Soc 132(33):11525–11538. doi:10.1021/ja102261m
47. Tuma C, Kerber T, Sauer J (2010) The tert-butyl cation in H-zeolites: deprotonation to isobutene and conversion into surface alkoxides. Angew Chem Int Ed 49(27):4678–4680. doi:10.1002/anie.200907015

Theor Chem Acc (2012) 131:1221
DOI 10.1007/s00214-012-1221-z

REGULAR ARTICLE

Effects of mutations on the absorption spectra of copper proteins: a QM/MM study

Antonio Monari · Thibaut Very · Jean-Louis Rivail · Xavier Assfeld

Received: 9 February 2012 / Accepted: 8 April 2012 / Published online: 25 April 2012

Abstract The ground and excited state properties of copper proteins are studied and analyzed using hybrid quantum mechanics/molecular mechanics technique. Wild-type plastocyanin, characterized by an intense blue color, and wild-type nitrosocyanin, a red protein, are considered. These proteins differ from some ligands of the copper containing chromophore; we also studied the effects of selective mutations of one of the active site residue in plastocyanin. It is shown that this mutation is able to strongly modify the UV/VIS spectrum continuously modifying the absorption spectrum of the protein that from blue becomes red. Electrostatic and polarization effects of the macromolecular environment on the chromophore are taken into account using original techniques. Principal transitions are analyzed by mean of natural transition orbitals.

Keywords QM/MM · TD-DFT · Absorption spectra · Copper protein

Dedicated to Professor Marco Antonio Chaer Nascimento and published as part of the special collection of articles celebrating his 65th birthday.

Electronic supplementary material The online version of this article (doi:10.1007/s00214-012-1221-z) contains supplementary material, which is available to authorized users.

A. Monari (✉) · T. Very · J.-L. Rivail (✉) · X. Assfeld
Equipe de Chimie et Biochime Théoriques, SRSMC-UMR 7565, Université de Lorraine-Nancy and CNRS, BP-70239, 54506 Nancy-Vandoeuvre, France
e-mail: antonio.monari@cbt.uhp-nancy.fr

J.-L. Rivail
e-mail: Jean-Louis.Rivail@cbt.uhp-nancy.fr

1 Introduction

Among the many metalloproteins, copper proteins occur in a wide variety of biological systems, ranging from bacteria to humans [1–3]. In particular, plastocyanins (see Fig. 1) play a very important role as electron carriers during the photosynthesis process in higher plants. They contain a copper II ion and are reduced by cytochrome f of photosystem II [1]. The reduced form in which the extra electron is mainly located on the copper ion can transfer an electron to photosystem I [4, 5]. This class of plastocyanins can be found in many superior plants; the difference between them comes from few structural modifications from species to species. More important mutations give rise to the class of azurine, found in bacteria. The latter proteins start to show a great interest as antitumoral agents [6].

It is striking to notice how much the physicochemical properties of these compounds differ from those of aqueous copper complexes [7–12]. In particular, they exhibit an anomalously high reduction potential and usually a very intense absorption band in the visible range, which shifts from a compound to another. The role of the protein appears to be crucial in these changes, and it is twofold. First, it constrains the geometry of the coordination arrangement of the copper ion, and second, the various residues entering this coordination sphere can induce important changes in these physicochemical properties. Indeed, several such plastocyanins differ from a parent one by a mutation in the coordination arrangement. Finally, artificial mutations can be imposed in order to tune some property of interest [13], or in some cases, equilibrium between green and blue form has been reported [14].

Many theoretical investigations on such systems have been performed [15–25] in the past, and recently some

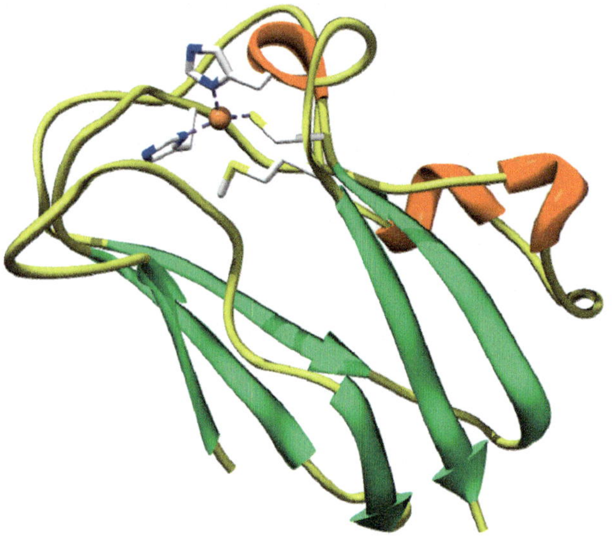

Fig. 1 The structure of plastocyanin, active site in ball and stick representation

mixed ab initio molecular dynamics and Car Parrinello dynamical studies have been reported [22–25].

In a previous paper [26], we have shown that a mixed quantum mechanics/molecular mechanics (QM/MM) approach of the spinach plastocyanin is quite appropriate to address the spectroscopic as well as the redox properties of this molecule and to analyze the influence of the various characteristics of the macromolecular surroundings on these properties. In the present paper, we focus our attention on the influence of the changes of the copper II fourfold coordination on the spectroscopic properties by considering the mutations of the methionine residue by the cysteine anion, cysteine, glutamate and homocysteine anion. We also considered the case of nitrosocyanin [27–30] (see Fig. 2), which is an example of a system in which the copper ion is surrounded by five ligands. Indeed, nitrosocyanin, the structure of which was resolved much more recently [27, 28], has a still unclear biological role, and contrary to plastocyanin, it exhibits an intense red color. The two proteins can be clearly related, and both show a quite rigid secondary structure, dominated by a β-sheet motif, with the chromophore being placed in a relative peripheral region. The active sites of both systems, see Fig. 3, show quite important similarities between them; in particular in both cases, two histidines are present as ligands, as well as one cysteine anion. The most important difference is the replacement of the relatively weak ligand methionine by a glutamate anion capable of a much more important interaction. As previously said, this reorganization of the active site leaves a free position in the coordination sphere of the copper II ion that is usually filled with a water molecule, therefore, leading to a pentacoordinate structure. Copper II ion bears, in both cases, an unpaired electron, mainly centered on metal but exhibiting quite an important delocalization over the different ligands in particular with the cysteine residue, as confirmed for instance by electron spin resonance (ESR) spectrum.

Recently, the effect of the mutation of the methionine ligand on the absorption spectrum of the blue-copper proteins has been underlined and proved experimentally [13], as well as the artificial mutation of the active site loop sequence of plastocyanin with the nitrosocyanin one giving rise to a red-copper protein [31].

In the present article, the absorption properties of the two native proteins, as well as the ESR and circular dichroism (CD) parameters, will be analyzed, at density functional theory (DFT) and time-dependent DFT (TD-DFT) level. The use of hybrid QM/MM approach will allow us to take into account the role of the macromolecular environment [32–37]. The evolution of the spectroscopic parameters with mutations will be also considered. The use of natural transition orbitals (NTOs) [26, 38] techniques will allow us to perform a systematic analysis of the nature and electronic properties of the excited states,

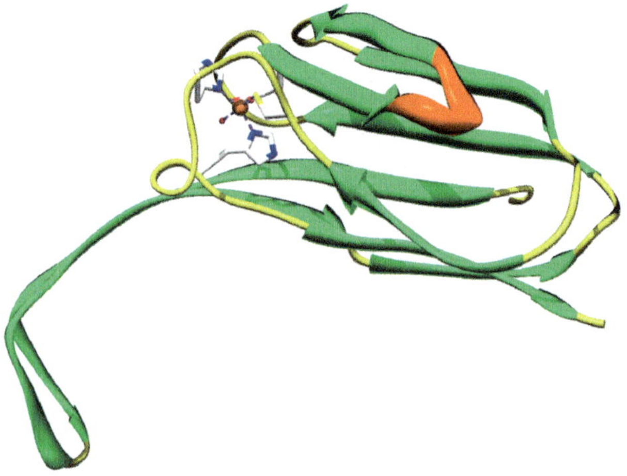

Fig. 2 The structure of nitrosocyanin, active site in ball and stick representation. Note the water molecule coordinating to copper

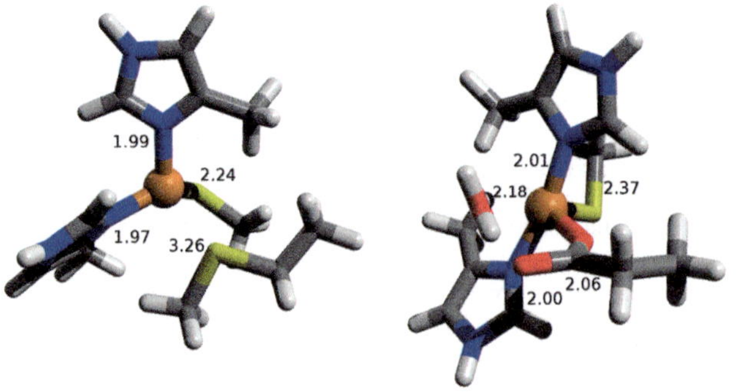

Fig. 3 The optimized active site of plastocyanin (*left*) and nitrosocyanin (*right*) in stick representation. The distances between the metal ion and the ligands are reported

allowing a systematic comparison of the different spectroscopic properties.

The present article is organized as follows: in Sect. 2, we present the computational strategy and details used in the present work, Results are analyzed and discussed in Sect. 3, and finally conclusion is drawn in Sect. 4.

2 Computational details and strategy

In order to correctly treat the macromolecular environment, we have partitioned our system in two regions: the active site treated at QM level and the outer region treated at MM level. Hybrid QM/MM computations have been performed using a local modified version of Gaussian 09 code [39], in which the local self-consistent field technique (LSCF) [40, 41] has been implemented. The coupling with MM has been performed by a local modified version of the Tinker code [42].

2.1 Computational strategy

Native proteins structures have been retrieved from the Protein Data Bank (PDB), blue-copper spinach plastocyanin is identified by the PDB code as 1AG6 (crystallographic resolution of 1.70 Å), while red-copper nitrosocyanin derived from the nitrifying bacterium *Nitrosomonas Europea* has the code 1IC0 (resolution of 1.65 Å). Nitrosocyanin crystallizes as a tetramer and, in order to simplify the computational problem, only one monomer of the protein has been considered. Mutated proteins have been constructed starting from the native plastocyanin by substitution of the Met91 residue, with cysteine, deprotonated cysteine, glutamate and the non-proteinogenic homocysteine aminoacid. The MM part of the system has been modeled by using CHARMM [43] force field parameters, as previously done for the study of plastocyanin. The QM/MM frontier has been located on the C_α–C_β bond of each of the four aminoacids directly interacting with the copper atom: His3, His86, Cys83, Met91 for plastocyanin and Glu60, Cys95, His98, His103 for nitrosocyanin.

The QM part was therefore composed of the copper II ion and of the lateral chains of the four aminoacids for plastocyanin. In the case of nitrosocyanin, an extra water molecule has been added to fill the vacancy in the coordination sphere. The effects of the MM point charges on the QM wavefunction have been directly taken into account by means of the so called electrostatic embedding, that is, by their inclusion into the QM Fock operator [40, 41].

In all the cases, the QM active site has been optimized, also allowing the geometry relaxation of the neighboring MM aminoacids. During the geometry optimization, a LANL2DZ [42–46] basis including pseudopotential for copper has been used together with B3LYP [47, 48] hybrid exchange and correlation functional. The frontier bonds have been treated by means of strictly localized bonding orbitals (SLBO) [40, 41] placing a doubly occupied local orbital, obtained from a model system (ethane), on each frontier bond. Subsequently, the LSCF technique [40, 41] was used to optimize the wavefunction under the constrain due to the SLBO, following a strategy that already proved efficiently in the calculations of biological systems.

Once the system has been optimized, excited states have been obtained at TD-DFT [38, 49] level in order to model absorption and circular dichroism spectra. In this case, we still used B3LYP exchange functional but the basis set was changed to a 6-31+G(d,p) set [50–52] in order to take into account the effect of diffuse functions. For these calculations, the frontier was treated using link-atoms (LA) technique [40, 41], that is, capping the dangling bond with an hydrogen atom, since the use of diffuse functions could produce unphysical transition on the frontier atoms when using SLBO techniques. Notice that when computing the UV/VIS spectra with the LANL2DZ basis, LA or SLBO techniques give absolutely superimposable results. CD spectra intensities [53, 54] have been obtained computing the rotatory strength by means of the "velocity" algorithm; the "length" formalism although showing a similar behavior gave the wrong sign for some of the less intense transitions. Finally, note that in order to match experimental results, vertical transitions have been convoluted with Gaussian functions of fixed half-length width of 0.2 eV. As previously reported [26] we underline that, although dealing with charge transfer transitions, in isolated model systems similar to our active site, the use of long range corrected functional gives practically indistinguishable spectra. To be assured of such an occurrence, we also performed more tests computing the absorption spectrum of the isolated plastocyanin active site (at the QM/MM B3LYP optimized geometry) using three more functionals: CAM-B3LYP [55], wB97XD [56], PBE0 [57]. The difference on the excitation energies of the most intense transitions with respect to the B3LYP results does not exceed 0.02 eV. This is probably due to the dimension of the active site and strongly justifies our choice to perform all the subsequent computations with B3LYP functional, only.

The ESR parameters, mainly the g-factors and the isotropic hyperfine A values, have also been obtained using the 6-31+G(d,p) basis set; rough g-factors tensor has been diagonalized in order to better compare the experimental data.

2.2 Electrostatic polarization of the MM part

When dealing with the modification induced on the UV/VIS spectra of a chromophore embedded in a macromolecular environment, one has to take into account three main effects: the geometric deformation induced by the rather rigid environment, the electrostatic effect modeled by the polarization of the wavefunction induced by the MM point charges, and finally the electronic polarization of the environment due to the modification of the electronic density of the environment to adapt itself to the new excited state electronic density.

The electrostatic effects can be easily taken into account by using electrostatic embedding in the QM/MM calculations. Conversely, the influence of the fixed charges on classical atoms can be easily evidenced by setting them to zero and performing another QM computation on the previously optimized fixed geometry. The electronic response of the surroundings (ERS) [26, 31–34] technique allows us to simulate the role of electronic polarization, that is, the modifications of the electronic distribution of the classical subsystem treated at MM level under the influence of the density modification in the QM part. Indeed, if within the framework of the Franck–Condon principle, the electronic transitions are vertical and do not involve any nuclear position rearrangement, the electronic cloud of the environment can, however, instantaneously adapt itself to the new electronic distribution of the chromophore. Since the electrons of the MM part are not treated explicitly, different possibilities exist to take this phenomenon into account. One solution consists in using polarizable force fields. Although rigorous, and able to take into account all the anisotropies of the environment, this choice suffers the drawback of a consequently augmented computational cost. Moreover, the use of polarizable force fields is nowadays not a standard task and can be quite problematic. In the framework of ERS technique, we model the electronic response of the environment by means of a polarizable continuum medium characterized by a dielectric constant extrapolated at infinite frequency. Technically, the chromophore is placed in a site cavity created in the continuum and the MM points charges of the protein do not interact with the continuum. ERS technique already proved its efficiency in modeling the important effects of electronic polarizability on the position and intensities of electronic transitions, at a very moderate additional computational cost [26, 31–34].

2.3 Natural transition orbitals

One of the most useful features of excited state computation rests in the possibility of making an accurate analysis of the excited states in terms of electron density rearrangement with respect to the ground state. Indeed, in linear response TD-DFT formalism, one can express excited states as a linear combination of single-electron excitation from occupied to virtual Kohn–Sham orbitals. If in some cases, such an analysis is straightforward, since the excited state is dominated by only one excitation, there are cases in which many couples of orbitals are participating to the expansion with a similar weight, therefore making the analysis quite cumbersome. To overcome such a problem, one can use the so called NTOs [26, 38, 58], based on a unitary transformation of the 1-electron transition density matrix $\mathbf{T}$. Indeed, NTOs are obtained by a singular value decomposition of the $\mathbf{T}$ matrix, leading to two matrices $\mathbf{U}$ and $\mathbf{V}$ storing occupied and virtual NTOs, respectively. The singular values $\mathbf{\Lambda}$ represent the weight of the transition from an occupied to the corresponding virtual orbital. In contrast with the molecular orbitals representation, only one, or in the worst cases two, transitions dominate the expansion and the description of the excited states, making the analysis much simpler. Speaking in more physically terms, one can think of the dominant occupied NTOs as of the orbital from which the electron is taken away upon the excitation, that is, representing a sort of hole density, while the virtual one represents the orbital that describes the electron rearrangement in the excited states. NTOs have been obtained by a postprocessing of gaussian 09 transition densities using a code specifically designed and produced in our laboratory. The code, licensed under gnu public license, is free and can be downloaded from http://nancyex.sourceforge.net [26].

3 Results and discussion

In this Section, we will analyze and compare the results obtained for plastocyanin and nitrosocyanin, in terms of chromophore structure, UV/VIS and circular dicroism spectrum and ESR parameter.

3.1 Structural parameters

The QM/MM optimized active site for plastocyanin and nitrosocyanin is represented in Fig. 3 (left and right panel, respectively); more structural parameters are reported in Supplementary Materials. One can immediately see that in the case of Copper II plastocyanin, the active site assumes a distorted tetrahedral arrangement with three ligands much more strongly bonded than methionine. Indeed, as already underlined computationally [22–26] and experimentally [1–6], the Cu–Met distance is quite large and close to 3.50 Å; conversely, the Cu–Cys bond length is quite shorter going to 2.20 Å. The angles comprising methionine also appears noticeably distorted with the Met–Cu–His angle being close to 90°.

The structure of nitrosocyanin appears quite different, and the geometry, when we include the water molecule, appears much closer to a trigonal bipyramid. Indeed, the two histidines act like the apical ligands, the angle between the two nitrogen atoms and copper ion being close to 180°, on the other hand, the anionic glutamate and cysteine, together with water, can be considered as the three equatorial ligands, roughly lying on the same plane. As the distances are concerned, we can see that the Cys–Cu bond length is only slightly perturbed but is lengthened to the value of 2.37 Å. On the contrary, the carboxylic group of the glutamate ion is much closer to the metal, the closest oxygen atom being at a distance of 2.06 Å only from copper. This aspect reveals a totally different role played by such a residue compared to methionine in plastocyanine. On the other hand, the distances between the histidine nitrogen atom and the copper ion are still close to 2.00 Å and not too much changed compared to the case of blue plastocyanin. Finally, as the water molecule is concerned, the distance between oxygen and copper is quite a common one of 2.18 Å. If we consider the angles between ligands, as already stated, the two histidines forming the apical ligands the N–Cu–N angle has a value of about 170°, the angles between the equatorial ligands are indeed much more distorted and vary from about 89° (water–Cu–Glu) to 160° (water–Cu–Cys). This deviation can also be ascribed to the possibility for the water molecule to come closer to the second oxygen of glutamate to be stabilized by a hydrogen bond. Indeed, the distance between the hydrogen of water and glutamate oxygen is of 1.62 Å.

Of course when mutations are introduced in the plastocyanin active site, structural deformations appear to be less important in magnitude than for the nitrosocyanin case. The distorted tetrahedral site is preserved, the most affected parameter being the distance between the mutated residue and the copper ion. In the case of protonated and deprotonated cysteine, the Cu–S distance for the mutated residue remains quite large and greater than 4.50 Å. This peculiarity can be explained also by the fact that cysteine has a lateral chain shorter than methionine. Therefore, the protein backbone geometric constrain keeps the cysteine residue quite far from the copper ion, even when deprotonation occurs. This is confirmed when one looks at the homocysteine case. This aminoacid has a lateral chain constituted by two CH_2 groups bonded to a SH function, its length is therefore comparable to that of methionine, but the presence of the anion centered on the S atom should lead to a much stronger interaction. Indeed, the Cu–S distance is reduced to 2.40 Å and becomes equivalent to the Cu–S distance of the native cysteine, leading to a much more symmetric arrangement. Finally, in the case of glutamate, the C–O distance is much shorter and goes down to 2.00 Å, indicating a much stronger interaction with the metal ion.

3.2 Ground state electronic structure and ESR parameters

Both plastocyanin and nitrosocyanin proteins host a copper II ion; therefore, they both are in a doublet spin state, bearing an unpaired electron formally attributed to the d^9 metal electronic configuration. However, the electron experiences quite an important delocalization over the ligands in both structures, even if the metallic character of the unpaired electron is much more pronounced in the case of nitrosocyanin, as confirmed by population analysis. This indicates that in the case of nitrosocyanin, one faces a less important, or less efficient, overlap between ligands and metal d orbitals. The computed ESR g-factors and hyperfine parameters reported in Table 1, which compare very well with experimental parameters [3, 29], show one dominant component of the diagonalized g-factor tensor and two smaller ones, similar between them, and greater than the free electron g-factor. This occurrence indicates a partial $d_{x^2-y^2}$ nature of the free electron in the ground state, coherent with DFT calculations that place such an orbital as the singly occupied molecular orbital. We can also notice that the value of the g-factor should be directly proportional to the amount of the d-character and inversely proportional to the ligand field energy difference between $d_{x^2-y^2}$ and d_{xy} orbitals [29]. Since the value of the computed g-factors are almost equal for both proteins, while the ligand field separation should be much higher, as confirmed spectroscopically by Basumalick et al. [29], we have another confirmation of the more pronounced metallic character of the ground state in nitrosocyanin.

3.3 Native protein UV/VIS and CD spectra

In Figs. 4 and 5, we report the TD-DFT computed spectra of wild-type plastocyanin and nitrosocyanin, respectively. One can immediately see that the UV/VIS spectrum of plastocyanin is dominated by the intense absorption at about 600 nm, responsible for the blue color of the protein; some weaker structures appear at shorter and longer

Table 1 Computed ESR g-factors and hyperfine A constants

	Plastocyanin	Nitrosocyanin
g_{xx}	2.059 (2.047)	2.037 (2.036)
g_{yy}	2.094 (2.059)	2.099 (2.059)
g_{zz}	2.185 (2.226)	2.225 (2.245)
A_y	30.3 G (|25| G)	51.5 G (|16| G)
A_z	−112.0 G (|177| G)	−138.7 G (|140| G)

Experimental results in parenthesis taken from [29]. Note that experimental techniques were not able to retrieve the sign of hyperfine A constant. Therefore, only the absolute value is given

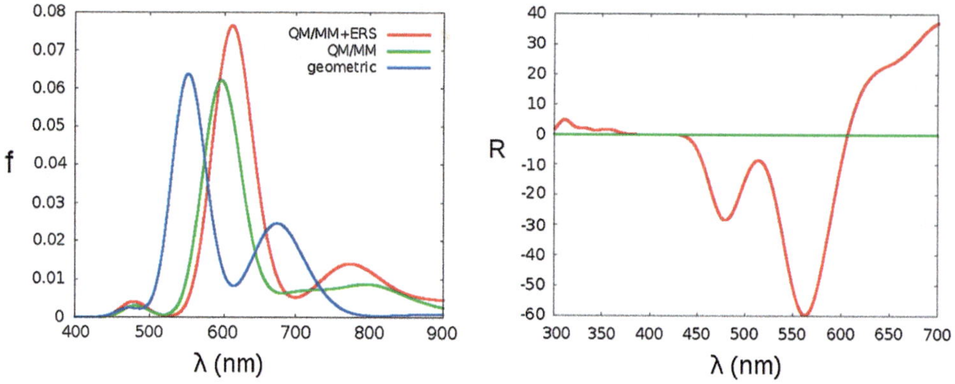

Fig. 4 The TD-DFT computed UV/VIS (*left*) and circular dichroism (*right*) spectra of plastocyanin. CD spectra are computed taking into account ERS polarization. Wavelengths in nm, intensities in arbitrary units. "QM/MM+ERS" stands for the QM/MM computation including polarization of the protein; "QM/MM" for the computation limited to electrostatic classical-quantum interactions and "geometric" means that the only geometric constrains are taken into account (the classical charges are switched off)

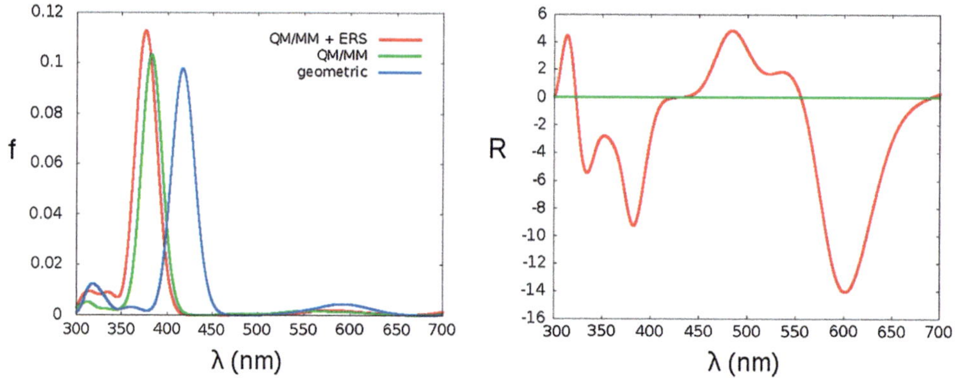

Fig. 5 The TD-DFT computed UV/VIS (*left*) and circular dichroism (*right*) spectra of nitrosocyanin. CD spectra are computed taking into account ERS polarization. Wavelengths in nm, intensities in arbitrary units. The notations are the same as for Fig. 4

wavelengths. As already stated on a previous paper [26], the UV/VIS spectrum matches well the experimental value [7] characterized by an absorption maximum at 598 nm. It also appears that the ERS effects, contrary to the electrostatic ones, do not have a very large magnitude although they enhance the intensity and modify the shape of the low-energy part of the spectrum. If one considers the CD spectrum, one can see that the presence of transitions in the region close to 550 nm is evidenced by a peak showing quite an important negative rotatory strength, and at the same time, the large band at 600 nm gives a very intense positive rotatory strength band. Notice also that now the region close to 450 nm is much better resolved than in the case of the UV/VIS spectrum. All these results are quite in agreement with the experimental ones (see for instance the works of Solomon [2]) reproducing all the main features of the CD spectrum, even if the intensities are in this case much more sensitive and, therefore, more difficult to reproduce at the same level of precision than for the UV/VIS spectra.

Contrarily to plastocyanin, nitrosocyanin exhibits a very strong absorption band centered at 380 nm, that compares very well with the experimental one centred at 390 nm [29], and which is responsible for the red color. We can also notice a very weak band at 500 nm, comparable to the experimental one centred at 491 nm, and two very large and weak bands, one around 570 nm (experimentally 568 nm) and a second one around 670 nm (experimental 666 nm). Again, the inclusion of ERS effect does not change too much the 390-nm band, even if the pure electrostatic treatment gives a peak red-shifted for about 20 nm, but the weak bands appear much more affected from the inclusion of polarization, in particular in the low-energy part of the spectrum. Finally, it must be emphasized that a treatment limited to the geometric deformation only is absolutely insufficient to quantitatively recover the spectrum features, the most intense band appearing, in this case, at 425 nm. As far as the CD spectrum is concerned, one can see an even more complicated structure when compared with the plastocyanin one. The 380-nm band gives rise to an intense CD band showing a negative rotatory strength. This compares well with experiment (see for instance Basumaick et al. [29]) although the calculated intensity for this band appears underestimated. Two bands with positive rotatory strength at 490 and 550 nm are again identified in agreement with the experiment, as well as with some of the low intensity features already evidenced for the UV/VIS spectrum. Finally, one can identify the band at 600 nm, which shows a very intense negative rotatory strength; the latter can be directly compared with the band responsible of the blue color in the case of plastocyanin.

The different behavior of the two proteins can easily be rationalized by considering the NTOs (Fig. 6) for the most intense transition occurring at 600 and 390 nm for plastocyanin and nitrosocyanin, respectively. In both cases, the transition can be characterized as mainly a ligand (cytosine) to metal charge transfer, with a non-negligible participation of the other ligands. In the case of plastocyanin, however, the occupied NTO has a very strong π character while the main feature of nitrosocyanin appears as a σ antibonding interaction. It looks obvious that the change from one color to the other one takes place by enhancing the π or the σ metal to ligand charge transfer, respectively, while the less active transition undergoes a quenching of its intensity.

3.4 Plastocyanin mutations

Following the experimental works recently published on azurine by Clark et al. [13], we decided to selectively mutate the Met91 residue in spinach plastocyanin to assess for the different spectroscopic properties. The results for some of the most important bands are reported in Table 2. All the spectra reported there have been computed at the QM/MM level taking into account the ERS technique to model the response of the macromolecular environment.

The data fit very well experimental results reported by Clark et al. [13], even if care should be taken since in that work azurine protein was used instead, our results being able to correctly reproduce the variation of the absorption spectrum.

Notice the important effect due to the protonated or deprotonated state of the mutated cysteine, see also Fig. 7, and hence the strong dependence of the spectrum on the pH of the solution, a result again observed in the experimental study of mutated azurine [13].

It is also worth mentioning that in the case of glutamate mutation TD-DFT calculated spectrum presents also a second quite intense band at about 500 nm, as well as the band listed in Table 2. Note also that glutamate could show the same dependence with pH already evidenced for the cytosine mutant. Unfortunately, experimental data reported from Clark et al. [13] are not totally sufficient to entirely clarify this feature, as well as to clarify the pH dependence of the UV/VIS spectrum of such mutant.

Notice anyway that, as expected from the analysis of the structural parameters, the simple mutation of methionine

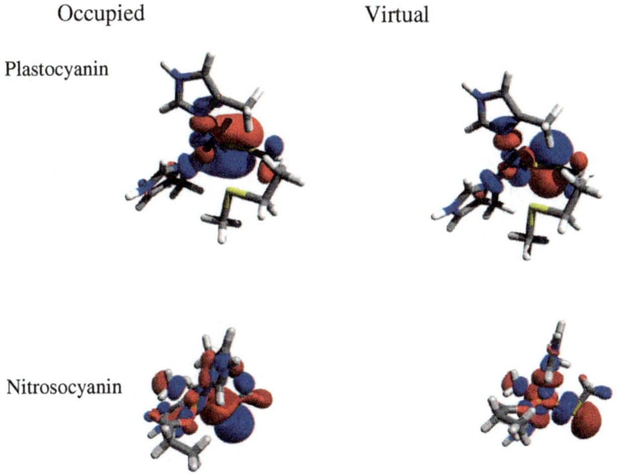

Fig. 6 NTOs for the most intense spectral band of plastocyanin (600 nm) and nitrosocyanin (390 nm)

Table 2 Computed TD-DFT principal UV/VIS absorption transitions for the mutated Plastocyanin

	First band (nm)	Second band (nm)	Intensity ratio
Cys	589	452 (451)	2.00
Cys^{-}	526	447 (441)	0.44
Glu^{-}	550	440 (450)	0.66
Hcy^{-}	452	403 (410)	0.84

The aminoacid used to substitute the Meth91 residue is indicated a "–" indicates an anionic form. The values have been computed at QM/MM level including electronic response of the environment. In parenthesis experimental value for azurine taken from [13]

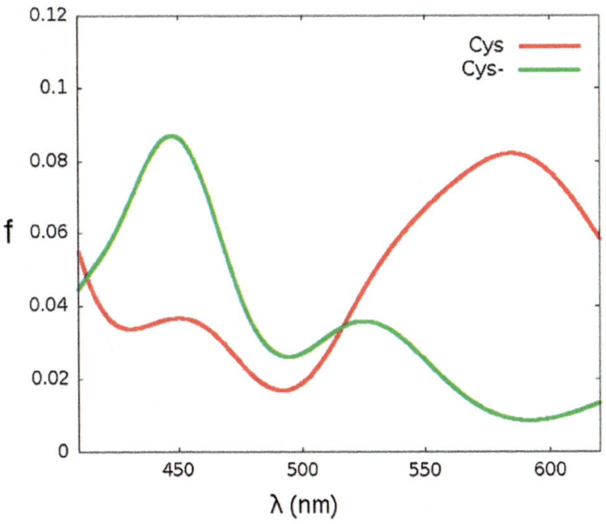

Fig. 7 The TDDFT spectrum of the mutated plastocyanin bearing a cysteine residue, in protonated and deprotonated state. Wavelengths in nm, intensities in arbitrary units

by a much stronger ligand like anionic cysteine is not sufficient to induce the change from a blue to a red protein, even if the spectrum is significantly altered and generally shifted to the short wavelengths. Indeed, the long distance between copper and the terminal sulfur atom imposed by the protein geometric constrain strongly limits the interaction strength. More flexible and extended ligands, such as glutamate and anionic homocysteine, on the other hand, produce a much more important shift. In particular with the mutation by the Hcy residue, this effect is so pronounced to be able to produce a red-copper protein.

4 Conclusions

We performed a systematic study of the structural and spectroscopic properties of two wild-type copper proteins: plastocyanin and nitrosocyanin and of mutated form of the first one.

In particular, optimized geometries, ESR parameters, UV/VIS, and CD spectra have been obtained, at QM/MM level, using DFT and TD-DFT techniques.

The different nature of the active site has been elucidated, as well as its influence on the spectroscopic properties, correctly reproducing experimental results, for the blue plastocyanin as well as for the red nitrosocyanin.

The UV/VIS spectra have also been analyzed to take into account the different effects produced by geometric deformation, electrostatic interactions, and polarization of the macromolecular moiety.

The nature of the excited states has been analyzed in terms of NTO showing how the main transition in the two proteins have a very different nature, one being based on a π interaction and the other on a σ one.

The effect of the mutation of the weak methionine ligand on the tetrahedral active site of plastocyanin has been studied, evidencing how the UV/VIS spectrum can be altered significantly by this mutation.

Again in coherence with experimental results on a similar azurine protein, we have shown how the mutation of only one ligand is able to induce the change from a blue to a red protein. The role of pH and of the protonation or deprotonation of the ligands has also been considered, as well as the influence of the lateral chain length of the mutated aminoacid. Indeed, because of the constrains by the protein backbone, when lateral chains are too short, they keep the electron donor atom too far away from the metal cation to produce a sufficiently strong interaction. The capability of reproducing such important and complex spectroscopic changes determined by a relative minor mutation can be considered as a significant proof of the robustness of our QM/MM methodology and strategy in the determination of excited state properties of biological molecules.

In the future, we plan to extend the present work in two directions: influence of solvation by a combined molecular dynamics (MD) and QM/MM study of the solvated proteins, and change in the properties of plastocyanin when all the amino acids of the active site are replaced by those of nitrosocyanin.

The first aspect is based on the run of a significantly long MD trajectory that will give access to the possible configuration assumed by the macromolecule and by its active site, and subsequently on the extraction of snapshots from the trajectory to compute the spectrum at the TD-DFT level. Even if the backbone of both copper proteins appears to be quite rigid, such a treatment will allow us to take into account many more effects, and in particular the role of local geometric modifications and the possible influence of the solvent.

The mutation of the whole active site aminoacid sequence will allow us to compare our findings with some very recent experimental data and to get even more insight on the role played by the macromolecular structure in finely tuning the rather particular and fascinating spectroscopic properties of these classes of proteins.

Acknowledgments Supports from Université de Lorraine and CNRS are gratefully acknowledged, also for the financing of the "chaire d'excellence" (A. M.). We also acknowledge support from the ANR project ANR-09-BLAN-0191-01 "PhotoBioMet".

References

1. Holm RH, Kennepohl P, Solomon EI (1996) Chem Rev 96:2239
2. Solomon EI, Hadt RG (2011) Coord Chem Rev 255:774
3. Solomon EI, Szilagyi RK, De Beer S, Basumallick G, Basumallick L (2004) Chem Rev 104:419
4. Gross EL (1993) Photosynth Res 37:103
5. Skyes AG (1991) Adv Inorg Chem 36:377
6. Mohamed MS, Fattah SA, Mostafa HM (2010) Curr Res J Biolog Sci 2:396
7. Book LD, Arnett DC, Hu H, Scherer NF (1998) J Phys Chem A 102:4350
8. Xue Y, Ökvist M, Hansson Ö, Young S (1998) Protein Sci 7:2099
9. Gewirth AA, Solomon EI (1998) J Am Chem Soc 110:3811
10. Nersissian AM, Nalbandyan RM (1988) Biochim Biophys Acta 957:446
11. Gray HB, Malström BG, Williams RJP (2000) J Biol Inorg Chem 5:551
12. Lockwood DM, Cheng Y-K, Rossky PJ (2001) Chem Phys Lett 345:159
13. Clark KM, YU Y, Marshall NM, Sieracki NA, Nilges MJ, Blackburn NJ, van der Donk WA, Lu Y (2010) J Am Chem Soc 132:10093
14. Ghosh S, Xie X, Dey A, Sun Y, Scholes CP, Solomon EI (2009) Proc Nat Acad Sci USA 106:4969
15. Ando K (2010) J Chem Phys 133:175101

16. Corni S, De Rienzo F, Di Felice R, Molinari E (2005) Int J Quant Chem 102:328
17. Ryde U, Olsson MHM, Pierloot K, Roos BO (1996) J Mol Biol 261:586
18. Olsson MHM, Ryde U (1999) J Biol Inorg Chem 4:654
19. Ryde U, Olsoon MHM (2001) Int J Quant Chem 81:335
20. Ando K (2004) J Phys Chem B 108:3940
21. Pierloot K, De Kerpel JOA, Ryde U, Olsson MHM, Roos BO (1998) J Am Chem Soc 120:13156
22. Sinnecker S, Neese F (2006) J Comput Chem 27:1463
23. Cascella M, Cuendet MA, Tavernelli I, Rohtlisberger U (2007) J Phys Chem B 111:10248
24. Le Bard DN, Matyushov DV (2008) J Phys Chem B 112:5218
25. Ando K (2008) J Phys Chem B 112:250
26. Monari A, Very T, Rivail J-L, Assfeld X (2011) Comp Theor Chem. doi:10.1016/j.comptc.2011.11.026
27. Lieberman RL, Arciero D, Hooper AB, Rosenzweig AC (2001) Biochemistry 40:5674
28. Arciero DM, Pierce BS, Hendrich MP, Hooper AB (2002) Biochemistry 41:1703
29. Basumalick L, Sarangi R, De Beer Giorgi S, Elmore B, Hooper AB, Hedmann B, Hodgson KO, Solomon EI (2004) J Am Chem Soc 127:3531
30. Besley NA, Robinson D (2011) Farady Discuss148:55
31. Berry SM, Bladhoim EL, Mostad EJ, Schenewerk AR (2011) J Biol Inorg Chem 16:473
32. Laurent AD, Assfeld X (2010) Interdiscip Sci Comput Life Sci 2:38
33. Loos P-F, Dumont E, Laurent AD, Assfeld X (2009) Chem Phys Lett 475:120
34. Jacquemin D, Perpète EA, Laurent AD, Assfeld X, Adamo C (2009) Phys Chem Chem Phys 11:1258
35. Dumont E, Laurent AD, Loos P-F, Assfeld X (2009) J Chem Theory Comput 5:1700
36. Ambrosek D, Loos P-F, Assfeld X, Daniel C (2010) J Inorg Biochem 104:893
37. Rodriguez-Ropero F, Zanuy D, Assfeld X, Aleman C (2009) Biomacromolecules 10:2338
38. Dreuw A, Head-Gordon M (2005) Chem Rev 105:4009
39. Frisch MJ, Trucks GW, Schlegel HB, Scuseria GE, Robb MA, Cheeseman JR, Scalmani G, Barone V, Mennucci B, Petersson GA, Nakatsuji H, Caricato M, Li X, Hratchian HP, Izmaylov AF, Bloino J, Zheng G, Sonnenberg JL, Hada M, Ehara M, Toyota K, Fukuda R, Hasegawa J, Ishida M, Nakajima T, Honda Y, Kitao O, Nakai H, Vreven T, Montgomery JA Jr, Peralta JE, Ogliaro F, Bearpark M, Heyd JJ, Brothers E, Kudin KN, Staroverov VN, Kobayashi R, Normand J, Raghavachari K, Rendell A, Burant JC, Iyengar SS, Tomasi J, Cossi M, Rega N, Millam NJ, Klene M, Knox JE, Cross JB, Bakken V, Adamo C, Jaramillo J, Gomperts R, Stratmann RE, Yazyev O, Austin AJ, Cammi R, Pomelli C, Ochterski JW, Martin RL, Morokuma K, Zakrzewski VG, Voth GA, Salvador P, Dannenberg JJ, Dapprich S, Daniels AD, Farkas Ö, Foresman JB, Ortiz JV, Cioslowski J, Fox DJ (2009) Gaussian 09, Revision A.1. Gaussian Inc., Wallingford, CT
40. Ferré N, Assfeld X, Rivail J-L (2002) J Comput Chem 23:610
41. Assfeld X, Rivail J-L (1996) Chem Phys Lett 266:100
42. Tinker code package: http://dasher.wustl.edu/tinker/
43. Brooks BR, Bruccoleri RE, Olafson BD, States DJ, Swaminathan S, Karplus M (1983) J Comput Chem 4:187
44. Hay PJ, Wadt WR (1985) J Chem Phys 82:270
45. Hay PJ, Wadt WR (1985) J Chem Phys 82:284
46. Hay PJ, Wadt WR (1985) J Chem Phys 82:299
47. Becke AD (1993) J Chem Phys 98:1372
48. Lee C, Yang W, Parr RG (1993) Phys Rev B 37:785
49. Runge E, Gross EKU (1984) Phys Rev Lett 52:997
50. Ditchfield R, Hehre WJ, Pople JA (1971) J Chem Phys 54:724
51. Hehre WJ, Ditchfield R, Pople JA (1972) J Chem Phys 56:2257
52. Rassolov VA, Ratner MA, Pople JA, Redfern PC, Curtiss LA (2001) J Comp Chem 22:976
53. Hansen E, Bak KL (1999) Enantiomer 4:455
54. Warnke I, Furche F, Wire Inter Rev. doi:10.1002/wcms.55
55. Yanai T, Tew D, Handy N (2004) Chem Phys Lett 393:51
56. Chai J-D, Head-Gordon M (2008) Phys Chem Chem Phys 10:6615
57. Adamo C, Barone VJ (1999) Chem Phys 110(110):6158
58. Martin RL (2003) J Chem Phys 118:4775

Theor Chem Acc (2012) 131:1214
DOI 10.1007/s00214-012-1214-y

REGULAR ARTICLE

Structure and electronic properties of hydrated mesityl oxide: a sequential quantum mechanics/molecular mechanics approach

Marcus V. A. Damasceno · Benedito J. Costa Cabral · Kaline Coutinho

Received: 14 February 2012 / Accepted: 24 March 2012 / Published online: 22 April 2012

Abstract The hydration of mesityl oxide (MOx) was investigated through a sequential quantum mechanics/molecular mechanics approach. Emphasis was placed on the analysis of the role played by water in the MOx *syn–anti* equilibrium and the electronic absorption spectrum. Results for the structure of the MOx–water solution, free energy of solvation and polarization effects are also reported. Our main conclusion was that in gas-phase and in low-polarity solvents, the MOx exists dominantly in *syn*-form and in aqueous solution in *anti*-form. This conclusion was supported by Gibbs free energy calculations in gas phase and in-water by quantum mechanical calculations with polarizable continuum model and thermodynamic perturbation theory in Monte Carlo simulations using a polarized MOx model. The consideration of the in-water polarization of the MOx is very important to correctly describe the solute–solvent electrostatic interaction. Our best estimate for the shift of the π–π^* transition energy of MOx, when it changes from gas-phase to water solvent, shows a red-shift of $-2{,}520 \pm 90\ \text{cm}^{-1}$, which is only $110\ \text{cm}^{-1}$ (0.014 eV) below the experimental extrapolation of $-2{,}410 \pm 90\ \text{cm}^{-1}$. This red-shift of around $-2{,}500\ \text{cm}^{-1}$ can be divided in two distinct and opposite contributions. One contribution is related to the *syn* → *anti* conformational change leading to a blue-shift of $\sim 1{,}700\ \text{cm}^{-1}$. Other contribution is the solvent effect on the electronic structure of the MOx leading to a red-shift of around $-4{,}200\ \text{cm}^{-1}$. Additionally, this red-shift caused by the solvent effect on the electronic structure can by composed by approximately 60 % due to the electrostatic bulk effect, 10 % due to the explicit inclusion of the hydrogen-bonded water molecules and 30 % due to the explicit inclusion of the nearest water molecules.

Keywords Solvent effect 1 · Theoretical calculations 2 · Absorption electronic spectrum 3

1 Introduction

A better understanding of solvent effects on molecular properties is of fundamental interest for explaining the mechanisms and dynamics of chemical and biochemical processes in solution [1, 2]. The knowledge of these mechanisms at the molecular level may have a strong impact on the design of biochemical and technological applications. Solvent effects may influence the properties of a solvated species in many different ways. They may affect, in particular, the conformational equilibrium relative to the gas-phase or low-polarity solvent [3, 4]. This effect can be explained, in general, by the free energy difference between conformations in solution, mainly in polar solvents, where it is expected that structures with a larger dipole moment are favored [5]. Solvents effects are also very important for understanding the vibrational and

Dedicated to Professor Marco Antonio Chaer Nascimento and published as part of the special collection of articles celebrating his 65th birthday.

Electronic supplementary material The online version of this article (doi:10.1007/s00214-012-1214-y) contains supplementary material, which is available to authorized users.

M. V. A. Damasceno · K. Coutinho (✉)
Instituto de Física, Universidade de São Paulo, CP 66318, São Paulo, SP 05314-970, Brazil
e-mail: kaline@if.usp.br

B. J. Costa Cabral
Grupo de Física Matemática da Universidade de Lisboa, Av. Professor Gama Pinto 2, 1649-003 Lisbon, Portugal

electronic spectra in solution [6]. It is well-known that the electronic absorption spectra in solution may, in comparison with gas-phase data, exhibit significant changes in the position, intensity and shape of the absorption bands [1, 6–8]. These changes are related, in general, to the solvent polarity although some specific local interactions (e.g. hydrogen bonding) are also important. The concept of solvatochromism was adopted for describing the change relative to gas phase in the position of the maximum absorption wavelength, λ_{max}, that can be related, in many cases, to the polarity of a particular solvent. In this context, hypsochromic [bathochromic] shifts are associated with a blue [red]-shift of λ_{max}. Solvatochromic effects are driven by solute–solvent interactions. Therefore, the understanding of these effects depends on the analysis of different contributions to the solvent–solute interactions, including specific intermolecular contributions [9]. The solvent polarity has been studied through the transition energy of solvatochromic dyes, which exhibit absorption bands in the visible that are shifted with respect to the solvent [10]. Therefore, in this way, it was possible to define empirical parameters for the solvent polarity through UV–vis measurements for a large number of solvatochromic species with the purpose of establishing an empirical scale of solvent polarities. This procedure, actually, led to the definition of different polarity scales [1, 11–13].

The main purpose of the present work is to investigate the role played by solvent effects on the conformational stability and electronic spectrum of the mesityl oxide (MOx) molecule [$(CH_3)_2C{=}CHCOCH_3$, 4-methyl-3-penten-2-one] through theoretical calculations. The MOx can be considered as a model for larger species of ketones characterized by the presence of a carbonyl (–C=O) group linked to other two organic groups with a general formula RC–C(=O)–CR′. It should be observed that the presence of low-wavelength (high-energy) electronic transitions sensitive to the solvent polarity in non-saturated α,β-ketones makes these systems very important for understanding solvatochromic effects. Five isomers of the MOx can exist. The conjugate acetonic form, characterized by the presence of the –C=O group in a conjugate C–C=C- array, with two isomers (*syn* and *anti* or *cis* and *trans*) is illustrated in Fig. 1 (top). These two isomers differ by the relative orientation (torsional angle) of the C=C–C=O atoms. This torsional angle is 0° for the *syn*-isomer (*cis*) and 180° for the *anti*-isomers (*trans*). In the non-conjugate acetonic form, the isomer of oxide mesityl [4-methyl-4-penten-2-one] is named *iso*-mesityl oxide (IMOx) that can be also found in the *syn*- and *anti*-isomers, as illustrated in Fig. 1 (bottom). An enolic form for species with at least a α-hydrogen atom can be also identified. However, it was experimentally verified through the analysis of the vibrational spectrum that this enolic form is not present in

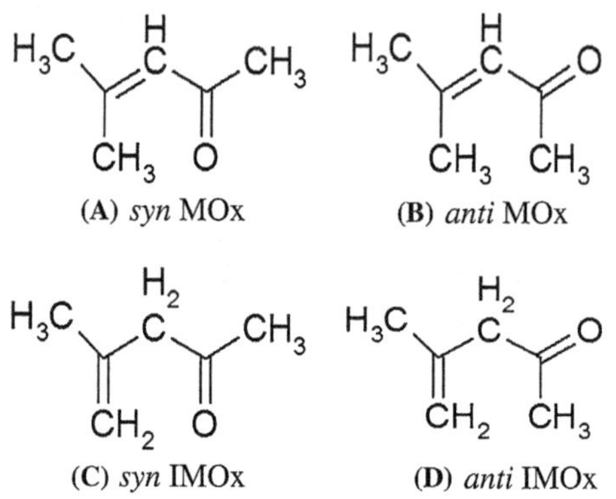

Fig. 1 Chemical structure of *syn*- and *anti*-isomers of mesityl oxide (MOx) and of iso-mesityl oxide (IMOx)

equilibrium at normal conditions, but it was found a composition of 91 % of conjugated isomer (MOx) and 9 % of unconjugated isomer (IMOx) [14]. Experimental results for the vibrational (IR) and electronic absorption (UV–vis) spectra, as well as their dependence on thermodynamic properties (boiling points) of MOx and IMOx, have been reported [14, 15]. In the non-polar solvent iso-octane, MOx (b.p. = 130 °C) exhibits an intense IR absorption band associated with the C=O double bond in a conjugate position [14] and two UV–vis absorption bands at 231 nm (strong, π–π^*) and 329 nm (weak, n–π^*) [15, 16]. In the same solvent, the IMOx molecule (b.p. = 121.5 °C) shows an IR absorption band typical of a C=O bond in a non-conjugate position [14] and only one UV–vis absorption band at 290 nm (weak, n–π^*) [15]. Therefore, the low-lying n–π^* absorption band in the region of 290–330 nm cannot be attributed to a specific isomer due to the presence of MOx and IMOx isomers in the experimental system. Additionally, for some solvents, like water and tetrafluoropropanol, the n–π^* band is submerged in the π–π^* band and its λ_{max} cannot be measured [16]. Then, the π–π^* absorption band in the region of 200–250 nm is used to characterize the MOx isomer [14, 15], like other molecules of the family of non-saturated α,β-ketones [17, 18] and α, β-aldehydes [19]. Kosower [16] reported the λ_{max} for the π–π^* band of the MOx in several solvents including iso-octane, $\lambda_{max} = 230.6 \pm 0.5$ nm ($43{,}365 \pm 90$ cm^{-1}), and water, $\lambda_{max} = 242.6 \pm 0.5$ nm ($41{,}220 \pm 90$ cm^{-1}). Therefore, the solvatochromic or bathochromic shift of the π–π^* band of MOx is well-known when the solvent changes from iso-octane to water, that is $-2{,}145 \pm 90$ cm^{-1} (-0.266 ± 0.011 eV). An interesting aspect cited by Kosower [16], as a private communication from Dr. M. C. Whiting from Oxford University, is that the equilibrium between *syn*- and *anti*-forms of the MOx is solvent dependent. In spite of this, no further investigation was conducted to confirm this aspect. As far as we know, there are no theoretical studies on this system which is a simple model for the large family of the conjugate ketones.

The present work is focused on the changes of the electronic properties of mesityl oxide (MOx) upon hydration using quantum mechanics (QM) and molecular mechanics (MM) in a sequential procedure [20, 21]. It is organized as follows. First, we present the theoretical approach and computational details. Gas-phase results for MOx are then reported. This is followed by the analysis of different properties of MOx in water including the structure of the solution (solvation shells), hydrogen bonding, polarization effects, dipole moments, Gibbs free energy difference between *syn–anti* MOx isomers, and electronic absorption of MOx. We conclude by stressing the importance of adopting quantum mechanics and molecular mechanics methods (QM/MM) to investigate the electronic properties of solvatochromic species in solution and also the different contributions of the solute–solvent interactions to the solvatochromic shift.

2 Computational methods and details

The gas-phase structure (see Fig. 1 top) and electronic properties of MOx were determined by carrying out calculations with different theoretical methods and basis-sets. Density functional theory (DFT) [22] with the B3LYP [23, 24] exchange–correlation functional, and Møller-Plesset second-order perturbation theory (MP2) [25, 26] were applied to investigate the gas-phase properties of MOx. These methods were used combined with the family of Pople basis-set, 6–31+G(d), 6–311+G(d,p) and 6-311++G(d,p) [27], and the family of Dunning correlated with consistent hierarchical basis-set, cc-pVDZ and aug-cc-pVxZ (x = D, T) [28]. The electronic excitation energies were determined through the application of time-dependent density functional theory (TDDFT) [29] with several approximations for the exchange–correlation functional including B3LYP and BHandHLYP as implemented in Gaussian program [30]. Some reference calculations for gas-phase excitation energies were carried out by using equation-of-motion-coupled cluster with single and doubles excitations (EOM-CCSD) [31].

The analysis of the properties of mesityl oxide in aqueous solution was based on the continuous and discrete models of the solvent. Here, we used the polarizable continuum model (PCM) [32, 33], and for the discrete model of the solvent, we performed the sequential use of quantum mechanics and molecular mechanics methods, S-QM/MM [20, 21]. In the S-QM/MM procedure, initially the liquid-phase configurations are sampled from molecular simulations, and after statistical analysis, only configurations with less than 10 % of statistical correlation are selected and submitted to quantum mechanical calculations. In our study, we used the Monte Carlo method (MC) with Metropolis sampling technique [34] to perform the liquid simulation. This sampling was carried out separately for the *syn*, *anti* and transition state (TS) structures of MOx in water. By adopting this approach, fully relaxed water structures around each different isomer and TS are generated for the statistical analysis. We assumed that the MOx structures in water are not significantly modified relative to the gas-phase ones (see discussion on gas-phase properties of MOx below). These structures (optimized at the B3LYP/6-31+G(d) level) were kept rigid during the MC sampling that was generated by the DICE program [35]. The MC sampling was carried out at normal conditions ($p = 1$ atm; $T = 298$ K) in the *NpT* ensemble for a system with one MOx molecule and $N = 500$ water molecules in a cubic box of $L \approx 24.9$ Å length and by applying periodic boundary conditions and image method. In the calculation of the pair-wise energy, the MOx interacts with all water molecules within a center-of-mass separation that is smaller than the cutoff radius $r_c = L/2$ (that is approximately 12.45 Å in this case). For separations larger than r_c, the long-range correction of the potential energy was calculated [36]. In each simulation, 15×10^6 MC steps were performed in the thermalization stage and 37.5×10^6 MC steps in the production stage.

Solute–water and water–water interactions were described by a Lennard-Jones intermolecular potential plus a Coulombic term described by the interactions between atomic point charges. Water was represented by the SPC model [37]. The choice of the Lennard-Jones parameters for the MOx was driven by the OPLS force field [38] (see Table 1). Although charges from the OPLS force field can also be used to describe the Coulombic contribution, different works pointed out the importance of taking into account the solute polarization by the solvent charges [39]. The Coulombic contribution was reparametrized by a polarization procedure taking into consideration the (SPC) charge background of the water molecules. However, to assess the importance of these effects in the specific case of MOx, three different charge distributions for the solute were considered: charges from the OPLS force field [38]; two other charges obtained from QM calculations with MP2/aug-cc-pVDZ level and using the fitting of the electrostatic potential with CHELPG procedure [40] for the isolated solute (gas-phase) and embedded in the electrostatic potential of the water background (polarized) using an iterative procedure in the sequential QM/MM scheme [39].

The Gibbs free energy of hydration, $\Delta G_x(aq)$, was calculated using the MC simulation coupled to the thermodynamic perturbation theory (TPT) [41–43], in which a series of MC runs is carried out. A system with one solute molecule and 1000 water molecules at normal conditions in the *NpT* ensemble was used for the TPT calculations. The

Table 1 Parameters of the Lennard-Jones intermolecular potential (ε in kcal/mol and σ in Å) and different sets of atomic charges (in e) and dipole moment (μ in D) for mesityl oxide at optimized geometry with B3LYP/6-31+G(d)

	OPLS			Gas phase		Polarized in water	
	ε	σ	q	q_{syn}	q_{anti}	q_{syn}	q_{anti}
C1	0.066	3.50	0.000	0.393	0.255	0.456	0.302
C2	0.066	3.50	−0.115	−0.585	−0.508	−0.634	−0.549
C3	0.066	3.50	0.500	0.746	0.716	0.868	0.837
O4	0.210	2.96	−0.500	−0.538	−0.532	−0.753	−0.769
C5	0.066	3.50	−0.180	−0.341	−0.299	−0.377	−0.311
C6	0.066	3.50	−0.180	−0.309	−0.185	−0.344	−0.168
C7	0.066	3.50	−0.180	−0.326	−0.238	−0.342	−0.319
H8	0.030	2.50	0.115	0.156	0.164	0.182	0.173
H9	0.030	2.50	0.060	0.085	0.076	0.116	0.102
H10	0.030	2.50	0.060	0.085	0.076	0.116	0.102
H11	0.030	2.50	0.060	0.084	0.079	0.088	0.086
H12	0.030	2.50	0.060	0.083	0.060	0.105	0.050
H13	0.030	2.50	0.060	0.087	0.059	0.099	0.073
H14	0.030	2.50	0.060	0.087	0.059	0.108	0.072
H15	0.030	2.50	0.060	0.080	0.074	0.095	0.110
H16	0.030	2.50	0.060	0.133	0.070	0.124	0.099
H17	0.030	2.50	0.060	0.080	0.074	0.093	0.110
μ			3.63/3.10 *syn*/*anti*	2.77	3.90	4.97 (3.85)	6.96 (5.25)

All the QM calculations were performed with MP2/aug-cc-pVDZ. For comparison, in parentheses, it is also presented the value of the dipole moment polarized with the water environment described with continuum model PCM

MOx–water interactions were the same as described above. In this case, we adopted the polarized representation of the MOx charges that takes into account the charge relaxation of MOx upon hydration. For each species, the Gibbs free energy of hydration was calculated, as usual, through a hypothetical process where the solute–water interactions are switched-off in several simulations using the double-wide sampling [42]. Four simulations were used to annihilate the Coulomb potential and six simulations to annihilate the Lennard-Jones potential. Therefore, in the total, 10 simulations were performed to vanish completely the solute molecule from the aqueous solution. Each simulation had 90×10^6 MC steps in the thermalization stage and 450×10^6 steps in production stage. More details about this procedure can be found in refs. [43, 44].

3 Results

3.1 Gas-phase properties of mesityl oxide

Full geometry optimization and vibrational frequency calculations in the gas phase of the *syn*-, TS- and *anti*-forms of the OM were performed at the B3LYP and MP2 level of theory with 6–31+G(d) and aug-cc-pVDZ basis-set. A true minimum was verified for the *syn*- and *anti*-forms, and the transition state was also confirmed with only one imaginary frequency. Theoretical data for the gas-phase-optimized geometry of MOx are presented in Table 2 for the *syn*- and *anti*-isomers (see Fig. 2). We are not aware of experimental data for the structure of mesityl oxide. As illustrated in Table 2, the optimized structures of MOx are not very dependent on the adopted theoretical methods and basis-sets. In addition, small differences are observed when we compare data for the *syn*- and *anti*-isomers. For example, at the MP2/aug-cc-pVDZ level, the C=O distances are quite similar, 1.237 Å of the *syn*-isomer and 1.240 Å for the *anti*-isomers. The geometry of the MOx was also optimized in water using the PCM model and one explicit water molecule (see Table 2). The MOx structure still not very dependent on the solvent model, but comparing to gas phase, the C=O distance stretched approximately 0.01 Å in the presence of water, as expected [45].

Results for the gas-phase dipole moment of MOx are presented in Table 3, where they are compared with experimental information [1, 16, 46, 47]. For the *syn*-isomers, our MP2/aug-cc-pVDZ value (2.80 D) practically coincides with experiment values (2.80 D [16], 2.79 D in benzene [46], 2.83 D in dioxane [46] and 2.85 D in carbon tetrachloride [47]). For the *anti*-isomer, our MP2/aug-cc-pVDZ value (3.97 D) is also in good agreement with the experimental estimations (~3.7 D [16] and 3.88 D [47]).

Table 2 Data for the optimized structures of the mesityl oxide in *syn-* and *anti*-isomers

	B3LYP/6-31+G(d)		B3LYP/aug-cc-pVDZ		MP2/aug-cc-pVDZ	
	Syn	*Anti*	*Syn*	*Anti*	*Syn*	*Anti*
Bond lengths						
C1=C2	1.354 (1.357)	1.354 (1.357)	1.355	1.354	1.364	1.364
C1–C6	1.509 (1.507)	1.512 (1.510)	1.507	1.510	1.510	1.512
C2–H8	1.090 (1.091)	1.089 (1.089)	1.093	1.093	1.097	1.097
C3=O4	1.228 (1.236) 1.235[a]	1.229 (1.240) 1.236[a]	1.226	1.227	1.237	1.240
H8...O4	3.307 (3.307)	2.423 (2.435)	3.311	2.427	3.332	2.452
C2...O4	2.405 (2.408)	2.324 (2.325)	2.404	2.324	2.419	2.330
Valence angles						
C2, C1, C6	119.9 (119.7)	118.9 (118.6)	119.9	118.9	119.8	119.0
C1, C2, C3	128.3 (128.7)	133.1 (132.5)	128.1	132.9	127.9	131.5
C2, C3, O4	124.7 (124.9)	117.5 (117.5)	124.7	117.5	124.7	117.8
C2, C3, C5	115.0 (114.9)	123.6 (123.9)	115.0	123.4	114.6	122.8
O4, C3, C5	120.3 (120.2)	119.0 (118.6)	120.3	119.0	120.8	119.4
Dihedral angles						
C6, C1, C2, C3	180.0 (180.0)	180.0 (180.0)	180.0	180.0	180.0	178.6
C6, C1, C2, H8	0.0 (0.0)	0.0 (0.0)	0.0	0.0	0.0	−1.7
C7, C1, C2, C3	0.0 (0.0)	0.0 (0.0)	0.0	0.0	0.0	−1.0
C1, C2, C3, O4	0.0 (0.0)	180.0 (180.0)	0.0	180.0	0.0	173.5
C1, C2, C3, C5	180.0 (180.0)	0.0 (0.0)	180.0	0.0	180.0	−7.8

Bond distances in Å, valence and dihedral angles in degrees. In parenthesis, it is shown the results for the MOx geometry optimized in water with polarizable continuum model, PCM

[a] Results using the MOx geometry optimized in a complex with one hydrogen-bonded water molecule

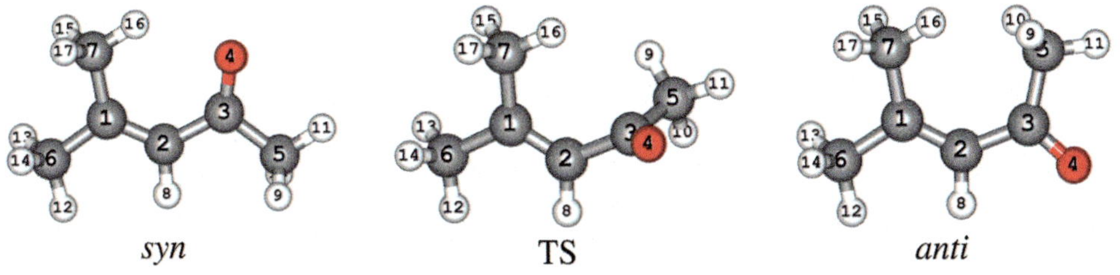

Fig. 2 Illustration of the optimized geometry of *syn-* and *anti*-isomers, and the transition state (TS) of mesityl oxide. The atomic labels are shown

These results are consistent with other relevant information concerning on the gas-phase Gibbs free energy differences associated with the conformational change between the *syn-* and *anti*-isomers and the transition state (TS). These results are reported also in Table 3 and show that $\Delta G_{x\rightarrow y}(gas)$ is dependent on the theoretical method. Thus, MP2 free energy differences ($\Delta G_{syn\rightarrow anti}(gas) = 1.34$ kcal/mol and $\Delta G_{syn\rightarrow TS}(gas) = 3.57$ kcal/mol) are significantly lower than B3LYP predictions (2.20 and 5.93 kcal/mol, respectively). Based on the calculated values of $\Delta G_{syn\rightarrow anti}(gas)$ with different method, the population of the *anti*-isomer in gas phase was calculated and reported in the parenthesis. All used methods predict that the *syn*-isomer is more stable in the gas phase and the population of this particular isomer is large than 90 % in agreement with experimental findings [18, 19, 46–49].

Using the optimized geometry with B3LYP/6-31+G(d), the results for the dipole moment and relative free energy are close to those obtained with MP2/aug-cc-pVDZ. The calculated MOx dipole moment of the gas phase is 2.77 D for the *syn*-isomer and 3.90 D for the *anti*-isomers, and the relative free energy of the *anti*-isomer and of the transition state compared with the *syn*-isomer is 0.95 and 3.36 kcal/mol, respectively. Therefore, all further results reported in this work will be obtained using the geometry optimized with B3LYP/6-31+G(d) level.

The electronic absorption spectra of mesityl oxide in different solvents with different polarities ranging from

Table 3 Gas-phase quantum mechanical dipole moments (μ in D) and Gibbs free energy changes ($\Delta G_{x \to y}$(gas) in kcal/mol) for the *syn*- and *anti*-isomers of mesityl oxide and for the transition state structure (TS)

Dipole moment, μ	*Syn*	TS	*Anti*
B3LYP/6-31+G(d)	3.14	3.30	4.38
B3LYP/aug-cc-pVDZ	3.00	3.22	4.29
MP2/aug-cc-pVDZ	2.80	3.00	3.97
MP2/aug-cc-pVDZ[a]	2.77	3.01	3.90
Experimental	2.80[b]; 2.79[c]; 2.83[d]		~3.7[b]; 3.88[d]
Relative free energy, $\Delta G_{x \to y}$(gas)		*Syn* → TS	*Syn* → *anti*
B3LYP/6-31+G(d)		5.93	2.20 (2 %)
B3LYP/aug-cc-pVDZ		6.80	2.00 (3 %)
MP2/aug-cc-pVDZ		3.57	1.34 (9 %)
MP2/aug-cc-pVDZ[a]		3.36	0.95

The experimental values for the dipole moment are also presented. The population of the *anti*-isomer in gas phase is reported in the parenthesis

[a] Geometry and thermal, enthalpy and entropic corrections at B3LYP/6-31+G(d) and electronic energy with MP2/aug-cc-pVDZ

[b] Ref. [16]

[c] In benzene [46]

[d] In carbon tetrachloride [47]

iso-octane (very low polarity, $E_T^N = 0.012$) to water (high polarity, $E_T^N = 1.000$) were determined by Kosower [16]. Two bands were well characterized in the experimental work: a strong π–π^* and a weak n–π^* band [16]. Here, we will focus on the long-wavelength π–π^* band and analyze its solvatochromic shift when the MOx change from gas phase to water. On the basis of the experimental transition energies measured for several solvent with different polarities, it is possible to obtain an experimental extrapolation to the gas-phase transition energy. Classifying the solvents used by Kosower [16] with the normalized Reichardt polarity scale [2], E_T^N, the solvent with lowest polarity is iso-octane ($E_T^N = 0.012$) and with largest polarity is water ($E_T^N = 1.000$). The λ_{max} measured for the MOx in iso-octane is 230.6 ± 0.5 nm (43,365 ± 90 cm^{-1}) and in water is 242.6 ± 0.5 nm (41,220 ± 90 cm^{-1}) [16]. In Fig. 3, we present a plot of the experimental values for the maximum absorption transition energy, E_{max}, of the MOx in several solvent [16] versus the normalized Reichardt solvent polarity scale, E_T^N [2]. Fitting the data with a linear regression, an extrapolation for the E_{max} of MOx in gas phase ($E_T^N = -0.111$) is obtained as 43630 cm^{-1} (229.2 nm). This is only 1.4 nm (265 cm^{-1} = 0.033 eV) lower than the iso-octane value. Therefore, the solvatochromic or bathochromic shift of the π–π^* band of MOx is measured as −2,145 ± 90 cm^{-1} (−0.266 ± 0.011 eV) when the solvent changes from iso-octane to water [16] and

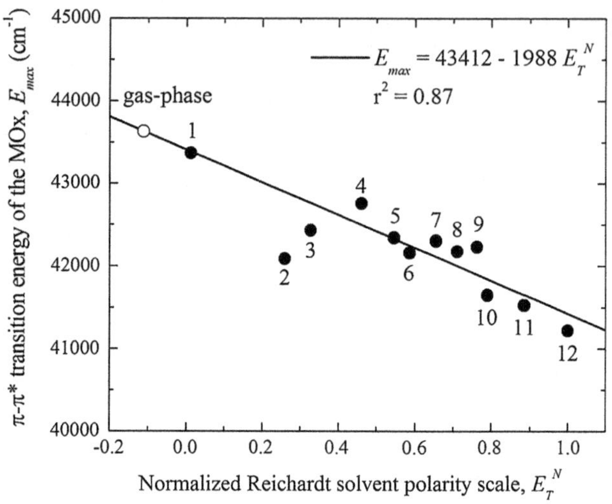

Fig. 3 Extrapolation of the maximum absorption transition energy, E_{max}, of the mesityl oxide in gas phase, $E_T^N = -0.111$, as 43,630 cm^{-1} (229.2 nm) using a linear regression in the plot of the experimental E_{max} [16] versus the normalized Reichardt solvent polarity scale, E_T^N [2]. The solvents and experimental values (E_{max} and E_T^N) are (1) iso-octane (43,365 cm^{-1} = 230.6 nm and 0.012), (2) chloroform (42,088 cm^{-1} = 237.6 nm and 0.259), (3) ethylene dichloride (42,427 cm^{-1} = 235.7 nm and 0.327), (4) acetonitrile (42,753 cm^{-1} = 233.9 nm and 0.460), (5) isopropyl alcohol (42,337 cm^{-1} = 236.2 nm and 0.546), (6) n-butyl alcohol (42,159 cm^{-1} = 237.2 nm and 0.586), (7) ethanol (42,301 cm^{-1} = 236.4 nm and 0.654), (8) 95 % ethanol (42,176 cm^{-1} = 237.1 nm and 0.710), (9) methanol (42,230 cm^{-1} = 236.8 nm and 0.762), (10) ethylene glycol (41,649 cm^{-1} = 240.1 nm and 0.790), (11) tetrafluoropropanol (41,528 cm^{-1} = 240.8 nm and 0.886) and (12) water (41,220 cm^{-1} = 242.6 nm = and 1.000)

extrapolated as −2,410 ± 90 cm^{-1} (−0.299 ± 0.011 eV) from gas phase to water.

TDDFT results for the gas-phase π–π^* electronic excitation energies (in cm^{-1}) are gathered in Table 4. The best agreement with the extrapolated experimental gas-phase value of 43,630 cm^{-1} (229.2 nm) is the *syn*-isomer of 43,685 cm^{-1} (228.9 nm) using the B3LYP/6-311+G(d,p), which is inside the experiment precision of 0.5 nm (90 cm^{-1} in this region). For the *anti*-isomer, with the same level of calculation, the electronic excitation was obtained as 44,895 cm^{-1} (222.7 nm). For both properties, dipole moment and excitation energy, the *syn*-isomer calculated values are in better agreement with the experimental values than the *anti*-isomer.

As discussed above assuming a *syn/anti* population of 91:9 % at MP2/aug-cc-pVDZ in the gas phase, the properties can be calculated as weighted averages of the *syn*- and *anti*-isomers. This leads to a calculated gas-phase dipole moment of 2.90 D with MP2/aug-cc-pVDZ and excitation energy of 43,794 cm^{-1} (228.3 nm) with B3LYP/6-311+G(d,p). This represents only a small change of 0.10 D in the dipole moment and of −0.6 nm in the excitation

Table 4 Gas-phase results for the π–π^* excitation energies (E in cm^{-1}) of the *syn-* and *anti*-isomers of the mesityl oxide and the difference between the excitation energies, $\delta E_{syn \to anti} = E_{anti}(gas) - E_{syn}(gas)$

	B3LYP			BHandHLYP			EOM-CCSD		
	Syn	*Anti*	$\delta E_{syn \to anty}$	*Syn*	*Anti*	$\delta E_{syn \to anty}$	*Syn*	*Anti*	$\delta E_{syn \to anty}$
6-31+G(d)	44,084	45,273	1,189	46,753	48,407	1,654	49,604	51,863	2,259
6-311+G(d,p)	43,783	45,031	1,248	46,414	48,114	1,700			
6-311++G(d,p)	43,685	44,895	1,210	46,352	48,065	1,713			
6-311++G(d,p)[a]	43,440	44,570	1,130	46,015	47,590	1,575			
6-311++G(d,p)[b]	43,580	44,590	1,010	46,205	47,675	1,470			
cc-pVDZ	44,998	46,290	1,292	47,651	49,307	1,656	51,056	53,476	2,420
aug-cc-pVDZ	43,543	44,735	1,192	46,194	47,920	1,726	48,556	50,814	2,258
aug-cc-pVTZ	43,510	44,725	1,215	46,134	47,858	1,724			
In iso-octane[c]	43,365 ± 90								
In gas phase[d]	43,630 ± 90								

[a] Results using the MOx geometry optimized in water with PCM model

[b] Results using the MOx geometry optimized in a complex with one hydrogen-bonded water molecule

[c] Experimental value of E_{max} measured in iso-octane [16]

[d] Extrapolated value of E_{max} in gas phase (see Fig. 3)

energy when compared to a population of 100 % of *syn*-isomer in gas phase. Therefore, in this work, for simplification, we assumed a composition of 100 % of MOx *syn*-isomer in gas phase and 100 % of MOx *anti*-isomer in aqueous solution.

The molecular orbitals involved in this π–π^* excitation are the highest occupied molecular orbital (HOMO) to the lowest unoccupied molecular orbital (LUMO). The graphic representations of these molecular orbitals are presented in Fig. 4. As it can be seen, there is a charge transfer character in this HOMO → LUMO excitation: from the $(CH_3)_2C$=C region to the CHCO region. This evidence corroborates the experimental interpretation of the π–π^* band for other molecules of the family of non-saturated α,β-ketones [18].

At this theoretical level (B3LYP/6-311++G(d,p)), the π–π^* excitation energy for the *anti*-isomer is blue-shifted by 1,210 cm^{-1} (0.15 eV) relative to the *syn*-isomer, $\delta E_{syn \to anti} = E_{anti}(gas) - E_{syn}(gas)$, as shown in Table 4. It should be observed that this value is quite similar to the one predicted by using the aug-cc-pVTZ basis-set (1,215 cm^{-1}). Therefore, the 6–311++G(d,p) basis-set can be a good choice for the calculations in the MOx–water solution since it is computationally less demanding. It should be noticed that gas-phase excitation energies of MOx and the $\delta E_{syn \to anti}$ exhibit some dependence on the theoretical method and significant deviations from experiment are observed for the B3LYP and BHandHLYP functionals. Using the aug-cc-pVDZ basis-set, the B3LYP predicts a $\delta E_{syn \to anti}$ of 1,192 cm^{-1} (0.15 eV) and the BHandHLYP predicts a $\delta E_{syn \to anti}$ of 1,726 cm^{-1} (0.21 eV). But the results also indicate that $\delta E_{syn \to anti}$ is not very dependent on the basis-set, ranging from 1,189 to 1,292 cm^{-1} using B3LYP and from 1,654 to 1,726 cm^{-1} to BHandHLYP.

In the absence of experimental data for $\delta E_{syn \to anti}$, it could be of interest to carry out calculations relying on a reference theoretical method to predict excitation energies. EOM-CCSD gas-phase excitation energies for the MOx *syn-* and *anti*-isomers are also reported in Table 4. The EOM-CCSD results indicate that DFT, in particular, the B3LYP exchange–correlation functional, underestimates $\delta E_{syn \to anti}$. However, the difference between EOM-CCSD (2,258 cm^{-1} = 0.28 eV) and BHandHLYP (1,726 cm^{-1} = 0.21 eV) calculations is quite small (0.07 eV).

The π–π^* excitation energies were also calculated for the MOx geometries optimized in water (using the PCM model and a cluster with one explicit water molecule). These values are also reported in Table 4. The results indicate that the C=O distance stretching induced by the presence of water leads to a small red-shift of the π–π^* excitation energies relative to the values for the isolated optimized geometry. In this case, the $\delta E_{syn \to anti}$ is 1,575 cm^{-1} (0.20 eV or −7.2 nm) using the optimized geometry with PCM and 1,470 cm^{-1} (0.18 eV or −7.7 nm) using the optimized geometry with one bonded water molecule, both results with BHandHLYP/6-311++G(d,p). Therefore, the inclusion of the MOx relaxation in the presence of the aqueous solution provides a small red-shift of around 100–300 cm^{-1} (0.01–0.04 eV). This result is in concordance with values previously reported for acetone in water [45].

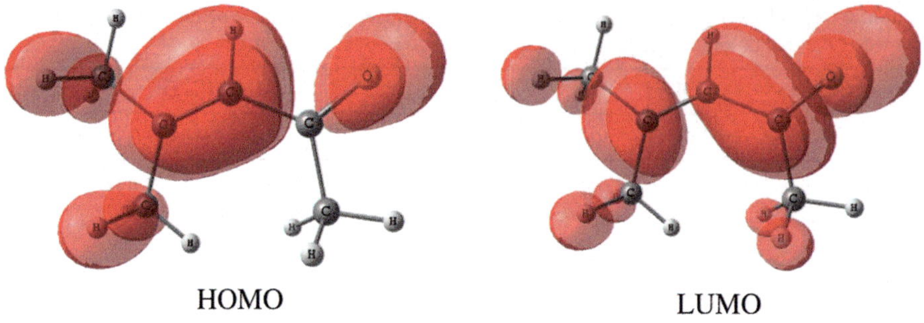

Fig. 4 Illustration of the frontier molecular orbitals involved in the π–π^* transition: highest occupied molecular orbital (HOMO) and lowest unoccupied molecular orbital (LUMO)

3.2 Properties of the mesityl oxide–water solution

3.2.1 Polarization effects of water on hydrated mesityl oxide

The gas-phase QM calculated dipole moment of the MOx *anti*-isomer (3.90 D) is larger than the *syn*-isomer (2.77 D). In contrast with this, the OPLS charges predict that the *syn*-isomer has a stronger dipole (3.63 D) than the *anti*-isomer (3.10 D), see Table 1. Therefore, these atomic charges are inadequate to be used in molecular simulations. The dipole moments from the polarized procedure using S-QM/MM in water are 4.97 D (*syn*) and 6.96 D (*anti*) and thus correspond to our estimate of the (effective) dipole moments for MOx in water. These three sets of atomic charges and the dipole moments are shown in Table 1. For comparison, the polarized dipole moments of the *syn*- and *anti*-isomers of the MOx were also calculated with the continuum model, PCM, for the water environment. These values are 3.85 D for the *syn*-isomer and 5.25 D for the *anti*-isomer that are around 30 % less than that obtained with the iterative procedure using S-QM/MM method. This is a reasonable result considering that the specific interactions, like the hydrogen-bonded water molecules, are included in the QM calculation as SPC point charges. Therefore, its effect seems to be stronger than that considered in the continuum model.

Figure 5 shows the convergence of the MOx dipole moment during the iterative polarization procedure. Each point in the Fig. 5 corresponds to a QM calculation of the MOx with an embedding represented by an electrostatic average of the solvent (ASEC) [49], where the solvent molecules are considered as point charges. In the present case, this ASEC includes the contribution of 56,250 (= 75 × 250 × 3) point charges from the superposition of 75 statistically uncorrelated configurations with 250 water molecules. As illustrated in Fig. 5, for both isomers, it is observed that an increase of 80 % of the dipole moment relative to the gas-phase value after convergence is attained. The set of atomic charges obtained with this polarization procedure are shown in Table 1.

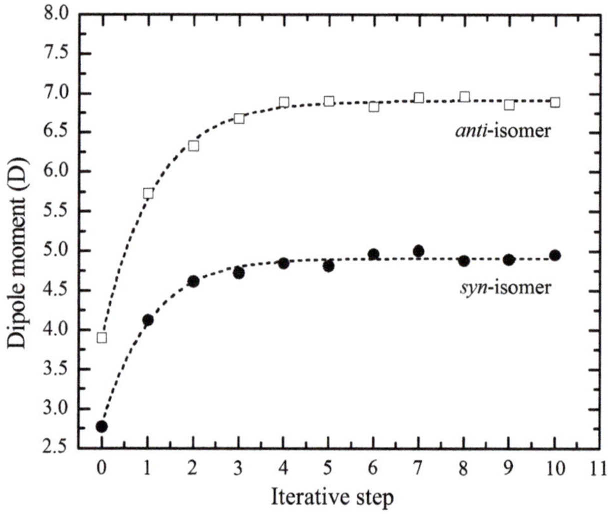

Fig. 5 Behavior of the mesityl oxide dipole moment in the iterative polarization procedure for the *syn*-isomers (*solid circle*) and *anti*-isomers (*open square*)

3.2.2 Structure and hydrogen bonding in the mesityl oxide–water solution

The analysis of the structure was carried out by using the configurations generated with the gas-phase atomic charges (non-polarized) and those generated in-water (polarized) atomic charges. The organization of the water around MOx can be discussed through the analysis of the radial distribution function or equivalently by the minimum distance distribution function (MDDF). This distribution corresponds to the histogram calculated with the minimum distance between the atoms of the solute MOx and the atoms of the solvent water molecules and normalized by a parallelogramic uniform distribution [50]. The MDDF is shown in Fig. 6 for the sampling with polarized (bottom) and non-polarized charges (top) and for the *syn*- and *anti*-isomers. The MDDF exhibits a shoulder with a peak at 1.8 Å, which is related to a micro-solvation shell and the hydrogen bond formation (see discussion on hydrogen bonding below). Integration of the MDDF shoulder up to 2.1 Å leads to the number of micro-solvation shell of

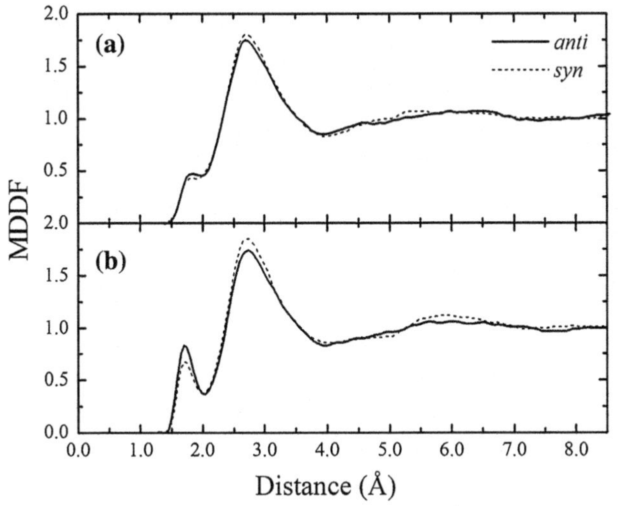

Fig. 6 Minimum distance distribution function (MDDF) for the *syn* and *anti* mesityl oxide isomers with polarized (**a**) and non-polarized (**b**) atomic charges

approximately 2–3 water molecules, which is also reported in Table 5. Integration of the MDDF up to the first minimum (4.0 Å), that defines the first hydration shell, leads to the first coordination number of 34–35 water molecules. The integration up to the second minimum (7.5 Å), defining a second shell, leads to a second coordination number of 141–145 water molecules. These structural data are gathered in Table 5 for different models and MOx isomers. The results indicate that the organization of the water molecules around the solute is not very dependent on the choice of the charges (non-polarized or polarized). Moreover, with the exception of micro-solvation shell, quite small differences are observed for the water structure around the MOx *syn*- and *anti*-isomers.

The analysis of the hydrogen bond (HB) can be performed using the micro-solvation shell. However, the small distance between the solute–solvent does not fully defines the HB formation. Therefore, the more accurate analysis of the HB should be carried out by defining the geometric and energetic criteria [51, 52]. In this work, we used a distance between the MOx oxygen (O_M) and water oxygen (O_w), $R(O_M \ldots O_W) \leq 3.25$ Å (value of the first minimum in the O_MO_W RDF), an angle between O_M and O_WH, $\theta(O_M \ldots O_WH) \leq 40°$ and a binding energy (calculated by the interaction model) $E_{ij} \leq -0.01$ kcal/mol. This energy criterion is used only to guarantee that the water molecule is binding to the MOx. Using these criteria, the results for the statistical analysis on HB distribution and the average properties of HB in the MOx–water solution and its standard deviation are reported also in Table 5. The following specific aspects are worth noticing. For both non-polarized and polarized charges, the percentage of water molecules with only one hydrogen bond (1 HB) to MOx is dominant. If we consider the *syn*-isomer, the numbers are 55.3 and 42.7 % for non-polarized and polarized charges, respectively. However, comparison between the *syn*- and *anti*-isomers shows that in contrast with the results for the non-polarized charges, polarized leads to quite similar populations for both isomers (42.7 %) and a significant change on the number of water molecules with two hydrogen bonds (2 HB) to MOx. The fraction of configurations with three hydrogen bonds (3 HB) is quite small.

Table 5 Number of water molecules in the micro-, first and second solvation shells and the hydrogen-bonded (HB) to the mesityl oxide, and the statistical analysis (average and standard deviation values) of HB properties for two different charge distribution models used in the MC simulations: non-polarized (gas phase) and polarized in water (see Table 1)

	Non-polarized		Polarized	
	Syn	*Anti*	*Syn*	*Anti*
Number of water molecules				
Micro-solvation shell (up to 2.1 Å)	1.9	1.9	2.4	2.8
1st shell (up to 4.0 Å)	34	34	35	35
2nd shell (up to 7.5 Å)	143	141	145	143
% of Configurations with				
0 HB	36.3	35.6	22.1	0.4
1 HB	55.3	44.0	42.7	42.7
2 HB	8.4	28.8	33.5	55.3
3 HB	0.0	1.6	1.7	1.6
HB average properties				
Number of water $\langle N_{HB} \rangle$	0.7	1.1	1.2	1.6
Distance $\langle R(O_M \ldots O_W) \rangle$ (in Å)	2.82 ± 0.15	2.82 ± 0.16	2.74 ± 0.15	2.75 ± 0.16
Angle $\langle \theta(O_M \ldots O_WH) \rangle$ (in °)	15 ± 9	15 ± 9	12 ± 7	12 ± 7
Energy $\langle E_{ij} \rangle$ (in kcal/mol)	−4.9 ± 1.2	−4.8 ± 1.2	−7.3 ± 1.5	−8.2 ± 1.5

These results are not surprising if we take into account the increase of the charges (and dipole moment) of MOx in water when a polarization procedure is adopted to estimate the charge distribution (see Fig. 5 and Table 1). By taking the average of the HB, $\langle N_{HB} \rangle$, for the different isomers, we conclude that they are of ~1 and ~1.5 for the *syn*- and *anti*-isomers, respectively. Therefore, the polarization effect does not increase significantly the quantity of HB (only 0.5 HB), but makes it more oriented, shorter and stronger, as it can be seen in Table 5. For the *anti*-isomer, the HB O…O distance decreases from 2.82 to 2.75 Å, the O…OH angle decreases from 15° to 12°, and the binding energy increases from 4.8 kcal/mol to 8.2 kcal/mol for the non-polarized and the polarized models, respectively. These results show the relevance of using a polarized model for the solute.

3.2.3 *Gibbs free energy and conformational change of mesityl oxide in water*

The conformational equilibrium in solution was investigated by calculating the Gibbs free energy change associated with the interconversions between different isomers (*syn* and *anti*) and with the transition state structure (TS). The Gibbs free energy change in aqueous solution $\Delta G_{x\to y}(\text{aq})$ can be determined through the definition of a thermodynamic cycle and estimated as:

$$\begin{aligned}\Delta G_{x\to y}(\text{aq}) &= \Delta G_{x\to y}(\text{gas}) + (\Delta G_x(\text{aq}) - \Delta G_y(\text{aq})) \\ &= \Delta G_{x\to y}(\text{gas}) + \Delta\Delta G_{x\to y}(\text{aq})\end{aligned} \quad (1)$$

where x and y represent the different solutes (*syn*, *anti* and TS), $\Delta G_x(\text{aq})$ is the Gibbs free energy of hydration associated with the species x (calculated by TPT), $\Delta\Delta G_{x\to y}(\text{aq})$ is the differential free energy of hydration, and $\Delta G_{x\to y}(\text{gas})$ is the gas-phase free energy difference that can be evaluated by QM methods (see Table 3). Moreover, it is possible to separate the different contributions (electrostatic and non-electrostatic) to the total $\Delta G_x(\text{aq})$ [43, 44]. Once, the $\Delta G_x(\text{aq}) = -\Delta G_{x\to 0}(\text{aq})$, where $\Delta G_{x\to 0}(\text{aq})$ is the free energy of annihilation of the species x in aqueous solution. This annihilation process is divided in two parts: first the Coulombic potential is annihilated then the Lennard-Jones potential. All the calculated values of these variations of free energy are presented in Table 6. For comparison, the free energies of hydration of the MOx were also calculated with the continuum model, PCM, for the water environment and are shown in Table 6. Only for these QM calculations of the solvation free energy, the PCM was

Table 6 Non-electrostatic and electrostatic contributions to the Gibbs free energy of hydration, $\Delta G_x(aq)$, the total hydration free energy, $\Delta\Delta G_{x\to y}(\text{aq})$, and the Gibbs free energy change in aqueous solution, $\Delta G_{x\to y}(\text{aq})$, of the MOx isomers (*syn* and *anti*) and transition state structure (TS) obtained with QM calculation with PCM-UAHF(HF/6-31G(d)) and with TPT in the MC simulations

	Syn	TS	*Anti*
QM continuum model, PCM-UAHF			
ΔG_x(aq) non-electrostatic contribution	2.19	2.73	2.22
ΔG_x(aq) electrostatic contribution	−5.96	−6.90	−7.49
Total ΔG_x(aq)	−3.77	−4.17	−5.27
		Syn → TS	*Syn* → *anti*
$\Delta\Delta G_{x\to y}$(aq)		−0.40	−1.50
		−0.20[a]; −0.39[b]	−1.27[a]; −1.43[b]
$\Delta G_{x\to y}$ (aq)[c]		3.17	−0.16
	Syn	TS	*Anti*
TPT in MC simulation			
ΔG_x (aq) non-electrostatic contribution	−0.64 ± 0.03	−1.12 ± 0.06	−1.26 ± 0.06
ΔG_x (aq) electrostatic contribution	−12.71 ± 0.64	−12.79 ± 0.63	−16.20 ± 0.80
Total ΔG_x (aq)	−13.35 ± 0.67	−13.91 ± 0.69	−17.46 ± 0.86
		Syn → TS	*Syn* → *anti*
$\Delta\Delta G_{x\to y}$ (aq)		−0.56 ± 0.96	−4.11 ± 1.09
$\Delta G_{x\to y}$ (aq)[c]		3.01 ± 0.96	−2.77 ± 1.09 (99 %)

All values are in kcal/mol, and the deviations are the statistical error. The population of the *anti*-isomer in water is reported in the parenthesis

[a] Obtained with PCM-UAHF(B3LYP/aug-cc-pVDZ)

[b] Obtained with PCM-UAHF(MP2/aug-cc-pVDZ)

[c] Obtained from the thermodynamic cycle (see Eq. 1) with $\Delta G_{x\to y}$ (gas) using MP2/aug-cc-pVDZ (see Table 3)

performed using the united atom topological model applied on atomic radii optimized for the HF/6-31G(d) level of theory (UAHF). This UAHF radii option is consistent with the non-electrostatic parametrized terms of the solvation free energy implemented in Gaussian program and gives the best performance when compared with the experimental results [53, 54].

It should be noticed that the free energy of hydration, ΔG_x(aq), of MOx conformations (*syn*, *TS* and *anti*) exhibits some dependence on the theoretical method. In Table 6, comparing the non-electrostatic contribution obtained with PCM solvent model (all positive values) and with TPT in simulation (all negative values), it seems that the PCM is underestimating the van der Waals and dispersion interactions (negative contributions to the free energy of hydration) or super estimating the cavitation contribution (positive contributions to the free energy of hydration), or both. Additionally, comparing the electrostatic contribution, we believe that the PCM solvent model is underestimating the in-water polarization of the MOx, as described in previous section. Therefore, all the electrostatic contribution values obtained with PCM are smaller than with TPT. However, for all used method, the *anti*-isomer solvates better in water solution than the *syn*-isomer by around 4 kcal/mol using TPT in simulations and 1.5 kcal/mol using QM calculations with PCM-UAHF. Using the QM calculations for $\Delta G_{x\rightarrow y}$(gas) reported in Table 3 (best result with MP2/aug-cc-pVDZ) and Eq. (1), the free energy change in water, $\Delta G_{x\rightarrow y}$(aq), can be estimated and the results are also reported in Table 6. Based on the results for $\Delta G_{x\rightarrow y}$(aq) calculated with TPT, the *anti*-isomer is dominant in the aqueous solution by more than 99 % of population. These results show that the *syn–anti* equilibrium of the MOx is solvent dependent and it is in agreement with the citation of Kosower [16].

3.2.4 Absorption spectrum of mesityl oxide in water

According with experimental data, the π–π^* band of MOx is shifted by $-2{,}145 \pm 90$ cm^{-1} (-0.266 ± 0.011 eV) when the solvent changes from iso-octane to water [16] and extrapolated as $-2{,}410 \pm 90$ cm^{-1} (-0.299 ± 0.011 eV) from gas phase to water using a linear regression in the a plot of the experimental values for the maximum absorption transition energy, $E_{\max}$, of the MOx in several solvent [16] versus the normalized Reichardt solvent polarity scale, E_T^N [2] (see Fig. 3). In order to discuss, the absorption spectrum of solute MOx (S) in water different theoretical representations of the solvent was adopted: (1) the polarizable continuum model (S + PCM), which provides a simplified representation of the solvent as a continuum dielectric medium; (2) the average solvent electrostatic configuration (S + ASEC), where 250 water molecules of

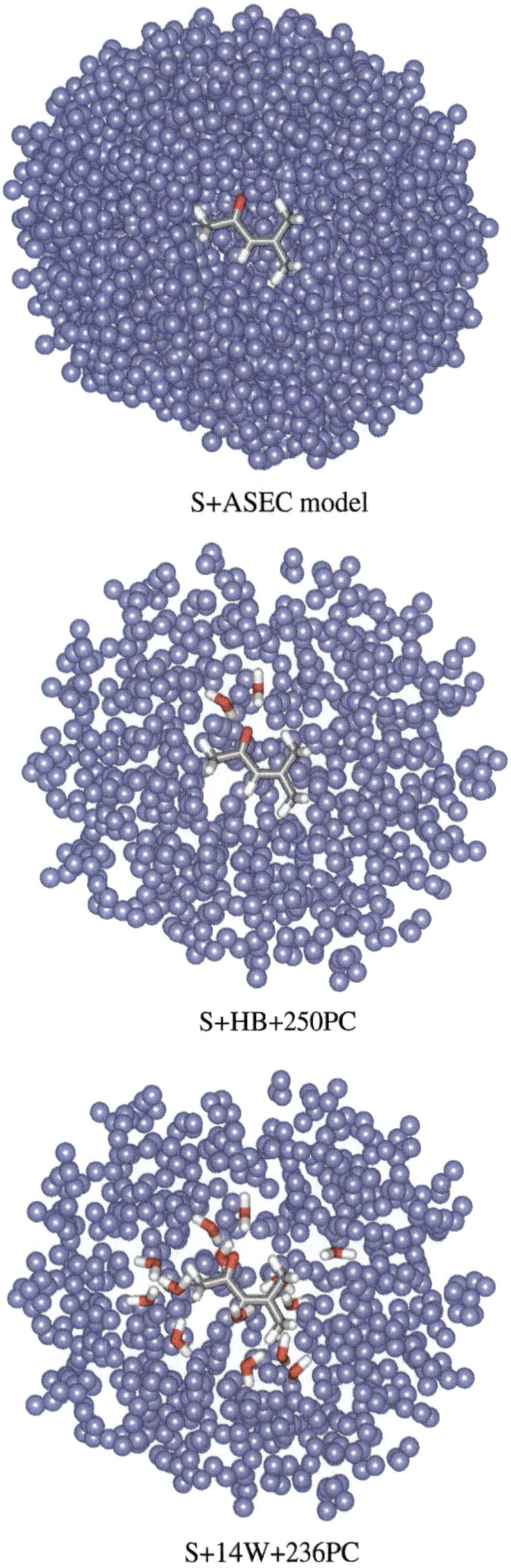

Fig. 7 An illustration of the models used in the QM calculation: S+ASEC, one configuration of S+HB+250PC and one configuration of S+14W+236PC. The point charges are represented by small spheres

Table 7 Solvent effect on the π–π^* excitation of mesityl oxide

Model	ΔE_{syn}		ΔE_{anti}		$\Delta E_{syn\rightarrow anti}$	
	B3LYP	BHandHLYP	B3LYP	BHandHLYP	B3LYP	BHandHLYP
S+PCM	−1,510	−1,551	−1,855	−2,088	−645	−375
S+ASEC	−1,645	−1,615	−2,385	−2,526	−1,175	−813
S+HB+250PC	−2,129 ± 59	−2,021 ± 60	−3,025 ± 85	−3,022 ± 70	−1,815 ± 85	−1,309 ± 70
S+8W+242PC	−3,039 ± 112	−2,644 ± 71	−4,077 ± 92	−3,979 ± 83	−2,867 ± 92	−2,266 ± 83
S+12W+238PC	−3,187 ± 85	−2,777 ± 72	−4,356 ± 101	−4,156 ± 87	−3,146 ± 101	−2,443 ± 87
S+14W+236PC	−3,256 ± 83	−2,810 ± 73	−4,459 ± 115	−4,232 ± 88	−3,249 ± 115	−2,519 ± 88
Experimental shift of iso-octane to water [16]						−2,145 ± 90
Extrapolated shift of gas phase to water (see Fig. 3)						−2,410 ± 90

$\Delta E = E(\text{aq}) - E(\text{gas})$ where E(aq) and E(gas) are the excitation energies in aqueous solution and gas phase, respectively. TDDFT calculations with the 6-311++G(d,p) basis-set. ΔE_x values assume both, E(aq) and E(gas), in the x-isomer, and the $\Delta E_{syn\rightarrow anti}$ values considered the excitation energy difference between the gas-phase MOx in the *syn*-isomer and the in-water MOx in the *anti*-isomers and represent the total solvent effect on the π–π^* excitation. The description of the models is as follows: *S* MOx solute molecule, *PCM* the polarizable continuum model, *ASEC* the average solvent electrostatic configuration, where the 250 water molecules of 75 configurations are represented as point charges, *HB* the hydrogen-bonded water molecules explicitly included in the QM calculation, *nW* the nearest *n* water molecules explicitly included in the QM calculation, where $n = 8$, 12 and 14, *mPC* remaining *m* water molecules described by point charges, where $m = 242$, 238 and 236. For all nW + mPC models, the nearest 250 water molecules were used ($n + m = 250$). All values are in cm^{-1}

75 configurations are superposed and represented as re-scaled point charges; (3) the hydrogen bond model (S+HB+250PC) that includes explicitly a few water molecules hydrogen-bonded to MOx (in average 1.2 for the *syn*-isomer and 1.6 for the *anti*-isomer, see Table 5) and this system is embedded in the electrostatic field of the remained 250 water molecules represented by point charges (250PC); and (*iv*) the explicit model (S+nW+mPC) where the nearest n water molecules to MOx (nW) are explicitly included in the calculation, where $n = 8$, 12 and 14, and this system is embedded in the electrostatic field of the remained solvent molecules represented by m point charges, where $m = 242$, 238 and 236. For all S+nW+mPC models, the nearest 250 water molecules were used ($n + m = 250$). In Fig. 7, we show an illustration of the S+ASEC model (top), one configuration of the S+HB+250PC (center) and S+14W+236PC (bottom) models. Note that for the S+PCM and S+ASEC models, only one QM calculation is performed to obtain the π–π^* excitation energy of the MOx in water because both models represent the solvent average behavior. But for the S+HB+250PC and S+nW+mPC models, the π–π^* excitation energy of the MOx in water is obtained as an average over 75 statistically uncorrelated configurations generated from the MC simulations.

The shift of the π–π^* excitation energy of the MOx upon hydration, considering the *syn* → *anti* isomerization change, is

$$\Delta E_{syn\rightarrow anti} = E_{anti}(\text{aq}) - E_{syn}(\text{gas}), \tag{2}$$

where $E_{anti}(aq)$ is the excitation energy of the hydrated MOx *anti*-isomer, and $E_{syn}(gas)$ is the excitation energy of the isolated MOx *syn*-isomer. This expression can be re-written as

$$\begin{aligned}\Delta E_{syn\rightarrow anti} &= (E_{anti}(\text{gas}) + \Delta E_{anti}) - E_{syn}(\text{gas}) \\ &= \delta E_{syn\rightarrow anti} + \Delta E_{anti},\end{aligned} \tag{3}$$

where $\Delta E_{anti} = E_{anti}(\text{aq}) - E_{anti}(\text{gas})$ is the excitation energy shift due to the solvent effect on electronic structure of the specific *anti*-isomer, and $\delta E_{syn\rightarrow anti} = E_{anti}(\text{gas}) - E_{syn}(\text{gas})$ is the excitation energy shift due to the isomerization change. The ΔE_x values are reported in Table 7, where x is the *syn*- and *anti*-isomers and the $\delta E_{syn\rightarrow anti}$ are reported in Table 4. Therefore, on the basis of Eq. (3), the analysis of the solvent effect on the MOx excitation spectrum should take into consideration two distinct and opposite contributions. The first one is related to the *syn* → *anti* conformational change leading to a blue-shift of 1,210 cm^{-1} [1,713 cm^{-1}] for B3LYP [BHandHLYP] calculations with the 6–311++G(d,p) basis-set (see Table 4). The second contribution is related to the solvent effect and leads to a red-shift of the excitation energy. The results reported in Table 7 for ΔE_x are in keeping with the experimentally observed trend that points out to a red-shift of the excitation energy when we move from a low- (iso-octane) to a high-polarity solvent (water). However, the magnitude of ΔE_x is clearly dependent on of the adopted procedure to represent the solvent effect. In the specific case of the *anti*-isomer and for BHandHLYP calculations, ΔE_{anti} changes from −2,088 cm^{-1} (PCM) to −4,232 ± 88 cm^{-1} for the model S+14W+236PC, where the solute is in interaction with 14 (explicit) water molecules embedded in the electrostatic field of 236 water molecules represented as point charges. Moreover, it appears that ΔE is also dependent on the number (mW) of water molecules explicitly included in the calculations. However, the difference between the results with 12 and 14 explicit water molecules (76 cm^{-1} using BHandHLYP and 103 cm^{-1} using B3LYP for the *anti*-isomer) indicates that ΔE_x is near convergence with 14 W. Therefore, it will be assumed that our best estimate for ΔE_x relies on the (S+14W+236PC) calculation.

According to Eq. (3) by adding to ΔE_{anti} the contribution from the gas-phase *syn* → *anti* isomerization ($\delta E_{syn \rightarrow anti}$ = 1,210 and 1,713 cm^{-1}, for B3LYP and BHandHLYP, respectively), the total solvatochromism on the MOx π–π^* excitation energy is a red-shift of -3,249 ± 115 cm^{-1} using B3LYP and −2,519 ± 88 cm^{-1} using BHandHLYP, whereas the experimental extrapolated value is −2,410 ± 90 cm^{-1}. Our theoretical estimate (using BHandHLYP) is red-shifted by only 109 cm^{-1} (0.014 eV) from the extrapolated experimental result and indicates that our best approximation for representing the solvent effect on the MOx π–π^*excitation is adequate.

4 Conclusions

Sequential QM/MM calculations were carried out to investigate the hydration of a model non-saturated α,β-ketone. Our main conclusions are the following. The *syn–anti* equilibrium of mesityl oxide (MOx) is solvent dependent. In gas-phase and low-polarity solvents, it exists dominantly in *syn*-form and in aqueous solution in *anti*-form. This conclusion is supported by Gibbs free energy calculations in gas phase and in-water by applying quantum mechanical calculations with polarizable continuum model and thermodynamic perturbation theory in Monte Carlo simulations using a polarized MOx model. The consideration of the in-water polarization of the MOx is very important to correctly describe the solute–solvent electrostatic interaction. We predict that the dipole moment of MOx in water is ~5 D for the *syn*-isomer and ~7 D for the *anti*-isomer. It represents an increase of 2.2 D in the *syn*-isomer and 3.1 D the *anti*-isomer compared to the gas-phase MOx. Statistical analysis of MOx–water hydrogen bonding and comparison between polarized and non-polarized models indicate that by using polarized charges on the solute, the number of configurations with two water molecules hydrogen-bonded to MOx (*anti*) is increased by ~26 % and the hydrogen bonds became ~70 % stronger relative to the non-polarized model. Our best estimate for the shift of the π–π^* transition energy of MOx, when it changes from gas-phase to water solvent, shows a red-shift of −2,520 ± 90 cm^{-1}, which is only 109 cm^{-1} (0.014 eV) below experimental extrapolation of −2,410 ± 90 cm^{-1}. Therefore, this red-shift of around −2,500 cm^{-1} can be divided in two distinct and opposite contributions. One contribution is related to the *syn* → *anti* conformational change leading to a blue-shift of ~1,700 cm^{-1}. Other contribution is the solvent effect on the electronic structure of the MOx leading to a red-shift of around −4,200 cm^{-1}. Additionally, this red-shift caused by the solvent effect on the electronic structure (~4,200 cm^{-1}) can be composed by approximately 60 % due to the electrostatic bulk effect (~2,500 cm^{-1}), 10 % due to the explicit inclusion of the hydrogen-bonded water molecules (~500 cm^{-1}) and 30 % due to the explicit inclusion of the nearest water molecules (~1,200 cm^{-1}). The set of information on the structure, thermodynamics and electronic properties of MOx in water stresses the interest of performing theoretical studies that combine statistical, molecular and quantum mechanics with explicit solvent molecules for investigating solvent effects.

Acknowledgments Work partially supported by FCT (Portugal), CNPq, CAPES, FAPESP, INCT-FCx and nBioNet (Brazil).

References

1. Reichardt C (1979) Solvent effects in organic chemistry. Verlag Chemie, Weinheim, New York
2. Reichardt C (1994) Chem Rev 94:2319
3. Maitland GC, Rigby M, Smith EB, Wakeham WA (1987) Intermolecular forces-their origin and determination. Oxford University Press, Oxford
4. Ratajczak H, Orville-Thomas WJ (1980–1982) Molecular Interactions. vol. 1-3 Wiley, New York
5. Timasheff SN (1970) Acc Chem Res 3:62
6. Suppan P, Ghoneim N (1997) Solvatochromism. The Royal Society of Chemistry, London
7. Sheppard SE (1942) Rev Mod Phys 14:303
8. Archer WL (1996) Industrial solvents handbook. Dekker, New York
9. Reichardt C (1965) Angew Chem 4:29
10. Novaki PL, El Soud OA (1997) Ber Bunsenges Phys Chem 101:902
11. Kosower EM (1968) An introduction to physical organic chemistry. Wiley, New York
12. Streitwieser A, Heathcock CH, Kosower EM (1992) Introduction to organic chemistry. McMillan, New York
13. Buckingham DA, Lippert E, Bratos E (eds) (1978) Organic liquids—structure, dynamics and chemical properties. Wiley, New York
14. Stross FH, Monger JM, Finch HV (1947) J Am Chem Soc 69:1627
15. Gray HF Jr, Rasmussen RS, Tunnicliff DD (1947) J Am Chem Soc 69:1630
16. Kosower EM (1958) J Am Chem Soc 80:3261
17. Woodward RB (1941) J Am Chem Soc 63:1123
18. Turner RB, Voitle DM (1951) J Am Chem Soc 73:1403
19. Forbes WF, Shilton R (1959) J Am Chem Soc 81:786
20. Coutinho K, Canuto S (2000) J Chem Phys 113:9132
21. Coutinho K, Canuto S, Zerner MC (2000) J Chem Phys 112:9874
22. Parr RG, Yang W (1994) Density functional theory of atoms and molecules. Oxford Science Publications, Oxford
23. Becke AD (1993) J Chem Phys 98:5648
24. Imamura Y, Otsuka T, Nakai H (2007) J Comput Chem 28:2067
25. Møller C, Plesset MS (1934) Phys Rev 46:618

26. Krishnan R, Pople JA (1978) Int J Quantum Chem 14:91
27. Ditchfield D, Hehre WJ, Pople JA (1971) J Chem Phys 54:724
28. Dunning Jr TH(1987) J Chem Phys 90:1007
29. Fiolhais C, Nogueira F, Marques M (eds) (2003) A primer in density functional theory. Chapter 4, Springer, New York
30. Frisch MJ, Trucks GW, Schlegel HB, Scuseria GE, Robb MA, Cheeseman JR, Montgomery JA, Vreven T, Kudin KN, Burant JC, Millam JM, Iyengar SS, Tomasi J, Barone V, Mennucci B, Coss M, Scalmani G, Rega N, Petersson GA, Nakatsuji H, Hada M, Ehara M, Toyota K, Fukuda R, Hasegawa J, Ishida M, Nakajima T, Honda Y, Kitao O, Nakai H, Klene M, Li X, Knox JE, Hratchian HP, Cross JB, Bakken V, Adamo C, Jaramillo J, Gomperts R, Stratmann RE, Yazyev O, Austin AJ, Cammi R, Pomelli C, Ochterski JW, Ayala PY, Morokuma K, Voth GA, Salvador P, Dannenberg JJ, Zakrzewski VG, Dapprich S, Daniels AD, Strain MC, Farkas O, Malick DK, Rabuck AD, Raghavachari K, Foresman JB, Ortiz JV, Cui Q, Baboul AG, Clifford S, Cioslowski J, Stefanov BB, Liu G, Liashenko A, Piskorz P, Komaromi I, Martin RL, Fox DJ, Keith T, Al-Laham MA, Peng CY, Nanayakkara A, Challacombe M, Gill PMW, Johnson B, Chen W, Wong MW, Gonzalez C, Pople JA(2004) Gaussian 03, Revision D.01. Gaussian, Inc., Wallingford CT
31. Stanton JF, Bartlett RJ (1993) J Chem Phys 98:7029
32. Miertus S, Scrocco E, Tomasi J (1981) J Chem Phys 55:117
33. Cancès MT, Mennucci B, Tomasi J (1997) J Chem Phys 107:3032
34. Metropolis N, Rosenbluth AW, Rosenbluth MN, Teller AH, Teller E (1953) J Chem Phys 21:1087
35. Coutinho K, Canuto S (2009) DICE: a monte carlo program for molecular liquid simulation, v. 2.9. University of São Paulo, São Paulo
36. Allen MP, Tildesley DJ (1987) Computer simulation of liquids. Clarendon, Oxford
37. Berendsen HJC, Postma JPM, van Gunsteren WF, Hermans J (1981) In: Pullman B (ed) Intermolecular forces. Reidel, Dordrecht
38. Jorgensen WL, Maxwell DS, Tirado-Rives J (1996) J Am Chem Soc 118:11225
39. Georg HC, Coutinho K, Canuto S (2006) Chem Phys Lett 429:119
40. Breneman CM, Wiberg KB (1990) J Comp Chem 11:361
41. Zwanzig RW (1954) J Chem Phys 22:1420
42. Jorgensen WL, Buckner KJ, Boudon S, Tirado-Rives J (1988) J Chem Phys 89:3742
43. Georg HC, Coutinho K, Canuto S (2005) Chem Phys Lett 413:16
44. Pasalic H, Aquino AJA, Tunega D, Haberhauer G, Gerzabek MH, Georg HC, Moraes TF, Coutinho K, Canuto S, Lischka H (2010) J Comp Chem 31:2046
45. Coutinho K, Saavedra N, Canuto S (1999) J Mol Struct (Theochem) 466(1999):69
46. Estok GK, Sikes JH (1953) J Am Chem Soc 75:2745
47. Bentley JB, Everard KB, Marsden RJB, Sutton LE (1949) J Chem Soc 2957. doi:10.1039/JR9490002957
48. Mecke R, Noack K (1960) Chem Ber 93:210
49. Coutinho K, Georg HC, Fonseca TL, Ludwig V, Canuto S (2007) Chem Phys Lett 437:148
50. Georg HC, Coutinho K, Canuto S (2007) J Chem Phys 126:34507
51. Mezei M, Beveridge DL (1981) J Chem Phys 74:622
52. Sceats MG, Rice SA (1981) J Chem Phys 72:3236
53. Barone V, Cossi M, Tomasi J (1997) J Chem Phys 107:3210
54. Ho J, Klamt A, Coote ML (2010) J Phys Chem A 114:13442

Theor Chem Acc (2013) 132:1301
DOI 10.1007/s00214-012-1301-0

REGULAR ARTICLE

Density functional and chemical model study of the competition between methyl and hydrogen scission of propane and β-scission of the propyl radical

Marc E. Segovia · Kenneth Irving · Oscar N. Ventura

Received: 20 June 2012 / Accepted: 7 November 2012 / Published online: 5 December 2012

Abstract In this work, we study the competence between the reactions of hydrogen and methyl scission during thermal cracking and combustion of propane, the emergence of the two isomers of the propyl radical, *n*-propyl and *i*-propyl, and their subsequent β-scission reaction to ethene and methyl radical. The purpose of the study was to analyze the accuracy of density functional (DFT) methods as applied on this relatively well-known subset of the reactions implied in the production of propylene oxide from propane and propene. Conventional (B3LYP, B3PW91) and state-of-the-art (PBE0, M06, BMK) DFT methods were employed, and their accuracy checked against experimental data and calculations performed using model chemistries (complete basis set CBS-4M, QB3, and APNO, and G4 methods) and ab initio methods (MP2, CCSD(T) with a large 6-311 ++G(3df,2pd) basis set). The results obtained at the BMK level for the thermodynamics of the reactions are closer to experimental data than those afforded by any other DFT method and very similar actually to CBS or CCSD(T) results, even if a medium size basis set is used. Activation energies determined using two- and three-parameter Arrhenius equations are also very good, but the preexponential factors are incorrect. Tunneling and internal rotation corrections must be applied to obtain semiquantitative results.

Dedicated to Professor Marco Antonio Chaer Nascimento and published as part of the special collection of articles celebrating his 65th birthday.

M. E. Segovia · K. Irving · O. N. Ventura (✉)
Computational Chemistry and Biochemistry Group (CCBG), DETEMA, Facultad de Química, UdelaR, CC1157 Montevideo, Uruguay
e-mail: onv@fq.edu.uy

Keywords Propane cracking · Density functional methods · BMK · Thermochemistry · Kinetics · CCSD(T) calculations · Composite models

1 Introduction

Hydrocarbon cracking is the name of the process by which higher molecular weight hydrocarbons are converted into lower molecular weight species, through carbon–carbon bond scission [1]. This process is central to obtain gasoline by petroleum refining, and several methods have been developed to make it more efficient.

Thermal cracking is the simplest and oldest method. It relies on the use of high temperatures (~450°–750°) and pressures (up to about 70 atm.) to produce the scission of C–C and C–H bonds, producing radicals that then react with other neutral or radical species. Olefins are produced during this thermal process, especially high yields of ethylene and smaller amounts of alpha olefins, like propene for instance. Normally, cracking processes involve many species and hundreds of reactions, a reason for which the experimental studies of cracking reaction kinetics are very difficult [2–4]. On the other side, optimizing the performance of these processes for different raw materials and process conditions requires detailed and accurate kinetic models and, thus, the knowledge of precise thermodynamic and kinetic data [5, 6].

Computational chemistry has advanced dramatically in the last decades, due to the increase in the availability of moderately priced hardware and software and may be an alternative or complementary method to study the thermodynamics and kinetics of cracking processes. It is well known that very accurate total and relative energies of molecular species can be obtained at the coupled cluster

CCSD(T) level when a sufficiently large basis set is used. Regretfully, this method is computationally very demanding and can be applied routinely only to small molecules. Therefore, simpler methods, which may be applied to larger molecules, need to be gauged against accurate CCSD(T) calculations and experimental data, when available.

In this paper, we describe a computational study of the cracking reactions of propane and the production of olefins, especially propene, as a preliminary step in our research of gas-phase production of propylene oxide. The purpose of the study is twofold. On the one hand, we aim to compare different computational schemes applied to a subset of reactions for which experimental data exist. On the other hand, we want to obtain precise estimates of the thermochemistry and kinetics of the radical chain initiation, propagation and termination reactions involved in the mechanism. Previous computational and experimental studies on this area of research have been performed by several authors, which results we will use to compare to our own.

Sabbe et al. [7] did the most recent theoretical calculations in 2007. The composite methods CBS-QB3 and G3B3, as well as density functionals MPW1PW91, BB1 K, and BMK, were used with several corrections for tunneling and internal rotation for the calculation of rate coefficients for carbon-centered radical addition and β-scission reactions. They found that the DFT-based values for β-scission rate coefficients deviate significantly from the experimental ones at 300, and the DFT methods do not accurately predict the equilibrium coefficient. Zheng and Blowers[8] published in 2006 an investigation of the propyl radical β-scission reaction kinetics and energetics using quantum chemical Gaussian-3 (G3) method and a specially designed complete basis set (CBS) composite energy method. Good agreement with experimental results was reported for the CBS calculations. Curran [9] in 2006 determined rate constant expressions for C1-C4 alkyl and alkoxy radicals decomposition via β-scission based on the reverse, exothermic reaction, the addition of a hydrogen atom or an alkyl radical to an olefin or carbonyl species with the decomposition reaction calculated using microscopic reversibility on the basis of experimental data. Hunter and East [10] studied theoretically, at the B3LYP, MP2, CCSD(T), and G2 levels, the bond strengths of the C–C bonds in alkanes up to $C_{20}H_{42}$. Bond strength was found to be very constant (88 kcal mol^{-1}), except for the α and β bonds (89 and 87 kcal mol^{-1}, respectively). For thermal cracking, the results suggest that the most favored initiation step is the breaking of the β bond of the alkane to create an ethyl radical. Xiao et al. [11] in 1997 performed UHF and DFT calculations to investigate the detailed kinetics and mechanisms of hydrocarbon thermal cracking. They studied specifically the bond dissociation energy (BDE) for C–C homolytic scission—a process occurring without transition states—finding values of about 95 kcal/mol at the MP2/6-31G* level and 89 kcal/mol at the B3LYP/6-31G* level. H-transfer reactions do exhibit transition states, as expected, with activation energies of about 15–17 kcal/mol at the MP2/6-31G* level and 10–12 kcal/mol at the B3LYP level. Finally, they found an activation energy for β-scission of about 30–33 kcal/mol, while the activation energy for the reverse addition reaction was found to be about 5 kcal/mol. The optimized transition state structure is product-like for the radical decomposition reaction and reactant-like for the addition reaction.

Bencsura et al. [12] in 1992 produced the most recent experimental data on the reaction kinetics of the propyl radical, based on previous work. Radicals were produced by pulsed laser photolysis and their unimolecular decay subsequently studied by photoionization mass spectrometry. Tsang [13] in 1988 produced a compilation of revised and evaluated data on the kinetics of reactions involving propane and the propyl radical, among other species. The general mechanism for the decomposition of propane was initially determined by Papic and Laidler [14, 15] who experimentally identified most of the products which are consistently predicted in the mechanism proposed in this paper. We followed their rationale for the development of our own model.

2 Methods

Ab initio, density functional (DFT) and model chemistry methods have been used in this paper. B3LYP [16–18], B3PW91 [16, 17, 19], PBE0 [20–22], M06 [23] and BMK [24] DFT geometry optimizations [25] and analytic frequency [26–28] calculations were performed using the 6-31G(d,p) (basis I), 6-311 +G(d,p) (basis II), and 6-311 ++G(3df,2pd) (basis III) Pople basis sets, as representative of small, medium, and large polarized basis sets. Post-Hartree–Fock second-order Møller–Plesset (MP2 [29–32]) geometry optimizations and frequency calculations were performed using only the larger basis set. For a small set of molecules, CCSD(T) [33, 34] frequency calculations were performed on the MP2 optimum geometries, using also the larger basis set. Complete basis set (CBS) model chemistry methods [35–41], namely CBS-4M [39], CBS-QB3 [39, 40], and CBS-APNO [41], were also employed to calculate thermodynamic properties of all stable reactants, products, and intermediate radicals, but not for transition states. Those calculations were also performed using Curtiss Gaussian-4 theory (G4) [42].

The choice of DFT methods employed in the calculations was not arbitrary. The default choice is the adiabatically corrected (or hybrid) B3LYP method, which gives fairly accurate results and has been widely used in the last decade. We ourselves prefer B3PW91—that is, with a different choice of the correlation energy potential—because we showed in a number of publications that both methods usually agree, but when they do not, then B3PW91 is closer to the experimental results than B3LYP. The reason is that the fitting parameters in B3LYP were actually not fitted but taken from B3PW91. The three other methods chosen represent the implementation of increasingly complex and accurate methodologies. PBE0 is the adiabatically coupled version of the non-empirical potential of Perdew, Burke, and Erzenhoff. M06 is a meta-hybrid GGA method which depends not only on the Laplacian of the density, but also on the spin kinetic energy density. The method depends on several parameters which are adjusted on a training set and is particularly suited to describe non-covalent interactions. Finally, the BMK DFT method includes also the spin kinetic energy density, but it is fitted to thermochemical and kinetic data, so that its main field of applicability is to study thermochemical and kinetic properties of chemical reactions.

Basis sets of different complexity were used with the DFT methods in order to assess their accuracy, since it is known that DFT methods do not depend so heavily on the basis set size as ab initio methods do. The most exact ab initio calculations performed in this paper are the CCSD(T)//MP2/6-311 ++G(3df,2pd) done on the smaller molecules. Regretfully, some of the species studied (depicted in the scheme in Fig. 1) were too large to follow this procedure. In those cases, the CBS-APNO results were considered as the nearer best and used for comparison to the presumably less accurate DFT and MP2 calculations.

Chemical models as CBS or Gn are also called composite energy methods or compound models and aim to reach high level of accuracy at a moderate cost through addition of energy corrections calculated at different levels. G4 is the latest in the series of Gn chemical models, an improvement over G3 theory. Geometries as well as zero point energies are determined using B3LYP DFT with the 6-31G(2df,p) basis set. Correlation level calculations are performed using Møller–Plesset perturbation theory up to fourth order and with coupled cluster theory. Large basis sets, including multiple sets of polarization functions, are used in the correlation calculations. The CBS methods, on the other side, rely on the extrapolation toward the complete basis set limit using the N^{-1} asymptotic behavior of N-configuration pair natural orbital (PNO) expansions. The more accurate method, CBS-APNO, includes up to QCISD(T)/6-311 ++G(2df,p) calculations for estimating the correlation energy.

Thermodynamic calculations were performed at 0 K, room temperature and 600 K, in all cases assuming atmospheric pressure. Rate constants were calculated using the standard canonical transition state theory with Eckart's correction for tunneling [43] and both with and without hindered rotor treatment of the internal rotation. Rate coefficients are calculated based on conventional transition state theory in the high-pressure limit. For the bimolecular radical addition, the rate coefficient is calculated according to Eq. 1:

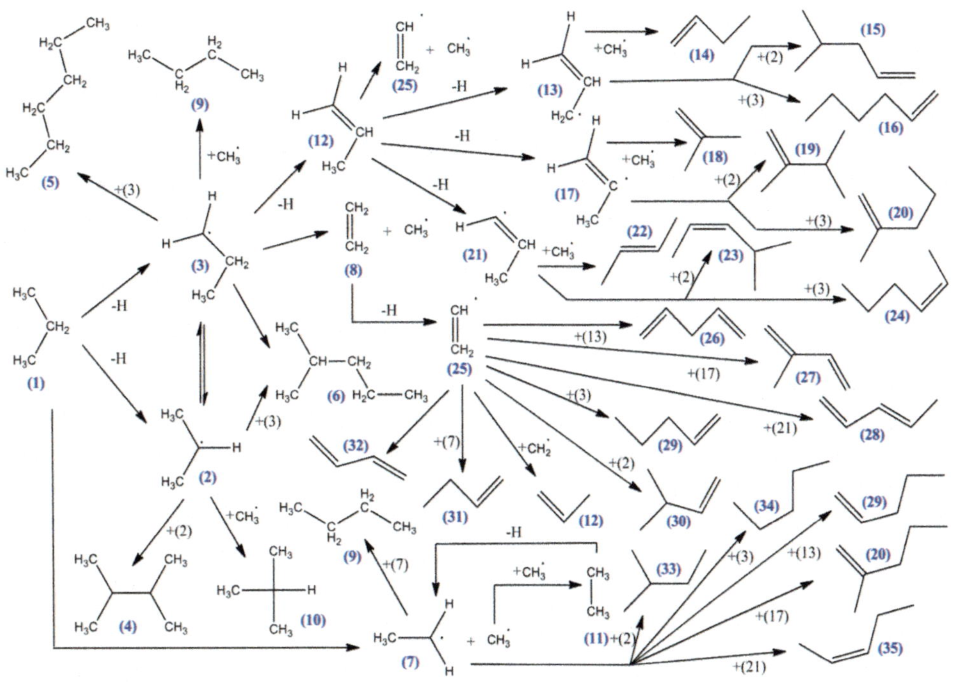

Fig. 1 Schematic depiction of the main reactants and products in the mechanism of cracking of propane

$$k\infty(T) = \kappa(T)(kBT/h) \left(n_{\text{opt},\#}q^{\#}/n_{\text{opt,A}}q^{\text{A}}n_{\text{opt,B}}q^{\text{B}}\right)\text{e}^{-\Delta\text{E(0K)/RT}}.$$

The monomolecular rate coefficients are calculated according to:

$$k\infty(T) = \kappa(T)(kBT/h)\left(n_{\text{opt},\#}q^{\#}/n_{\text{opt,P}}q^{\text{P}}\right)\text{e}^{-\Delta\text{E(0K)/RT}}$$

where q is the total molar partition function per unit volume, $\kappa(T)$ is the tunneling coefficient, and ΔE(0 K) is the activation barrier at 0 K including zero point vibrational energy (ZPVE). The number of optical isomers n_{opt} enters the equation because the partition functions are calculated for a single optical isomer whereas all configurations, including those that are not directly thermally accessible from the reference configuration, should be accounted for. External and internal symmetry numbers are contained within the partition functions.

Standard Arrhenius parameters A and E_a were derived from a regression of ln(k) against $1/T$ in the way

$$\ln k = \ln A - E_a/RT$$

Energies, geometries, frequencies, and other properties were calculated using the Gaussian 09 computer code [44]. The thermochemical and kinetic analysis were performed using the program TAMkin [45].

Figure 1 shows a general scheme of the species and reactions investigated in this paper. Forty-one chemical species were combined into 100 chemical reactions, to describe approximately the reaction system related to thermal cracking of propane in the absence of oxygen. Although possible from a theoretical point of view, we have not included in this study the CC and CH bond-breaking reactions of hydrocarbons larger than C_3H_8, generated by radical chain termination reactions. This decision was taken on the ground that the concentration of larger olefins is much smaller than those of propane, ethylene and related C_2 and C_3 species. Therefore, we expect that radicals generated from C_n species ($n > 3$) will be of marginal importance and were not included in the research.

3 Results and discussion

3.1 Initial steps

Modeling of thermal cracking of propane was performed by Sundaram and Froment [46] on the basis of experimental work by Van Damme et al. [47]. The kinetics of mercury-photosensitized decomposition of propane was studied in a pair of papers by Papic and Laidler [14, 15] in the seventies.

C–C and C–H bond breaking are possible decomposition reactions for propane, according to the following mechanisms

$$C_3H_8 \rightarrow C_2H_5\cdot + CH_3\cdot \quad (1)$$

$$C_3H_8 \rightarrow iC_3H_7\cdot + H\cdot \quad (2)$$

$$C_3H_8 \rightarrow nC_3H_7\cdot + H\cdot \quad (3)$$

$C_2H_5\cdot$ has only one structural conformer, while $C_3H_7\cdot$ is the global formula for two structural isomers, the 1-propyl ($nC_3H_7\cdot$) and 2-propyl ($iC_3H_7\cdot$) radicals, whose fates differ, as seen in Fig. 1. 2-propyl radical is more stable than 1-propyl radical at all levels, because of the stabilization produced by hyper-conjugation of the radical with the neighbor methyl groups. Both the hydrogen atom and the methyl radical produced in this chain initiation reactions can participate in dehydrogenation propagation reactions, in the form

$$C_3H_8 + H\cdot \rightarrow iC_3H_7\cdot + H_2 \quad (4)$$

$$C_3H_8 + H\cdot \rightarrow nC_3H_7\cdot + H_2 \quad (5)$$

$$C_3H_8 + CH_3\cdot \rightarrow iC_3H_7\cdot + CH_4 \quad (6)$$

$$C_3H_8 + CH_3\cdot \rightarrow nC_3H_7\cdot + CH_4 \quad (7)$$

We can use these first seven reactions to make a primary analysis of the performance of different theoretical methods. Since we are interested primarily in the comparison among methods, we will limit ourselves to the comparison of standard enthalpies of reaction at 298.15 K, ΔH°_{298}, for each one of the seven reactions, a property for which experimental data are readily available. These numbers are collected in Table 1.

Notice in the first place the effect of basis set extension in the results afforded by DFT methods. The most important effect is produced by the extension of the valence and diffuse part of the basis set, not by the addition of polarization functions. Extension from 6-31G(d,p) to 6-311+G(d,p) modifies the results for different reactions by 2–3 kcal/mol. Further extension of the polarization space from (d,p) to (3df,2pd) produces a tenfold smaller effect. The maximum absolute deviation (MAD) and root mean square error (RMSE) with respect to the experimental results obtained using reported standard enthalpies of formation show for almost all methods the counterintuitive result that an improvement in the basis set actually worsens the agreement with experiment. Only in the case of the τ-dependent BMK DFT method, developed specifically for thermochemical and kinetic studies, the increase in the basis sets results in better agreement with the experiment.

All DFT methods perform as well as or better than composite model chemistries employed. In particular, the results obtained with the M06 and BMK methods, independently of the basis sets, are better than even the more elaborated post-Hartree–Fock CBS-APNO and G4 calculations, where multiple steps are combined to obtain a good estimation of correlation energy. The pure MP2

Table 1 Enthalpies of reaction (at 298.15 K, in Kcal/mol) for reactions (1)–(7)

Method	Basis[a]	(R1) $C_3H_8 \rightarrow C_2H_5\cdot + CH_3\cdot$	(R2) $C_3H_8 \rightarrow iC_3H_7\cdot + H\cdot$	(R3) $C_3H_8 \rightarrow nC_3H_7\cdot + H\cdot$	(R4) $C_3H_8 + H\cdot \rightarrow iC_3H_7\cdot + H_2$	(R5) $C_3H_8 + H\cdot \rightarrow nC_3H_7\cdot + H_2$	(R6) $C_3H_8 + CH_3\cdot \rightarrow iC_3H_7\cdot + CH_4$	(R7) $C_3H_8 + CH_3\cdot \rightarrow nC_3H_7\cdot + CH_4$	MAD	RMSE
B3LYP	I	85.5	96.9	101.1	−9.3	−5.1	−8.7	−4.5	4.2	2.5
	II	81.6	94.8	98.7	−9.7	−5.9	−8.3	−4.4	6.6	3.8
	III	81.5	94.7	98.7	−10.0	−6.0	−8.5	−4.5	6.7	3.9
B3PW91	I	86.4	95.6	99.9	−7.5	−3.3	−8.7	−4.5	3.4	2.1
	II	83.2	93.8	97.7	−7.9	−3.9	−8.3	−4.4	5.2	3.3
	III	83.0	93.7	97.7	−8.1	−4.1	−8.6	−4.6	5.3	3.5
PBE0	I	88.8	95.4	99.5	−5.1	−0.9	−8.5	−4.4	3.6	2.0
	II	85.6	93.4	97.2	−4.7	−0.8	−8.1	−4.3	5.6	3.0
	III	85.4	93.2	97.2	−4.7	−0.7	−8.4	−4.4	5.8	3.2
M06	I	90.4	97.1	101.6	−4.9	−0.4	−9.0	−4.5	3.3	2.0
	II	87.2	95.2	98.7	−4.3	−0.8	−8.4	−4.9	3.8	2.3
	III	86.2	95.2	98.7	−4.2	−0.7	−8.7	−5.1	3.8	2.4
BMK	I	92.0	98.9	102.6	−3.5	0.2	−7.5	−3.8	3.8	2.2
	II	88.4	97.2	100.6	−4.5	−1.1	−7.1	−3.7	2.1	1.2
	III	88.6	97.3	100.8	−4.6	−1.1	−7.3	−3.8	2.1	1.2
CBS-4 M		90.4	99.1	102.2	4.0	7.0	−6.3	−3.2	10.2	5.3
CBS-QB3		89.7	98.9	102.0	−0.6	2.4	−6.5	−3.4	5.6	2.8
CBS-APNO		89.7	99.1	102.1	−0.3	2.7	−6.3	−3.3	5.9	3.0
G4		88.0	97.9	100.7	0.3	3.1	−6.6	−3.8	6.3	3.2
MP2	III	92.0	97.6	100.0	−0.4	2.1	−5.0	−2.5	5.3	3.7
CCSD(T)	III	87.0	96.5	99.7	−6.4	−3.1	−6.8	−3.6	2.5	1.3
Experimental[b]		88.2 ± 0.5	99.0 ± 0.5	101.0 ± 0.5	−5.1 ± 0.5	−3.2 ± 0.5	−5.7 ± 0.5	−3.8 ± 0.5		
Other experimental[c]		88.7[d]								

Maximum absolute deviations (MAD) and root mean square errors (RMSE) calculated with respect to experimental values estimated using data in Ref. [48]

[a] Definition of the basis sets: 6-31G(d,p) (basis I), 6-311 +G(d,p) (basis II), and 6-311 ++G(3df,2pd) (basis III)

[b] Experimental values were obtained using the data in Ref. [48]

[c] Alternative experimental data were obtained from individual sources

[d] Tsang [13]

calculation, which uses a large basis set but lacks the correlation energy correction contained in the composite methods, exhibits a larger RMSE, about three times larger than the BMK RMSE, with a considerably larger computational effort. As one could expect, the CCSD(T)//MP2/6-311 ++G(3df,2pd) procedure affords values closest to the experimental data, with an RMSE of 0.9 kcal/mol. This is not significantly better than the 1.2 kcal/mol registered at the BMK level using basis sets II or III.

Thus, using this restricted set of reactions, we can conclude that BMK/6-311 ++G(3df,2pd) is the best level of calculation we can reach from the thermochemical point of view. Moreover, the use of the extended basis set is not really necessary. The results are equally good resorting to the cheaper BMK/6-311 +G(d,p) procedure, therefore, allowing the calculation of enthalpies for larger molecules.

Zheng [49] performed composite method calculations on several cracking reactions. His results are collected in Table 2, together with those obtained by us. As expected, the G4 results obtained in this paper are closer on average to the experimental results than either G2 or G3. This is particularly noticeable for the abstraction of H from the $CH_3C(CH_3)_2$ radical (eighth row in Table 2) for which a 5 kcal/mol difference with G3 makes the G4 result much nearer to the experimental enthalpy of reaction. Neither the complete basis set methods nor the CCSD(T)//MP2 procedure is able to give an accurate enthalpy of reaction for this hydrogen abstraction. In fact, all composite methods behave in a similar way. The specially designed CBS-RAD(MP2) [50] method affords worse results than CBS-QB3 and the more complete CBS-APNO method is only marginally better. All of them give results of comparable accuracy to that of CCSD(T)//MP2, including the failure in the above-mentioned dehydrogenation reaction. The much cheaper BMK/6-311 +G(d,p) DFT method results in an maximum absolute deviation and rms error similar to G4 and better than those of the other methods, with the extra advantage that it can be applied to much larger molecules. A peculiar deviation from experiment is noticed for the CCSD(T) decomposition of the ethyl radical into the methyl radical and methylene carbene $CH_2CH_3\cdot \rightarrow :CH_2 + CH_3\cdot$. The CCSD(T)//MP2 value of 105.0 kcal/mol is approximately 6 kcal/mol larger than the experimental 98.8 kcal/mol. All the other methods, including MP2//MP2 which is not shown in the table, afford values differing less than 2 kcal/mol from experiment. This may be related to the lack of optimization of the geometry at the CCSD(T) level or a drawback of the method. Geometry optimization of the three species at the CCSD(T) level does afford the required correction, which can be linked almost completely to the better representation of the triplet CH_2 species. The agreement of CCSD(T) with experiment is now as good as usual.

Bond-breaking reactions (1)–(3) proceed without transition states. Theory and experiment agree well in this case; about 10 kcal/mol more is needed to break any of the C–H bonds than the C–C bond. Reactions (4)–(7) proceed instead through transition states. Further insight into these elementary reactions can then be obtained from the analysis of the reaction paths. Transition states were located for those reactions at the DFT and MP2 level of theory. Important geometrical parameters for these structures are shown in Fig. 2.

Transition states are more product-like when the hydrogen is abducted from the primary than from the secondary carbon, both for the abstraction by H· and $CH_3\cdot$. The structures obtained at the BMK/6-311 ++G(3df,2pd) level are in all cases looser than those obtained using the MP2 method and the same basis set. There are obviously no experimental data to compare with, but the differences among the methods are anyway small enough as to be of no substantial consequence with respect to the reaction mechanism. Abstraction reactions proceed in a simple way, as expected, and no extraordinary geometrical features are to be noticed. Barrier heights at different theoretical levels are collected in Table 3.

There are in these cases several sources of experimental data. The more consistent energies of activation for the direct and reverse reactions are those of Tsang [13], who reassessed previous experimental work and fitted modified Arrhenius equations in most cases. Nonetheless, we have considered also the other experimentally derived Arrhenius data for the direct reactions. In this case, all temperature dependence is built only in the exponential part. Therefore, the activation energies are larger than the exponential factors in Tsang results. Errors in the activation energies are higher than the ones for the enthalpies of reaction. In all cases, the theoretically derived energies of activation agree worse with Tsang values than with those from other sources. The influence of the inclusion of higher levels of correlation on the activation energy is quite obvious, comparing for instance the G4 values for reaction (4) to the MP2 ones. Results obtained using BMK are not in this case better than those obtained with other DFT methods, M06 being the best method overall, judging from the agreement with the experimental data available. Differences among methods are not so important, however, as to prefer definitely one over the other.

Since the four reaction paths involve proton transfers, it is possible that the disagreement between the theoretical and experimental data may be explained by tunneling effects. Moreover, the temperature dependence of the preexponential factor may be also a cause for the discrepancies between theoretical and experimental data. Therefore, we calculated the parameters in the Arrhenius and modified Arrhenius equations for the forward reactions using the BMK/6-311 ++G(3df,2pd) method and the Eckart correction for tunneling. They are collected in Table 4.

The activation energies are in reasonable agreement with the experimental data for Arrhenius equations. The preexponential factors, however, are not so good. A better agreement is obtained with the modified Arrhenius equation. Now, the preexponential factors and temperature dependence are reasonable, and activation energies are not far from experimental value (MADs are under 1 kcal/mol). From a kinetic point of view, however, the description is semiquantitative at best. If one plots the rate constants obtained theoretically and experimentally for these reactions, the picture shown in Fig. 3 is obtained. There is a quite a difference between experimental and theoretical data at each temperature, which could be adjusted by a simple multiplicative factor for each reaction. The calculations predict that the ratio of formation of the 1-propyl radical to the 2-propyl radical would be about 5 times faster theoretically than observed experimentally. Both theoretically and experimentally, one observes that increase in the temperature equalizes the rate of formation of both radicals, but experimentally this happens faster than what the theoretical calculations predict. This failure is not corrected even considering internal rotation to perform more precise thermochemical calculations.

3.2 Propyl radicals

Reactions (2)–(7) have in common the production of the two isomers of the propyl radical, *i*-propyl, **2**, and *n*-propyl, **3**. About 100 kcal/mol is necessary to obtain these products by hydrogen loss through reactions (2) and (3), without transition states. Reaction (1) requires about 12 kcal/mol less energy and should then proceed faster. However, after the methyl radicals start to accumulate in the reactor because of reaction (1), hydrogen abstraction through reactions (6) and (7) become more favorable, due to an activation energy of only about 10 kcal/mol. The process would lead then to the emergence of $C_2H_5\cdot$ and the propyl radicals, which would later react further.

Table 2 Comparison of computed enthalpies of reaction (at 298.15 K, in Kcal/mol, kcal/mol) for reactions in Ref. [49]

Reaction	G2[a]	G3[a]	G4[b]	CBS-QB3[a]	CBS-RAD(MP2)[a]	CBS-APNO[b]	BMK[b,c]	CCSD(T)//MP2[d]	Exp.[e]
$\cdot CH_2CH_3 \rightarrow CH_2CH_2 + H\cdot$	33.63	34.51	35.90	35.00	34.17	36.36	37.43	36.16	36.24
$CH_3CH_3 \rightarrow \cdot CH_2CH_3 + H\cdot$	101.34	99.65	100.68	100.20	100.49	101.80	100.25	99.40	100.50
$CH_3CH_2CH_2 \rightarrow CH_3CH{\cdot}CH_2 + H\cdot$	30.73	31.58	34.50	31.08	31.33	32.58	34.08	33.33	33.08
$CH_3CH{\cdot}CH_3 \rightarrow CH_3CHCH_2 + H\cdot$	33.67	34.25	37.33	34.56	34.31	35.58	37.48	36.58	34.98
$CH_3CH_2CH_3 \rightarrow \cdot CH_2CH_2CH_3 + H\cdot$	101.77	100.40	100.68	101.43	101.02	102.13	100.61	99.73	101.02
$CH_3CH_2CH_2CH_3 \rightarrow \cdot CH_2CH_2CH_2CH_3 + H\cdot$	103.63	99.99	100.65	100.94	100.69	102.29	101.01		101.56
$CH_3CH_2CH_2CH_3 \rightarrow CH_3CH{\cdot}CH_2CH_3 + H\cdot$	99.06	97.63	97.83	98.17	98.16	99.40	97.45		98.13
$CH_3C(CH_3)_2 \rightarrow \cdot CH_2C(CH_3)_2 + H\cdot$	33.77	34.20	39.67	34.51	34.43	34.88	40.75		41.10
$\cdot CH_2CH_3 \rightarrow CH_2 + \cdot CH_3$	98.69	97.49	99.05	100.79	98.53	100.27	99.92	105.04 (97.34)	98.77
$\cdot CH_2CH_2CH_3 \rightarrow CH_2CH_2 + \cdot CH_3$	24.21	20.96	23.19	22.77	21.76	23.97	25.25	23.42	23.46
$\cdot CH_2CH_2CH_2CH_3 \rightarrow CH_2CH_2 + \cdot CH_2CH_3$	21.25	21.10	22.36	21.85	22.16	22.95	22.83		21.67
$\cdot CH_2CH(CH_3)_2 \rightarrow \cdot CH_3 + CH_3CHCH_2$	19.27	20.19	23.86	22.21	21.20	22.14	23.14		22.70
$CH_3CH{\cdot}CH_2CH_3 \rightarrow \cdot CH_3 + CH_3CHCH_2$	21.30	21.05	24.82	23.43	22.32	23.34	25.29		23.70
Maximum absolute deviation	7.33	6.90	2.30	6.59	6.67	6.22	2.50	6.27 (1.60)	
Root mean square error	2.67	2.46	1.04	2.05	2.16	1.94	1.19	2.53 (1.04)	

MADs and RMSEs have been recalculated for the data present in Ref. [49], taking into account the experimental data in Ref. [48]

[a] From Ref. [49]

[b] This work

[c] BMK/6-311 +G(d,p)

[d] Values in parenthesis are calculated at the CCSD(T)//CCSD(T)/6-311 ++G(3df,2pd) level

[e] Experimental results taken from Ref. [48], except for $CH_2CH_2CH_2CH_3$; in this case, data from Ref. [49] were taken instead

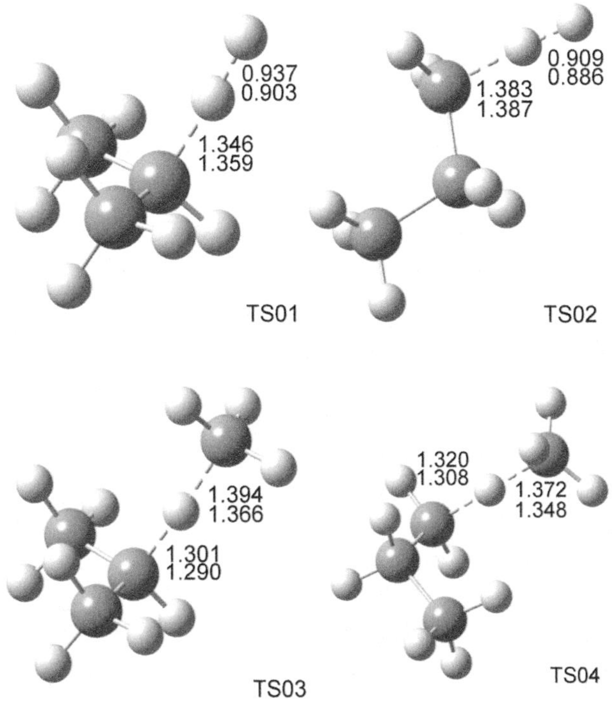

Fig. 2 Structure of the transition states for reactions (4)–(7), TS01 to TS04, respectively. Some important bond distances are shown. The upper entries correspond to BMK calculations, while the lower entries correspond to MP2 calculations. The 6-311 ++G(3df,2pd) basis set was employed in both cases. Distances are expressed in Å

Direct interconversion of the propyl radical isomers is not an easy step. The transition state for reaction (8) involves a three-atom 1,2-hydrogen transfer whose transition states TS05 are shown in Fig. 4.

$$iC_3H_7 \cdot (\mathbf{2}) \leftrightarrows nC_3H_7 \cdot (\mathbf{3}) \quad (8)$$

The barrier for this reaction is about 41 kcal/mol at room temperature (all methods agree to within one kcal/mol), but easy to overcome at the temperatures at which the more energy-consuming C–C or C–H bond breakings in propane take place. In a previous study of this reaction published by Matheu et al. [53], they derived the expression $k(\mathrm{T})\ s^{-1} = 5.36 \times 10^{12}\ \mathrm{T}^{0.88} \exp(-30.13/\mathrm{RT})$ with the temperature expressed in K and the exponential factor in kcal/mol. From our own BMK/6-311 ++G(3df,2pd), we obtained a somewhat similar expression $k(\mathrm{T})\ s^{-1} = 5.35 \times 10^{12}\ \mathrm{T}^{0.86} \exp(-46.82/\mathrm{RT})$ with a 50 % higher activation energy. The presence of propane lowers the barrier, according to experimental data by Berkley et al. [54], but we did not calculate the corresponding transition state, since even in the most unfavorably hypothesis, the equilibrium would be established without difficulties.

As would be expected, **2** is more stable than **3** because of a better stabilization of the unpaired electron. Papic and Laidler [14, 15] determined the energy difference between these isomers as −3.5 kcal/mol in the seventies. More recent work by Tsang in 1996 gauged the enthalpy

Table 3 Energies of activation for reactions (4)–(7) in kcal/mol

Reaction		B3LYP[a]	B3PW91[a]	PBE0[a]	M06[a]	BMK[a]	MP2[a]	CBS 4 M	CBS QB3	CBS APNO	G4	Experimental
(4)	$C_3H_8 + H\cdot \rightarrow iC_3H_7\cdot + H_2$	2.4	3.5	4.3	6.7	8.5	12.6	8.3	7.4	7.0	7.3	7.96[b], 4.47[c]
	$iC_3H_7\cdot + H_2 \rightarrow C_3H_8 + H\cdot$	13.4	12.6	11.0	14.9	14.2	14.0	15.0	14.8	13.4	14.9	8.67[c]
(5)	$C_3H_8 + H\cdot \rightarrow nC_3H_7\cdot + H_2$	5.3	6.6	7.6	8.7	11.5	15.6	11.3	10.3	9.9	10.3	9.37[b], 6.76[c]
	$nC_3H_7\cdot + H_2 \rightarrow C_3H_8 + H\cdot$	12.1	11.6	10.2	12.6	13.3	14.4	14.7	14.5	13.2	15.0	9.13[c]
(6)	$C_3H_8 + CH_3\cdot \rightarrow iC_3H_7\cdot + CH_4$	9.2	10.2	9.0	10.2	11.5	11.9	11.7	11.5	11.2	11.9	9.60[d], 5.48[c]
	$iC_3H_7\cdot + CH_4 \rightarrow C_3H_8 + CH_3\cdot$	18.1	19.0	17.6	19.2	19.1	17.2	18.2	18.3	17.6	18.8	10.80[c]
(7)	$C_3H_8 + CH_3\cdot \rightarrow nC_3H_7\cdot + CH_4$	12.2	12.4	11.2	13.2	14.5	15.6	14.2	13.9	13.4	14.2	11.50[d], 7.15[c]
	$nC_3H_7\cdot + CH_4 \rightarrow C_3H_8 + CH_3\cdot$	17.7	17.1	15.8	18.0	18.3	18.2	17.5	17.4	16.7	18.1	10.87[c]
MAD[e]		5.6/8.7	4.5/8.2	3.6/6.8	1.7/8.4	3.0/8.3	6.3/8.7	2.7/7.4	2.4/7.5	1.9/6.8	2.7/8.0	
RMSE[f]		3.5/5.0	2.7/4.7	2.0/3.6	1.1/5.4	2.1/6.1	4.5/7.1	2.0/6.0	1.6/5.8	1.3/5.1	1.9/6.1	

[a] Calculated in this using the 6-311 ++G(3df,2pd) basis set, at 1 atm and 298.15 K

[b] Baldin and Walker [51], 300–753 K

[c] Tsang [13] recommended value, 400–900 K

[d] Kerr and Parsonage [52], 550–750 K

[e] First entries are the MADs for the four direct reactions considering the experimental data from Refs. [51] and [52], and second entries are MADs for the eight direct and reverse reactions with respect to the experimental data in Ref. [13]

[f] First entries are the RMSEs for the four direct reactions considering the experimental data from Refs. [51] and [52], and second entries are RMSEs for the eight direct and reverse reactions with respect to the experimental data in Ref. [13]

Table 4 Comparison of theoretical and experimental Arrhenius and modified Arrhenius parameters

Reaction	Theoretical[a] A	Theoretical[a] E	Experimental[a,b] A	Experimental[a,b] E	Theoretical[a,c]	Experimental[a,c,d]
(4) $C_3H_8 + H\cdot \rightarrow iC_3H_7\cdot + H_2$	4.98×10^6	7.09	9.76×10^7	7.96	$3.63 \times 10^4 (T/T_{ref})^{3.7} \exp(-4.15/RT)$	$1.13 \times 10^4 (T/T_{ref})^{2.40} \exp(-4.47/RT)$
(5) $C_3H_8 + H\cdot \rightarrow nC_3H_7\cdot + H_2$	3.95×10^6	9.77	1.32×10^6	9.38	$3.72 \times 10^4 (T/T_{ref})^{3.6} \exp(-7.03/RT)$	$2.55 \times 10^4 (T/T_{ref})^{2.54} \exp(-6.76/RT)$
(6) $C_3H_8 + CH_3\cdot \rightarrow iC_3H_7\cdot + CH_4$	1.62×10^7	11.11	1.99×10^5	9.61	$1.04 \times 10^3 (T/T_{ref})^{7.4} \exp(-5.41/RT)$	$5.48 \times 10^0 (T/T_{ref})^{3.46} \exp(-5.48/RT)$
			8.57×10^4	13.25		
(7) $C_3H_8 + CH_3\cdot \rightarrow nC_3H_7\cdot + CH_4$	2.91×10^6	12.49	1.01×10^4	11.74	$1.70 \times 10^2 (T/T_{ref})^{7.6} \exp(-6.80/RT)$	$9.71 \times 10^0 (T/T_{ref})^{3.65} \exp(-7.15/RT)$
			4.52×10^5	14.92		
			4.37×10^4	17.51		

Theoretical calculations were performed using the BMK/6-311 ++G(3df,2pd) method including the Eckhart tunneling correction

[a] Preexponential factors in $m^3mol^{-1}s^{-1}$, activation energies in kcal/mol, $T_{ref} = 298$ K, T in K

[b] Data from Refs. [51], [52], [57] and [64]

[c] Preexponential factors in $m^3molecule^{-1}s^{-1}$, and exponential parameters in kcal/mol

[d] Data from Ref. [13]

difference at room temperature at -1.9 ± 1.0 kcal/mol [5]. Our own values at several theoretical levels are collected in Table 5. The best calculated results (BMK/6-311 ++G(3df,2pd), CBS-QB3, CBS-APNO) give a lower value, which can be expressed as -3.0 ± 0.5 kcal/mol.

The two isomers of the propyl radical can react with themselves or each other, in the following chain termination reactions

$$nC_3H_7\cdot(\mathbf{3}) + nC_3H_7\cdot(\mathbf{3}) \rightarrow nC_6H_{14}(\mathbf{5}) \quad (9a)$$

$$nC_3H_7\cdot(\mathbf{3}) + nC_3H_7\cdot(\mathbf{3}) \rightarrow C_3H_8(\mathbf{1}) + C_3H_6(\mathbf{12}) \quad (9b)$$

$$iC_3H_7\cdot(\mathbf{2}) + iC_3H_7\cdot(\mathbf{2}) \rightarrow (CH_3)_2CHCH(CH_3)_2(\mathbf{4}) \quad (10a)$$

$$iC_3H_7\cdot(\mathbf{2}) + iC_3H_7\cdot(\mathbf{2}) \rightarrow C_3H_8 + C_3H_6(\mathbf{12}) \quad (10b)$$

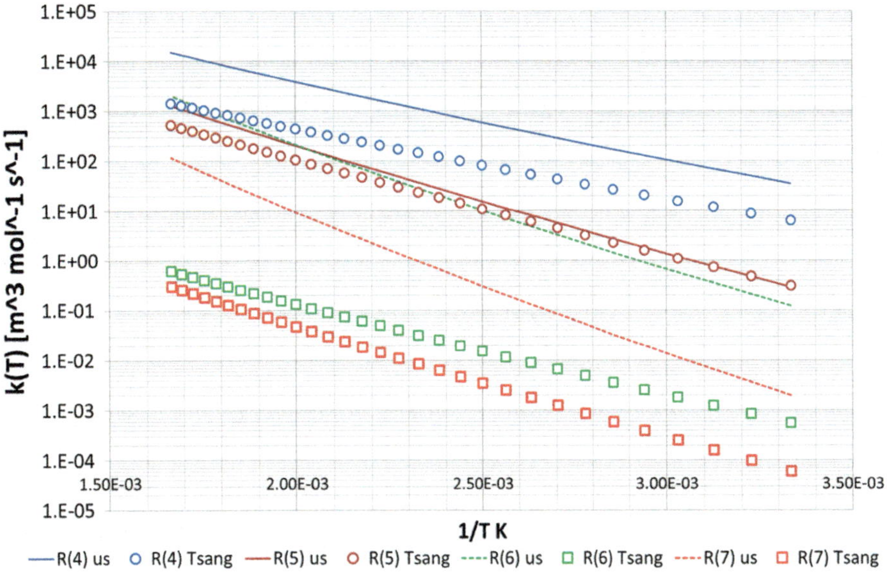

Fig. 3 Variation of the rate of reactions (4)–(7) with the inverse of the temperature, plotted in a logarithmic scale. *Circles* and *squares* identify fitted experimental data for hydrogen abstraction by the H and CH_3 radicals, respectively. *Lines* represent the theoretical data. Reactions are *color coded*, using the same color for the experimental and theoretical data

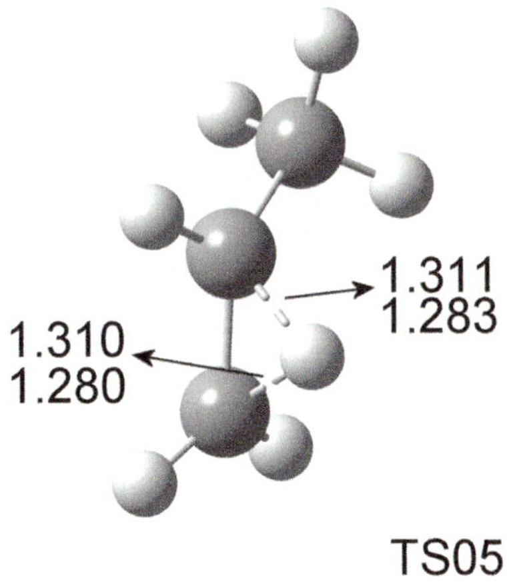

Fig. 4 Transition state TS05 for the interconversion between the *i*-propyl and *n*-propyl radicals

$$iC_3H_7 \cdot (\mathbf{2}) + nC_3H_7 \cdot (\mathbf{3}) \rightarrow (CH_3)_2CH(CH_2)_2CH_3(\mathbf{6}) \quad (11a)$$

$$iC_3H_7 \cdot (\mathbf{2}) + nC_3H_7 \cdot (\mathbf{3}) \rightarrow C_3H_8 + C_3H_6(\mathbf{12}) \quad (11b)$$

Experimental evidence for the existence of species **4–6** was found by Papic and Laidler [14, 15] who experimentally identified them as minor products for this reaction. Enthalpies in Table 5 show that reactions (9a)-(11a) liberate approximately the same energy that is required to produce C–C scission in propane, a fact that would help the reaction to proceed further. These are spontaneous reactions according to the free energy change and proceed without transition states.

Reactions (9b)-(11b) are somehow more difficult to treat. In this case, the two doublet species have to position with respect to each other in the collision complex in such a way their radical centers are separated, least they will end up giving the same products as in reactions (9a)-(11a). They can couple to form either a singlet or a triplet state complex, but the first one would be preferred since the final species are in their ground, closed-shell singlet states. Reaction in these cases would proceed through a hydrogen atom transfer like in the case of reactions (4)–(7). However, in this case, we have two radicals whose internal structures rearrange during the transfer. A electron density redistribution is produced in both participating monomers, due to the pairing of the preexisting lone electron with the one newly formed after H-atom transfer.

The correct treatment of such an open-shell complex should be performed using a multireference wave function. However, less accurate methods can give an idea of the structure of the complex. A semiempirical PM3 geometry optimization was first attempted for the complex between *n*-propyl and *i*-propyl radicals. The process converged toward a structure shown in Fig. 5a where it is very clear how the carbons bearing the lone electrons avoid each other. The spin distribution of this open-shell singlet state, Fig. 5c, shows the presence of one electron with alpha spin and one with beta spin localized on the primary and secondary carbons of each fragment, respectively. A single point BMK calculation was performed on this complex at the PM3 geometry using the larger basis sets, and the spin distribution of the singlet open shell is shown in Fig. 5d. It is similar to the PM3 spin distribution, only that the p-character of the orbitals is less marked. The open-shell singlet radical calculated at the DFT level is heavily spin-contaminated by the triplet, $\langle S^2 \rangle = 1.00$, $S = 0.61$, which is about 0.4 kcal/mol less stable. As must be obvious, a closed-shell calculation using this geometry lies at much higher energies (about 39 kcal/mol over the open-shell singlet).

Table 5 Reaction enthalpies calculated at 298.15 K for reactions 8 to 11a, 11b in this work compared to experimental results

Method	Basis	R8	R9a	R9b	R10a	R10b	R11a	R11b	MAD	RMSE
B3LYP	I	−4.2	−82.7	−63.5	−73.0	−55.1	−78.3	−59.3	13.5	8.0
	II	−3.9	−78.7	−63.1	−69.7	−55.3	−74.7	−59.2	16.8	9.9
	III	−4.0	−78.4	−63.5	−69.2	−55.5	−74.3	−59.5	17.3	10.0
B3PW91	I	−4.2	−83.6	−61.7	−73.9	−53.2	−79.2	−57.4	12.5	8.3
	II	−3.9	−80.4	−61.3	−71.5	−53.4	−76.4	−57.3	15.0	9.6
	III	−4.0	−80.1	−61.5	−71.0	−53.4	−76.0	−57.5	15.5	9.8
PBE0	I	−4.2	−86.5	−60.7	−77.8	−52.4	−82.6	−56.6	11.7	7.4
	II	−3.9	−83.2	−60.3	−75.4	−52.6	−79.8	−56.4	11.6	8.4
	III	−3.9	−83.0	−60.5	−74.8	−52.6	−79.4	−56.6	11.7	8.5
M06	I	−4.5	−89.0	−63.8	−81.8	−54.8	−85.9	−59.3	9.3	5.1
	II	−3.5	−84.4	−62.0	−79.3	−55.0	−82.4	−58.5	9.2	6.2
	III	−3.6	−83.9	−62.2	−78.5	−55.0	−81.6	−58.6	9.1	6.4
BMK	I	−3.7	−90.4	−67.0	−83.3	−59.5	−87.2	−63.2	4.6	2.7
	II	−3.4	−86.7	−66.5	−80.5	−59.7	−83.9	−63.1	6.0	3.4
	III	−3.5	−86.8	−67.2	−81.3	−60.2	−83.8	−63.7	5.2	3.1
CBS-4 M		−3.1	−90.3	−67.4	−86.0	−61.3	−88.6	−64.3	2.9	1.7
CBS-QB3		−3.0	−89.6	−68.1	−85.6	−62.0	−87.9	−65.1	2.1	1.3
CBS-APNO		−3.0	−89.7	−69.5	−85.6	−63.5	−88.1	−66.5	2.0	1.2
G4		−2.8	−87.5	−66.2	−84.0	−60.5	−86.1	−63.4	3.6	2.2
MP2	III	−2.5		−70.8		−65.9		−68.3	2.9	2.1
CCSD(T)	III	−3.3		−66.4		−59.9		−63.1	4.3	2.8
Experimental		−1.9 ± 1.0 −3.5	−87.7 ± 1.0	−67.9 ± 1.0	−86.5 ± 1.0	−64.1 ± 1.0	−87.6 ± 1.0	−66.0 ± 1.0		

Enthalpies in kcal/mol

No transition state was located for these reactions at the DFT level. Geometry optimizations using the BMK functional with either the medium or large basis sets led to the transfer of a hydrogen atom, H14 in Fig. 5, giving the optimum structure depicted in Fig. 5b. However, a priori one could not rule out the possibility of a conical intersection between the open-shell and closed-shell singlet potential energy surfaces. Multireference calculations, outside the scope of this paper, should be performed to rule out this path. However, this avenue of research was not pursued further in this paper because the thermochemistry of reactions (9b)-(11b) shows that they are less favorable than reactions (9a)-(11a) and less exothermic. Therefore, although these (b) channels might be open, they are anyhow probably less important than the (a) channels, the result of head-on collision of the radical centers in each one of the molecules.

3.3 β-scission reaction

An important difference between isomers **2** and **3** of the propyl radical is that the latter can produce ethylene and the methyl radical, which would increase the concentration of this species, arising from reaction (1). The process is known as a β-scission reaction and in the case of the propyl radical is represented by reaction (12)

$$nC_3H_7\cdot(\mathbf{3}) \rightarrow C_2H_4(\mathbf{8}) + CH_3\cdot \quad (12)$$

This reaction has been studied experimentally by Bencsura et al. [12] and theoretically using G3 and CBS models by Zheng and Blowers [8]. A drawback of this latter theoretical study is that the geometry optimization necessary to obtain the transition state was performed at the low MP2/6-31G(d) level. As shown in Fig. 6, the structure varies considerably when better levels of theory are applied. In particular, MP2 calculations tend to give a too tight transition state, while BMK and other DFT methods give a looser, more product-like transition state. Therefore, the barriers obtained with MP2 are higher than those obtained with DFT methods. To an extent, the drawback of the MP2 method is partially corrected by error compensation when a small basis set is used as in [8]. Higher levels of correlation energy, as added in G3 or CBS methods, may help correct this high activation energies obtained at the MP2 level.

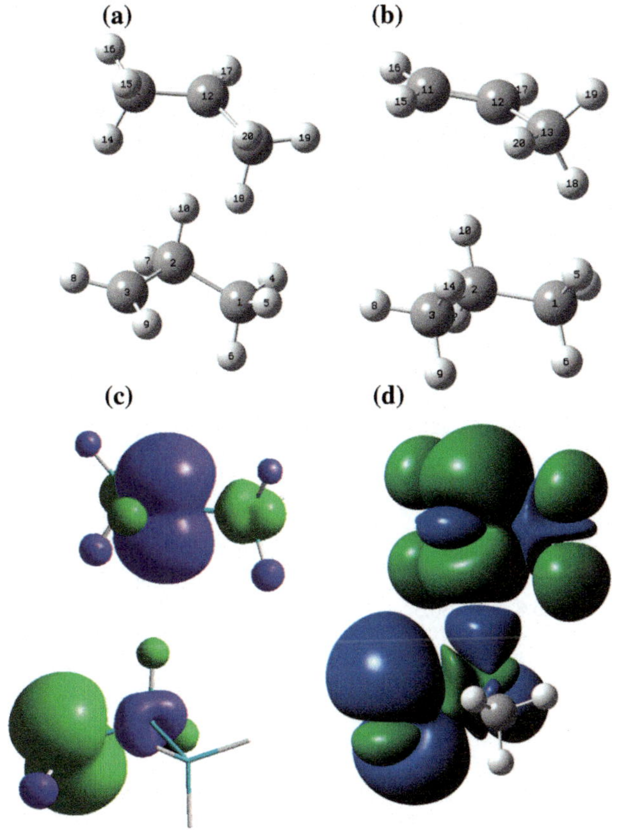

Fig. 5 **a** Optimum structure of the complex between *i*-propyl and *n*-propyl radicals, obtained at the semiempirical PM3 level of theory; **b** final optimized structure obtained at the DFT BMK/6-311 +G(d,p) level starting from the optimum PM3 geometry; **c** PM3 spin density at the PM3 optimum structure; **d** DFT BMK/6-311 +G(d,p) spin density at the PM3 optimum geometrical structure

Enthalpies of reaction and activation energies at room temperature are collected for several theoretical levels of calculation in Table 6 and compared to the experimental data available. With respect to the heat of reaction, DFT methods are reasonably close, in general, but tend to be outside the error bars of the experimental determination, except in the case of B3PW91 and BMK. Basis set extension is provoking a noticeable effect in the same way as mentioned before. BMK calculations with the more extended basis sets gave an enthalpy of reaction similar to that afforded by the CBS-4M or MP2 methods, while the other CBS and G4 composite methods are nearer to experiment. As expected, the CCSD(T) result is the nearest to the experimental value.

Thermal corrections are important, as one can see from the difference between the zero point-corrected energy barriers and the enthalpy barriers at 298 K. The barriers are in general several kcal/mol larger than the experimental activation energies reported in the literature, proportionally more significant for the reverse than for the forward reaction. To make a precise comparison, however, it is necessary to try two- and three-parameter Arrhenius fits and include corrections for tunneling and hindered internal rotations. The results obtained using the BMK method with the large basis set between 600 and 1,000 K, including tunneling and hindered rotor corrections, are shown in Table 7.

The agreement between the activation energies for both the forward and reverse reactions using the generalized Arrhenius equation is very good. Preexponential factors depend on the temperature in both cases, but the effect is more marked in the case of the reverse equation. A two-parameter Arrhenius equation does not represent well the curvature of k(T) with respect to T, and therefore, the activation energy is about 40 % too large. The extended Arrhenius parameters calculated at the BMK/6-311 ++G(3df,2pd) level are, however, in excellent agreement with the experimental results. Tunneling does not contribute markedly to the barrier, as expected, but internal rotation in both the *n*-propyl radical and the transition state does. Calculations were not repeated with the CBS, G3, or G4 methods, but it is to be expected that activation energies calculated with them will be too small, since the energy barriers are already in the same range as the experimentally observed activation energies.

3.4 Chain initiation reactions

Reactions (1)–(3) are not the only possible initiation elementary steps. Lower molecular weight hydrocarbons, like methane, ethane, ethene and propene, as well as higher molecular weight species are produced, as shown in Fig. 1. It is highly unlikely that higher molecular weight species are available in large enough concentrations to contribute

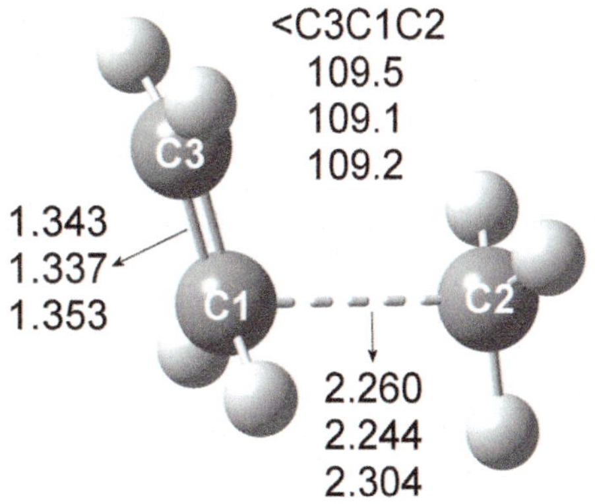

Fig. 6 Transition state TS06 for the β-scission reaction. Bond lengths in Å and bond angle in degrees. First entry corresponds to MP2/6-31G(d) optimization in Ref. [8], and second and third entries correspond to MP2 and BMK calculations using the 6-311 ++G(3df,2pd) basis set performed in this work

Table 6 Energy characterization of the β-scission reaction

Method	Basis	ΔH^{o}_{react}	$\Delta H^{\#}_{direct}$	$\Delta H^{\#}_{inverse}$	Energy barrier (direct reaction)	Energy barrier (inverse reaction)
B3LYP	I	24.8	30.2	5.5	34.3	6.6
	II	20.9	28.0	7.1	31.9	8.2
	III	20.4	27.7	7.3	31.7	8.4
B3PW91	I	27.6	33.1	5.5	37.2	6.6
	II	24.6	31.3	6.7	35.2	7.8
	III	24.2	31.0	6.8	35.0	7.9
PBE0	I	30.9	35.0	4.1	39.1	5.2
	II	27.8	33.2	5.4	37.0	6.5
	III	27.5	33.0	5.5	36.8	6.6
M06	I	29.3	32.9	3.5	37.2	4.8
	II	26.7	31.5	4.8	35.0	6.0
	III	26.2	31.3	5.1	34.9	4.0
BMK	I	28.7	35.0	6.3	38.7	7.4
	II	25.2	32.9	7.6	36.3	8.7
	III	24.9	32.7	7.8	36.2 [33.8][15]	9.0 [6.88][15]
CBS-4 M		24.9	30.6	5.8	30.5	7.0
CBS-QB3		23.2	29.4	6.2	29.2	7.3
CBS-APNO		24.0	30.3	6.4	30.2	7.6
G4		23.2	30.4	7.2	30.2	8.3
MP2	III	24.5	39.2	14.6	42.2	19.1
CCSD(T)	III	23.4				
Zheng G3[1]		20.0			30.0	10.0
Zheng CBS[1]		21.2			29.6	8.4
Other Theoretical					31.43[10], 30.07[10]	11.11[10], 7.43[10]
Experimental[16]		23.5 ± 1.0[2]			30.40[3], 32.99[4], 27.82[5], 32.59[6], 32.59[7], 31.40[8], 34.58[9]	7.89[6], 7.83[11], 7.35[12], 7.71[13,14]

Energies in kcal/mol

[1] Ref. [49, 50]; [2] Ref. [48]; [3] Ref. [55]; [4] Ref. [56]; [5] Ref. [57]; [6] Ref. [58]; [7] Ref. [15]; [8] Ref. [59]; [9] Ref. [60]; [10] CBS-QB3 Ref. [61]; [11] Ref. [62]; [12] Ref. [63]; [13] Ref. [52]; [14] Ref. [5]; [15] Values in brackets are energies of activation, see text..
[16] Experimental heat of reaction and energies of activation

Table 7 Energies of activation in kcal/mol obtained at the BMK/6-311 ++G(3df,2pd) level compared to experimental data

Method	E + ZPE Barrier	Two-parameter Arrhenius equation[a]			Three-parameter Arrhenius Equation[b]	Experimental
		Without corrections	Tunneling Correction	Tunneling + Hindered Rotation		
Direct reaction	36.2	34.4	34.3	33.8	31.8 ($n = 1.17$)	30.46 ($n = 0.87$)[c]
Reverse reaction	9.0	13.7	13.6	10.6	6.88 ($n = 2.43$)	6.13 ($n = 2.48$)[c]

[a] Expressed as A $\exp(-E_a/RT)$; [b] expressed as A $(T/298)^n \exp(-E_a/RT)$; [c] Ref. [9]

significantly to the mechanism. But it cannot be excluded that the simpler hydrocarbons do decompose at a similar or larger rate than propane, a question to be investigated in this section of the paper.

Eleven chain initiation reactions were considered in this paper. On one side, we have reactions (1)–(3) described previously for propane. In addition, we have also the following eight reactions

$$C_3H_6(\mathbf{12}) \rightarrow CH_2 = CHCH_2 \cdot (\mathbf{13}) + H\cdot \quad (13a)$$

$$C_3H_6(\mathbf{12}) \rightarrow CH_2C \cdot CH_3(\mathbf{17}) + H\cdot \quad (13b)$$

$$C_3H_6(\mathbf{12}) \rightarrow \cdot CH = CHCH3(\mathbf{21}) + H\cdot \quad (13c)$$

$$C_3H_6(\mathbf{12}) \rightarrow C_2H_3 \cdot (\mathbf{25}) + CH_3\cdot \quad (14)$$

$$C_2H_6(\mathbf{11}) \rightarrow C_2H_5 \cdot (\mathbf{7}) + H\cdot \quad (15)$$

$$C_2H_6(\mathbf{11}) \rightarrow CH_3 \cdot + CH_3\cdot \quad (16)$$

$$C_2H_4(\mathbf{8}) \rightarrow C_2H_3 \cdot (\mathbf{25}) + H\cdot \quad (17)$$

$$CH_4 \rightarrow CH_3 \cdot + H\cdot \quad (18)$$

Bold numbers identify the species in the global reaction mechanism depicted in Fig. 1. All of them occur without identifiable transition states, and the comparison among theoretical and experimental results is shown in Table 8. In this initiation reactions appear the H· and CH_3· radicals, as in reactions (1)–(3), but also other small radicals that can participate in chain elongation reactions, namely C_2H_3·, C_2H_5·, and different isomers of C_3H_5·.

The results for the CCSD(T) method are very good, as expected. The computed values are within the experimental error margins. CBS-APNO results are as good or better than the CCSD(T). In both cases, the results are not completely coherent because the optimization of geometry and the different contributions to the correlation energy are calculated using different methods. BMK results with the extended basis, on the contrary, are as good as either of the former, while now the geometry optimization as well as the calculation of all the energies and derivatives is performed at the same computational level.

According to the BMK results, reactions (13a), the generation of the propenyl radical, and (16), the cracking of ethane, occur at temperatures comparable to those necessary for the cracking of propane. Propene and ethane are both generated along the reaction path and are, therefore, not additional starting points. They just add to the full mechanism to proceed further toward products.

3.5 Chain propagation reactions

Reactions (4)–(8) and (12) are chain propagation reactions, where a new radical species is formed from an already existing one, adding to the pool of radicals present in the combustion chamber. Reaction of any of the radicals with any of the neutral molecules present in the reactor give either a new species or one already present from other branches of the reaction scheme. The reactions considered in this paper are shown in Table 9, where the values of the enthalpies of reaction obtained at each level with the largest basis set are reported. Experimental values, as well as MADs and RMSEs, are also included in the table.

About two-thirds of the propagation reactions in the table are exothermic, and one-third thermoneutral or endothermic. In all cases, the energy needed is much lower than the one necessary to initiate the chain of reactions. Again as before, BMK results are of comparable quality to G4 or CCSD(T), although the CBS-APNO calculations exhibit smaller deviations from the experimental values for this set of reactions. Needless to stress, the BMK calculations are faster and cheaper than the composite or coupled cluster methods, especially when larger molecules are considered.

3.6 Termination reactions

Reactions (9a), (9b)–(11a), and (11b) are termination reactions. Several more can be written, and some of them are collected in Table 10. Some of the experimental results are not available for these reactions. Thus, the theoretical values should be taken as estimates of the enthalpies of reaction.

The main difficulty in representing these reactions is that they are neither isodesmic nor isogyric. Errors do not compensate then, and only an equally accurate description of the radicals and the neutral molecules afford reasonable values for the reactions. As observed in the table, B3LYP is nearly useless for the description, since errors up to 17 kcal/mol occur. The BMK method exhibits both a MAD and RMSE of almost one-fourth those of B3LYP. The results are comparable to G4 and better than MP2, but not as good as CCSD(T) or CBS-APNO.

4 Discussion

Let us assume that the combustion takes place at a temperature where reaction (1) is feasible, but not (2) or (3). Since about 88 kcal/mol is needed to break the CC bond, there will be also enough energy to overcome the barriers of reactions (6) and (7). Therefore, once there are enough methyl radicals in the system, an equilibrium between the *n*-propyl, *i*-propyl, ethyl and methyl radicals will exist.

The calculations shown in this paper do not present any noticeable discrepancy with the known experimental values. Therefore, it is inmaterial whether the whole kinetic system is solved resorting to one type of data or the other. One could deduce from the BMK values of the energies for the reactions that the main processes can be schematized as in Fig. 7. The main products, therefore, will be methane, ethene, *n*-butane, and *i*-butane, which can in turn react forward in other chain reactions. The reduced mechanism presented in Fig. 7 amounts to the global reaction

Table 8 Comparison between theoretical and experimental enthalpies of reaction for chain initiation reactions, in kcal/mol

Method	Basis	R1	R2	R3	R13a	R13b	R13c	R14	R15	R16	R17	R18	MAD	RMSE
B3LYP	I	85.5	96.9	101.1	84.2	104.5	109.7	97.1	100.8	88.6	109.9	105.5	3.9	1.3
	II	81.6	94.8	98.7	82.6	102.9	108.0	93.5	98.5	84.5	108.1	103.1	7.5	2.0
	III	81.5	94.7	98.7	82.6	103.0	108.2	93.5	98.5	84.5	108.3	103.3	7.4	2.0
B3PW91	I	86.4	95.6	99.9	83.5	103.2	108.5	98.0	99.6	89.6	108.6	104.4	4.7	1.4
	II	83.2	93.8	97.7	82.0	101.7	106.9	95.1	97.5	86.2	107.0	102.1	6.1	2.1
	III	83.0	93.7	97.7	82.0	101.7	107.0	95.1	97.5	86.0	107.1	102.2	6.1	2.1
PBE0	I	88.8	95.4	99.5	83.1	102.9	108.1	100.3	99.2	91.7	108.2	103.9	5.1	1.4
	II	85.6	93.4	97.2	81.5	101.3	106.4	97.4	97.0	88.2	106.5	101.5	6.6	2.0
	III	85.4	93.2	97.2	81.4	101.2	106.5	97.3	96.9	88.1	106.5	101.6	6.7	2.0
M06	I	90.4	97.1	101.6	85.0	103.9	109.0	100.5	101.1	93.1	109.0	106.1	3.5	1.3
	II	87.2	95.2	98.7	83.5	102.3	107.4	97.4	99.0	89.4	107.3	103.6	4.6	1.6
	III	86.2	95.2	98.7	83.5	102.4	107.5	97.2	98.4	89.2	107.6	103.8	4.6	1.6
BMK	I	92.0	98.9	102.6	87.6	107.2	111.9	103.9	102.2	94.1	110.8	106.4	4.4	1.4
	II	88.4	97.2	100.6	86.1	105.9	110.6	100.6	100.2	90.3	109.4	104.3	2.0	0.9
	III	88.6	97.3	100.8	86.3	106.2	110.9	101.0	100.5	90.6	109.7	104.6	1.8	0.8
CBS-4 M		90.4	99.1	102.2	84.9	106.2	110.5	101.0	101.8	91.4	111.0	105.4	3.3	1.1
CBS-QB3		89.7	98.9	102.0	85.9	106.2	110.4	100.0	101.7	90.7	110.6	105.4	2.3	1.1
CBS-APNO		89.7	99.1	102.1	88.2	108.8	112.9	102.5	101.8	90.7	111.1	105.4	1.6	0.9
G4		88.0	97.9	100.7	85.8	105.4	109.5	98.6	100.7	89.0	110.0	104.5	2.4	1.0
MP2	III	92.0	97.6	100.0	91.9			110.0	99.6	91.9	114.7	102.5	9.1	4.0
CCSD(T)	III	87.0	96.5	99.7	87.4			100.8	99.4	88.1	110.7	103.3	2.6	1.4
Experimental[a]		88.2 ± 0.5	99.0 ± 0.5	101.0 ± 0.5	88.1 ± 0.5			100.9 ± 0.5	100.5 ± 0.5	89.6 ± 0.5	110.6 ± 0.5	104.8 ± 0.5		

[a] Ref. [13]

Table 9 Enthalpies of reaction at 298 K in kcal/mol for the chain propagation reactions

Reaction		B3LYP	B3PW91	PBE0	M06	BMK	MP2	APNO	G4	CCSD(T)	Experimental
4	$C_3H_8 + H\cdot \rightarrow iC_3H_7\cdot + H_2$	−10.0	−8.1	−4.7	−4.2	−4.6	−0.4	−0.3	0.3	−6.4	−5.1
5	$C_3H_8 + H\cdot \rightarrow iC_3H_7\cdot + H_2$	−6.0	−4.1	−0.7	−0.7	−1.1	2.1	2.7	3.1	−3.1	−3.2
6	$C_3H_8 + CH_3\cdot \rightarrow iC_3H_7\cdot + CH_4$	−8.5	−8.6	−8.4	−8.7	−7.3	−5.0	−6.3	−6.6	−6.8	−5.7
7	$C_3H_8 + CH_3\cdot \rightarrow nC_3H_7\cdot + CH_4$	−4.5	−4.6	−4.4	−5.1	−3.8	−2.5	−3.3	−3.8	−3.6	−3.8
8	$iC_3H_7\cdot \rightarrow nC_3H_7\cdot$	−4.0	−4.0	−3.9	−3.6	−3.5	−2.5	−3.0	−2.8	−3.3	−1.9
12	$nC_3H_7\cdot \rightarrow C_2H_4 + CH_3\cdot$	20.4	24.2	27.5	26.7	24.9	24.5	24.0	23.2	23.4	23.5
19	$C_3H_8 + C_2H_3\cdot \rightarrow nC_3H_7\cdot + C_2H_4$	−9.6	−9.4	−9.3	−8.8	−8.9	−14.7	−9.0	−9.3	−11.0	−9.5
20	$C_3H_8 + C_2H_3\cdot \rightarrow iC_3H_7\cdot + C_2H_4$	−13.6	−13.4	−13.3	−12.4	−12.4	−17.1	−12.0	−12.1	−14.2	−11.4
21	$C_3H_8 + C_2H_3\cdot \rightarrow C_3H_6 + C_2H_5\cdot$	−12.1	−12.1	−11.9	−11.0	−12.4	−18.0	−12.8	−10.7	−13.8	−12.7
22	$C_3H_8 + C_2H_5\cdot \rightarrow nC_3H_7\cdot + C_2H_6$	0.2	0.2	0.3	0.3	0.4	0.5	0.3	0.0	0.3	0.5
23	$C_3H_8 + C_2H_5\cdot \rightarrow iC_3H_7\cdot + C_2H_6$	−3.8	−3.8	−3.7	−3.3	−3.1	−2.0	−2.7	−2.8	−2.9	−1.4
24	$C_3H_8 + C_3H_5\cdot \rightarrow nC_3H_7\cdot + C_3H_6$	16.1	15.7	15.7	15.2	14.5	8.1	13.9	14.9	12.3	12.9
25	$C_3H_8 + C_3H_5\cdot \rightarrow iC_3H_7\cdot + C_3H_6$	12.1	11.7	11.8	11.6	11.0	5.6	10.9	12.1	9.0	11.0
26	$C_3H_8 + nC_3H_7\cdot \rightarrow iC_3H_7\cdot + C_3H_8$	−4.0	−4.0	−3.9	−3.6	−3.5	−2.5	−3.0	−2.8	−3.3	−1.9
27	$nC_3H_7\cdot \rightarrow C_3H_6 + H\cdot$	35.2	36.2	36.7	36.6	33.6	29.2	32.6	34.5	33.3	33.1
28	$iC_3H_7\cdot \rightarrow C_3H_6 + H\cdot$	39.2	40.2	40.6	40.1	37.1	31.7	35.6	37.3	36.6	35.0
29	$C_3H_6 + H\cdot \rightarrow C_3H_5\cdot + H_2$	−22.1	−19.8	−17.4	−18.6	−15.6	−6.1	−16.5	−18.8	−15.4	−16.1
30	$C_3H_6 + H\cdot \rightarrow C_2H_3\cdot + CH_4$	−9.7	−7.1	−4.3	−6.6	−3.6	7.5	−2.9	−5.9	−2.5	−3.9
31	$C_3H_6 + CH_3\cdot \rightarrow C_3H_5\cdot + CH_4$	−20.6	−20.3	−20.2	−20.3	−18.2	−10.6	−17.2	−18.7	−15.9	−16.7
32	$C_3H_6 + CH_3\cdot \rightarrow C_2H_3\cdot + C_2H_6$	9.1	9.1	9.2	8.0	10.4	18.1	11.8	9.6	12.7	11.3
33	$C_3H_6 + C_2H_3\cdot \rightarrow C_3H_5 + C_2H_4$	−25.7	−25.1	−25.0	−24.1	−23.4	−22.8	−22.9	−24.2	−23.3	−22.4
34	$C_3H_6 + C_2H_5\cdot \rightarrow C_3H_5\cdot + C_2H_6$	−15.9	−15.5	−15.5	−14.9	−14.1	−7.7	−13.6	−14.9	−12.0	−12.4
35	$C_3H_6 + C_3H_5\cdot \rightarrow C_2H_3\cdot + C_4H_8$	28.3	27.9	27.7	26.2	28.2		26.5	25.8		25.2
36	$C_3H_6 + nC_3H_7\cdot \rightarrow C_2H_3\cdot + nC_4H_{10}$	11.9	11.9	11.7	10.6	12.1		12.4	10.5		12.2
37	$C_3H_6 + iC_3H_7\cdot \rightarrow C_2H_3\cdot + iC_4H_{10}$	15.0	15.0	14.4	12.3	14.2		13.4	11.3		12.1
38	$C_2H_6 + H\cdot \rightarrow C_2H_5\cdot + H_2$	−6.2	−4.3	−2.0	−3.7	−1.5	1.6	−2.9	−3.9	−3.4	−3.7
39	$C_2H_6 + CH_3\cdot \rightarrow C_2H_5\cdot + CH_4$	−4.8	−4.8	−4.7	−5.4	−4.1	−3.0	−3.6	−3.8	−3.9	−4.3
40	$C_2H_6 + C_2H_3\cdot \rightarrow C_2H_5\cdot + C_2H_4$	−9.8	−9.6	−9.6	−9.1	−9.3	−15.1	−9.3	−9.3	−11.3	−10.1
41	$C_2H_4 + H\cdot \rightarrow C_2H_5\cdot$	−37.7	−38.9	−39.2	−38.7	−37.1	−32.5	−36.4	−35.9	−36.2	−36.2
42	$C_2H_4 + CH_3\cdot \rightarrow C_2H_3\cdot + CH_4$	5.1	4.8	4.9	3.7	5.2	12.2	5.7	5.5	7.4	5.7
	MAD	6.0	5.2	5.6	5.2	2.9	11.3	1.4	2.7	2.8	
	RMSE	2.8	2.4	2.3	2.2	1.3	5.0	0.7	1.4	1.2	

The 6-311 ++G(3df,2pd) results are shown for DFT and ab initio calculations. Experimental data from Ref. [48]

$$4C_3H_8 \rightarrow 2CH_4 + C_2H_4 + nC_4H_{10} + iC_4H_{10}$$

which is not immediately obvious from the mechanism in Fig. 1. Experimentally, the enthalpy of this reaction is 14.7 kcal/mol, while the BMK method affords 17.8 kcal/mol. The three CBS methods result in 16.9, 14.9, and 15.9 kcal/mol, respectively, and the G4 result is 14.3 kcal/mol. The CBS-APNO and G4 results are closer to experiment for this non-elementary reaction.

In relation to the accepted mechanism, our main difference is the route by which the propyl radical is obtained. The calculations show that the direct H abstraction is an unfavorable process, because C–C bond scission requires less energy to proceed. However, once the methyl radicals are produced in this process, the reaction with propane is feasible, and both the n- and *i*-propyl radicals are formed. While this modification thus affords a mechanistic explanation, it does not modify greatly the whole dynamics of the process.

5 Conclusions

DFT, ab initio and model chemistry calculations have been performed on the chain initiation, propagation and termination reactions for the cracking of propane. From the methodological point of view, it was shown that the DFT BMK results of enthalpies of formation and energies of activation are very similar to the experimental values, with rms error of about 2 kcal/mol. For reactions involving

Table 10 Enthalpies of reaction at 298 K in kcal/mol for some of the termination reactions

Reaction		B3LYP	B3PW91	PBE0	M06	BMK	MP2	APNO	G4	CCSD(T)	Experimental
9a	$nC_3H_7\cdot + nC_3H_7\cdot \rightarrow nC_6H_{14}$	−78.4	−80.1	−83.0	−83.9	−86.8		−89.7	−87.5		−87.7
9b	$nC_3H_7\cdot + nC_3H_7\cdot \rightarrow C_3H_8 + C_3H_6$	−63.5	−61.5	−60.5	−62.2	−67.2	−70.8	−69.5	−66.2	−66.4	−67.9
10a	$iC_3H_7\cdot + iC_3H_7\cdot \rightarrow (CH_3)_2CHCH(CH_3)_2$	−69.2	−71.0	−74.8	−78.5	−81.3		−85.6	−84.0		−86.5
10b	$iC_3H_7\cdot + iC_3H_7\cdot \rightarrow C_3H_8 + C_3H_6$	−55.5	−53.4	−52.6	−55.0	−60.2	−65.9	−63.5	−60.5	−59.9	−64.1
11a	$iC_3H_7\cdot + nC_3H_7\cdot \rightarrow (CH_3)_2CH(CH_2)_2CH_3$	−74.3	−76.0	−79.4	−81.6	−83.8		−88.1	−86.1		−87.6
11b	$iC_3H_7\cdot + nC_3H_7\cdot \rightarrow C_3H_8 + C_3H_6$	−59.5	−57.5	−56.6	−58.6	−63.7	−68.3	−66.5	−63.4	−63.1	−66.0
43	$H\cdot + H\cdot \rightarrow H_2$	−104.7	−101.8	−98.9	−102.2	−101.9	−98.0	−104.7	−104.5	−102.8	−104.2
44	$CH_3\cdot + H\cdot \rightarrow CH_4$	−103.3	−102.2	−101.6	−103.8	−104.6	−102.5	−105.4	−104.5	−103.3	−104.8
45	$CH_3\cdot + CH_3\cdot \rightarrow C_2H_6$	−84.5	−86.0	−88.1	−89.2	−90.6	−91.9	−90.7	−89.0	−88.1	−89.6
46	$C_2H_3\cdot + H\cdot \rightarrow C_2H_4$	−108.3	−107.1	−106.5	−107.6	−109.7	−114.7	−111.1	−110.0	−110.7	−110.6
47	$C_2H_3\cdot + CH_3\cdot \rightarrow C_3H_6$	−93.5	−95.1	−97.3	−97.2	−101.0	−110.0	−102.5	−98.6	−100.8	−100.9
48	$C_2H_3\cdot + C_2H_3\cdot \rightarrow CH_2 = CHCH = CH_2$	−108.9	−110.3	−112.6	−111.4	−114.7		−116.2	−114.0		−116.0
49	$C_2H_5\cdot + H\cdot \rightarrow C_2H_6$	−98.5	−97.5	−96.9	−98.4	−100.5	−99.6	−101.8	−100.7	−99.4	−100.5
50	$C_2H_5\cdot + CH_3\cdot \rightarrow C_3H_8$	−81.5	−83.0	−85.4	−86.2	−88.6	−92.0	−89.7	−88.0	−87.0	−88.2
51	$C_2H_5\cdot + C_2H_3\cdot \rightarrow C_4H_8$	−92.1	−93.9	−96.3	−95.7	−98.8		−104.7	−98.6		−102.0
52	$C_2H_5\cdot + C_2H_5\cdot \rightarrow C_4H_{10}$	−78.4	−80.0	−82.7	−83.3	−86.5		−88.9	−87.1		−86.8
53	$CH_2CHCH_2\cdot + H\cdot \rightarrow CH_3CHCH_2$	−82.6	−82.0	−81.4	−83.5	−86.3	−91.9	−88.2	−85.8	−87.4	−88.1
54	$CH_3CHCH\cdot + H\cdot \rightarrow CH_3CHCH_2$	−108.2	−107.0	−106.5	−107.5	−110.9		−112.9	−109.5		
55	$CH_3CCH_2\cdot + H\cdot \rightarrow CH_3CHCH_2$	−103.0	−101.7	−101.2	−102.4	−106.2		−108.8	−105.4		
56	$CH_2CHCH_2\cdot + CH_3\cdot \rightarrow CH_3CH_2CH = CH_2$	−65.2	−67.2	−69.6	−71.0	−72.8		−76.0	−72.8		−75.9
57	$CH_3CHCH\cdot + CH_3\cdot \rightarrow CH_3CH = CHCH_3$	−92.7	−94.4	−96.6	−96.7	−98.9		−103.9	−97.8		
58	$CH_3CCH_2\cdot + CH_3\cdot \rightarrow (CH_3)_2C = CH_2$	−87.2	−88.8	−91.3	−92.0	−94.4		−101.2	−94.3		
56	$CH_2CHCH_2\cdot + C_2H_3\cdot \rightarrow CH_2 = CHCH_2CH = CH_2$	−94.1	−95.5	−151.8	−97.3	−101.1		−108.3	−101.6		
57	$CH_3CHCH\cdot + C_2H_3\cdot \rightarrow CH_3CH = CHCH = CH_2$	−83.3	−85.4	−87.8	−87.5	−90.1		−93.4	−90.1		−93.8
58	$CH_3CCH_2\cdot + C_2H_3\cdot \rightarrow CH_2 = CH\text{-}C(CH_3) = CH_2$	−100.8	−102.3	−105.1	−104.9	−108.0		−111.1	−108.7		
59	$CH_2CHCH_2\cdot + C_2H_5\cdot \rightarrow CH_3CH_2CH_2CH = CH_2$	−62.2	−64.2	−66.9	−68.1	−70.6		−75.2	−72.1		−74.3
60	$CH_3CHCH\cdot + C_2H_5\cdot \rightarrow CH_3CH = CHCH_2CH_3$	−89.3	−90.9	−93.5	−93.2	−95.6		−102.9	−96.6		
61	$CH_3CCH_2\cdot + C_2H_5\cdot \rightarrow CH_3CH_2C(CH_3) = CH_2$	−84.4	−85.8	−88.8	−89.5	−92.9		−96.8	−94.8		
	MAD	17.3	15.5	11.7	9.1	5.2	9.1	2.7	3.7	4.2	
	RMSE	8.3	7.6	6.3	5.0	2.4	4.2	1.2	2.0	1.9	

The 6-311 ++G(3df,2pd) results are shown for DFT and ab initio calculations. Experimental data from Ref. [48]

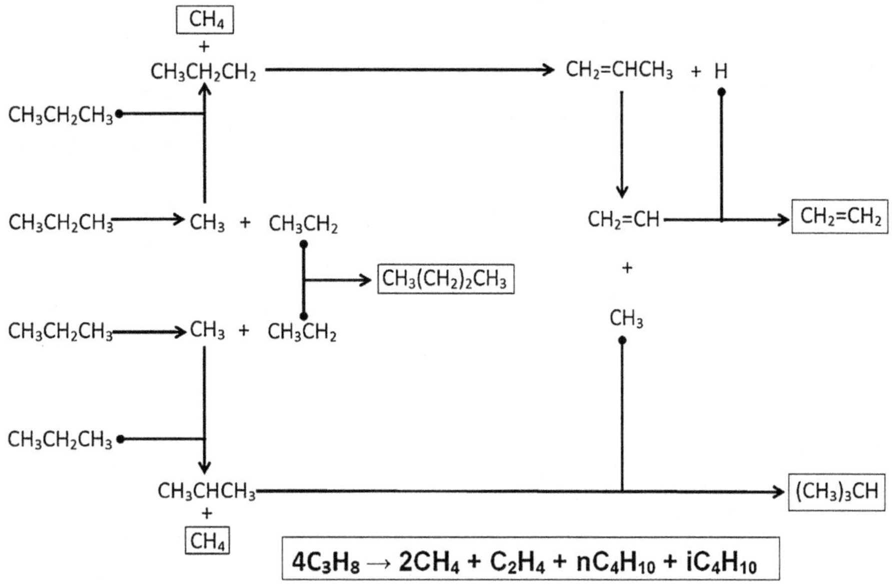

Fig. 7 Final scheme for the reduced mechanism (see *text*)

transition states, two- and three-parameter Arrhenius equations were calculated. It was shown that tunneling and internal rotation effects are important. Activation energies are then reproduced in good agreement with experimental ones. However, the preexponential factors calculated at the BMK level are too high, and the experimental relative concentration of radicals is then not reproduced accurately.

From the point of view of the reaction mechanism, calculations agree with the experimental result that the CC bond in propane is easier to break than the CH bonds. However, it was also shown that methyl radicals produced this way can abstract hydrogens from other propane molecules forming the *n*-propyl and *i*-propyl radicals. Although no attempt was performed in this paper to solve the full kinetic problem, a reduced mechanism is suggested, which proceeds through the formation of propene and ends up giving methane and butane, as observed experimentally.

Acknowledgments The authors acknowledge the financial support of CSIC, ANII, and Pedeciba through multiple grants.

References

1. Olah GA, Molnar A (2003) Hydrocarbon chemistry. Wiley, New York
2. Clymans PJ, Froment GF (1984) Comput Chem Eng 8:137
3. Hillewaert LP, Dierickx JL, Froment GF (1988) AIChE J 34:17
4. Ranzi E, Dente M, Plerucci S, Biardi G (1983) Ind Eng Chem Fund 22:132
5. Tsang W, Hampson RF (1986) J Phys Chem Ref Data 15:1087
6. Tsang W (1987) J Phys Chem Ref Data 16:471
7. Sabbe MK, Vandeputte AG, Reyniers M-F, van Speybroeck V, Waroquier M, Marin GB (2007) J Phys Chem A 111:8416
8. Zheng X, Blowers P (2006) Ind Eng Chem Res 45:530
9. Curran HJ (2006) Int J Chem Kinet 38:250
10. Hunter KC, East ALL (2002) J Phys Chem A 106:1346
11. Xiao Y, Longo JM, Hieshima GB, Hill RJ (1997) Ind Eng Chem Res 36:4033
12. Bencsura A, Knyazev VD, Xing S-B, Slagle IR, Gutman D (1992) Proc Combust Inst 24:629
13. Tsang W (1988) J Phys Chem Ref Data 17:887
14. Papic MM, Laidler KJ (1971) Can J Chem 49:535
15. Papic MM, Laidler KJ (1971) Can J Chem 49:549
16. Becke AD (1988) Phys Rev A 38:3098
17. Becke AD (1993) J Chem Phys 98:5648
18. Lee C, Yang W, Parr RG (1988) Phys Rev B 37:785
19. Perdew JP, Burke K, Wang Y (1996) Phys Rev B 54:16533
20. Perdew JP, Burke K, Ernzerhof M (1996) Phys Rev Lett 77:3865
21. Perdew JP, Burke K, Ernzerhof M (1997) Phys Rev Lett 78:1396
22. Adamo C, Barone V (1999) J Chem Phys 110:6158
23. Zhao Y, Truhlar DG (2008) Theor Chem Acc 120:215
24. Boese AD, Martin JML (2004) J Chem Phys 121:3405
25. Pople JA, Gill PMW, Johnson BG (1992) Chem Phys Lett 199:557
26. Johnson BG, Frisch MJ (1993) Chem Phys Lett 216:133
27. Johnson BG, Frisch MJ (1994) J Chem Phys 100:7429
28. Stratmann RE, Burant JC, Scuseria GE, Frisch MJ (1997) J Chem Phys 106:10175
29. Head-Gordon M, Pople JA, Frisch MJ (1988) Chem Phys Lett 153:503
30. Frisch MJ, Head-Gordon M, Pople JA (1990) Chem Phys Lett 166:275
31. Frisch MJ, Head-Gordon M, Pople JA (1990) Chem Phys Lett 166:281
32. Head-Gordon M, Head-Gordon T (1994) Chem Phys Lett 220:122
33. Bartlett RJ, Purvis GD III (1978) Int J Quantum Chem 14:561
34. Pople JA, Head-Gordon M, Raghavachari K (1987) J Chem Phys 87:5968
35. Nyden MR, Petersson GA (1981) J Chem Phys 75:1843
36. Petersson GA, Bennett A, Tensfeldt TG, Al-Laham MA, Shirley WA, Mantzaris J (1988) J Chem Phys 89:2193
37. Petersson GA, Al-Laham MA (1991) J Chem Phys 94:6081
38. Petersson GA, Tensfeldt TG, Montgomery JA Jr (1991) J Chem Phys 94:6091

39. Montgomery JA Jr, Frisch MJ, Ochterski JW, Petersson GA (2000) J Chem Phys 112:6532
40. Montgomery JA Jr, Frisch MJ, Ochterski JW, Petersson GA (1999) J Chem Phys 110:2822
41. Ochterski JW, Petersson GA, Montgomery JA Jr (1996) J Chem Phys 104:2598
42. Curtiss LA, Redfern PC, Raghavachari K (2007) J Chem Phys 126:084108
43. Eckart C (1930) Phys Rev 35:1303
44. Gaussian 09, Revision A.1, Frisch MJ, Trucks GW, Schlegel HB, Scuseria GE, Robb MA, Cheeseman JR, Scalmani G, Barone V, Mennucci B, Petersson GA, Nakatsuji H, Caricato M, Li X, Hratchian HP, Izmaylov AF, Bloino J, Zheng G, Sonnenberg JL, Hada M, Ehara M, Toyota K, Fukuda R, Hasegawa J, Ishida M, Nakajima T, Honda Y, Kitao O, Nakai H, Vreven T, Montgomery JA Jr, Peralta JE, Ogliaro F, Bearpark M, Heyd JJ, Brothers E, Kudin KN, Staroverov VN, Kobayashi R, Normand J, Raghavachari K, Rendell A, Burant JC, Iyengar SS, Tomasi J, Cossi M, Rega, Millam NJ, Klene M, Knox JE, Cross JB, Bakken V, Adamo C, Jaramillo J, Gomperts RE, Stratmann O, Yazyev AJ, Austin R, Cammi C, Pomelli JW, Ochterski R, Martin RL, Morokuma K, Zakrzewski VG, Voth GA, Salvador P, Dannenberg JJ, Dapprich S, Daniels AD, Farkas O, Foresman JB, Ortiz JV, Cioslowski J, Fox DJ (2009) Gaussian, Inc., Wallingford
45. Ghysels A, Verstraelen T, Hemelsoet K, Waroquier M, Van Speybroeck V (2010) J Chem Inf Model 50:1736
46. Sundaram KM, Froment GF (1979) Chem Eng Sci 34:635
47. Van Damme PS, Narayanan S, Froment GF (1975) AIChE J 21:1065
48. Linstrom PJ, Mallard WG, Eds (2003) NIST Chemistry WebBook, NIST Standard Reference Database Number 69, National Institute of Standards and Technology, Gaithersburg MD, 20899, http://webbook.nist.gov
49. Zheng X (2006) A computational investigation of hydrocarbon cracking: gas phase and heterogeneous catalytic reactions on zeolites, PhD Dissertation, University of Arizona, Tucson, Arizona
50. Zheng X, Blowers P (2005) Mol Simul 31:979
51. Baldwin RR, Walker RW (1979) J Chem Soc Faraday Trans 75:140
52. Kerr JA, Parsonage MJ (1976) Evaluated kinetic data on gas phase hydrogen transfer reactions of methyl radicals. Butterworths, London
53. Matheu DM, Green WH, Grenda JM (2003) Int J Chem Kinet 35:95
54. Berkley RE, Woodall GNC, Strausz OP (1969) Gunning HE 47:3305
55. Tsang W (1985) J Am Chem Soc 107:2872
56. Dean AM (1985) J Phys Chem 89:4600
57. Mintz KJ, Le Roy DJ (1978) Can J Chem 56:941
58. Camilleri P, Marshall RM, Purnell H (1975) J Chem Soc Faraday Trans I(71):1491
59. Lin MC, Laidler KJ (1966) Can J Chem 44:2927
60. Kerr JA, Calvery JG (1961) J Am Chem Soc 83:3391
61. Saeys M, Reyniers MF, Marin GB, Van Speybroeck V, Waroquier M (2004) AIChE J 50:426
62. Hogg AM, Kebarle P (1964) J Am Chem Soc 86:4558
63. Baulch DL, Cobos CJ, Cox RA, Esser C, Frank P, Just Th, Kerr JA, Pilling MJ, Troe J, Walker RW, Warnatz J (1992) J Phys Chem Ref Data 21:411
64. Leathard DA, Purnell JH (1968) Proc R Soc Lond A 306:553

Theor Chem Acc (2012) 131:1271
DOI 10.1007/s00214-012-1271-2

REGULAR ARTICLE

CompASM: an Amber-VMD alanine scanning mutagenesis plug-in

João V. Ribeiro · Nuno M. F. S. A. Cerqueira · Irina S. Moreira · Pedro A. Fernandes · Maria João Ramos

Received: 28 March 2012 / Accepted: 20 August 2012 / Published online: 30 September 2012

Abstract Alanine scanning mutagenesis (ASM) of protein–protein interfacial residues is a popular means to understand the structural and energetic characteristics of hot spots in protein complexes. In this work, we present a computational approach that allows performing such type of analysis based on the molecular mechanics/Poisson–Boltzmann surface area method. This computational approach has been used largely in the past and has proven to give reliable results in a wide range of complexes. However, the sequential preparation and manual submission of dozens of files has been often a major obstacle in using it. To overcome these limitations and turn this approach user-friendly, we have designed the plug-in CompASM (computational alanine scanning mutagenesis). This software has an easy-to-use graphical interface to prepare the input files, run the calculations, and analyze the final results. CompASM was built in TCL/TK programming language to be included in VMD as a plug-in. The CompASM package is distributed as an independent platform, with script code under the GNU Public License from http://compbiochem.org/Software/compasm/Home.html.

Dedicated to Professor Marco Antonio Chaer Nascimento and published as part of the special collection of articles celebrating his 65th birthday.

Electronic supplementary material The online version of this article (doi:10.1007/s00214-012-1271-2) contains supplementary material, which is available to authorized users.

J. V. Ribeiro · N. M. F. S. A. Cerqueira ·
I. S. Moreira · P. A. Fernandes · M. J. Ramos (✉)
REQUIMTE, Departamento de Química e Bioquímica, Faculdade de Ciências, Universidade do Porto, Rua Campo Alegre s/n, 4169-007 Porto, Portugal
e-mail: mjramos@fc.up.pt

Keywords Protein–protein interactions · Amber · VMD · MMPBSA · Software

1 Introduction

The association of proteins and the way they bind are a crucial topic in the study of living organisms. This importance stems from the fact that protein–protein interactions play a crucial role in the molecular recognition and cellular function. Mapping these interactions at the interface and revealing the key-stone residues will provide important insight into how these structures combine and how it is possible to manage them, as well as improving or inhibiting their association [1, 2].

One of the key features of these protein–protein interfaces is their sensitivity to mutations. This means that if we mutate a key interface residue by a residue alanine, there will be a significant variation in the protein–protein complex binding or association free energy. It has been defined in the literature that if the increase in the binding free energy is above 4 kcal/mol, then the mutated residue is extremely important and it is called a hot spot; if the energy increase upon mutation is between 2 and 4 kcal/mol, then this residue is relatively important for the protein–protein association and it is denominated a warm spot, and finally, if the mutations originate a binding free energy variation below 2 kcal/mol, then the residue is not particularly relevant for the interaction and it is termed a null spot [3–5].

Moreira et al. [3] have developed a protocol (schematized in Fig. 1), with low computational cost and high success rate that reproduces the quantitative free energy differences obtained from experimental mutagenesis procedures. This computational approach is transferable to any macromolecular complex and is a predictive model capable of anticipating the experimental results of mutagenesis, thus capable of guiding new experimental investigations. It is based on the all-atom methodology MMPBSA (molecular mechanics/Poisson–Boltzmann surface area) [6] to probe protein–protein interactions by calculating free energies combining molecular mechanics and a continuum solvent.

There are several web servers already available to compute this protein–protein interaction [7]. Despite the large variety of available possibilities to study this kind of interactions, our method still proved to be better from the point of view of returning quantitative values of the binding free energy differences with molecular dynamics as the sampling method. The features contrast with the qualitative values and the analysis of only a few structures of other methods available. At the end of the Sect. 4, values from other approaches obtained for our case studies are also presented for comparison.

2 Methodology

The first stage of CompASM involves the relaxation and equilibration of the wild-type complex that is being analyzed. This can be accomplished by a minimization procedure only or by a molecular dynamics simulation in a continuum medium, using the Generalized Born model. Subsequently, only the relaxed wild-type complex is divided in several alanine mutated complexes that were previously defined by the user, depending on the study that is intended to be performed. To the wild-type and mutated complexes, it is then applied the MMPBSA script to calculate the respective binding free energy differences.

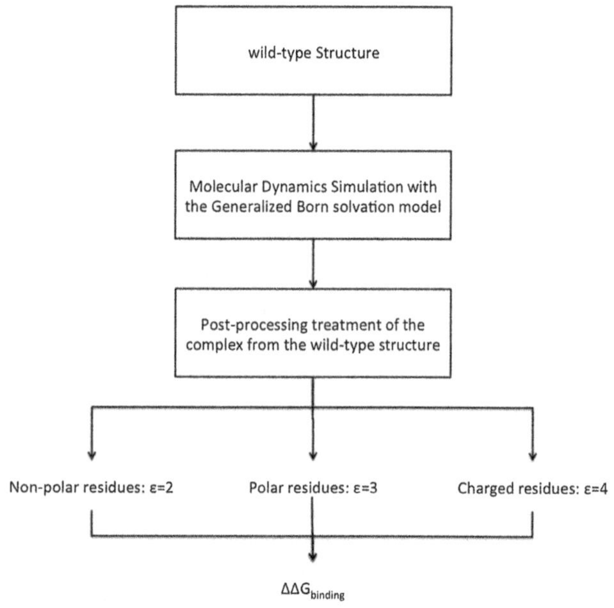

Fig. 1 General algorithm of the CompASM procedure [3]

To generate the structure of the mutant complex, a simple truncation of the mutated side chain is carried out, replacing carbon atom C_γ with a hydrogen atom, and setting the C_β–H bond direction to that of the former C_β–C_γ. The corresponding binding free energy can be calculated using the thermodynamic cycle shown in Fig. 2, in which ΔG_{gas} is the binding free energy between the two interacting partners in the gas phase, and $\Delta G_{\text{solv}}^{\text{lig}}$, $\Delta G_{\text{solv}}^{\text{recep}}$, and $\Delta G_{\text{solv}}^{\text{complex}}$ are the solvation free energy differences of the two binding partners and the complex, respectively. The binding free energy difference between an alanine mutant complex and a wild-type complex is defined as Eq. (1):

$$\Delta\Delta G_{\text{binding}} = \Delta G_{\text{binding-mutant}} - \Delta G_{\text{binding-wildtype}} \quad (1)$$

The binding free energy of two molecules in a complex is defined as the difference between the free energy of the complex and those of the respective monomers. On the other hand, the free energy of a protein–protein complex and its respective monomers can be calculated by Eq. (2), that is, by summing the internal energy (bond, angle, and dihedral), E_{internal}; the electrostatic and the van der Waals interactions, $E_{\text{electrostatic}}$ and E_{vdW}; the free energy of polar solvation, G_{polar} solvation; the free energy of nonpolar solvation, G_{nonpolar} solvation; and the entropic TS contribution for the molecule free energy.

$$G_{\text{molecule}} = E_{\text{internal}} + E_{\text{electrostatic}} + E_{\text{vdW}} + G_{\text{polar solvation}} + G_{\text{nonpolar solvation}} - TS \quad (2)$$

The first three terms in Eq. (2) are calculated using the Cornell force field [8] with no cutoff. The electrostatic solvation free energy is calculated by solving the Poisson–Boltzmann equation with the Delphi software [9, 10], which has been shown to constitute a good compromise between accuracy and computing time.

For the energy calculations, CompASM attributes specific values to three internal dielectric constant values, which depend exclusively on the type of amino acid that is mutated. Therefore, for the charged amino acids (aspartic acid, glutamic acid, lysine, arginine, and histidine), a constant of 4 should be used, for the remaining polar residues (asparagine,

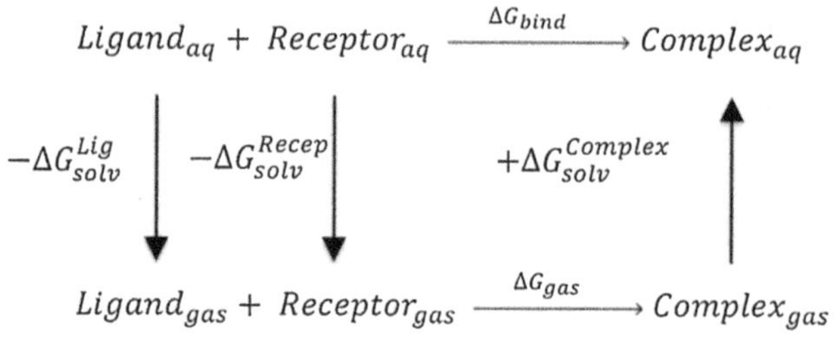

Fig. 2 Thermodynamic cycle used to calculate the binding free energy in the CompASM protocol

glutamine, cysteine, tyrosine, serine, and threonine) not ionized at physiological pH, the internal dielectric constant should be 3, and for the nonpolar amino acids (valine, leucine, isoleucine, phenylalanine, methionine, and tryptophan), the internal dielectric constant should be 2 [3]. The different internal dielectric constants account for the different degree of relaxation of the interface when different types of amino acids are mutated for alanine; the stronger the interactions these amino acids establish, the more extensive the relaxation should be, and the greater the internal dielectric constant value must be to mimic these effects.

However, for the sake of flexibility, in order to allow the user to set other values that he might see fit to the dielectric constants, we have added a feature in which the values of any number of dielectric constants can be changed from the default ones mentioned above or even added if necessary. This introduces flexibility to the plug-in and allows the user to improve the quality of the results, if considered necessary.

The nonpolar contribution to solvation free energy due to van der Waals interactions between the solute and the solvent and cavity formation was modeled as a term that is dependent on the solvent-accessible surface area of the molecule. It was estimated using empirical relation (3),

$$\Delta G_{\text{nonpolar}} = A+ \quad (3)$$

where A is the solvent-accessible surface area that was estimated using the MolSurf program, which is based on the idea primarily developed by Michael Connolly. Constants α and β are empirical, taking the values 0.005 42 kcal Å^{-2} mol^{-1} and 0.92 kcal mol^{-1}, respectively. The entropy term, obtained as the sum of translational, rotational, and vibrational components, was not calculated because it was assumed, on the basis of previous work [11, 12], that its contribution to $\Delta\Delta G_{\text{binding}}$ is negligible.

The calculation of $\Delta\Delta G_{\text{binding}}$ is achieved applying several modules of the AMBER program [13], and despite the apparent simplicity, this kind of study can easily become cumbersome. Beyond the repetitive tasks, such as the generation of mutation structures and the input for the MMPBSA [12] calculation, the handling of a large amount of files can be a tricky job. Regarding these difficulties and combining the visual facilities provided by visual molecular dynamics (VMD) [14] with the extremely intuitive graphical user interface (GUI) and the AMBER molecular dynamics calculations, we propose a new VMD-AMBER plug-in, named CompASM, which allows even the non-expert user to perform easily Alanine Scanning Mutagenesis calculations.

3 Software description

CompASM is a versatile tool created to study protein–protein interfaces, allowing the user to skip the repetitive task of creating input files and generating all necessary structures, as well as providing new options and procedures to perform a computational alanine scanning mutagenesis experiment. The input file is based on an AMBER-format file. To maintain the functionality in almost all situations, we have divided the software in two main structures: a VMD plug-in GUI and an independent CORE.

The CompASM GUI was designed to drive the user through the different steps of ASM, allowing simultaneously all the freedom needed to treat all kind of structures that require different specifications. Beyond the ligand/receptor differentiation and mutation selection, this GUI allows the user to exclude or include non-protein structures (known as heteroatoms) as well as to insert their parameterization files (.mol2 and/or.frcmod files). Another facility presented by CompASM GUI is the molecular minimization/dynamics simulation set-up tab. Here, the user can set all variables to the values needed to submit a molecular dynamics simulation or just a minimization to AMBER. The user can add more variables directly in the input file following the instructions that pop up as one proceeds in the calculation. Another useful tool is the mutation selection by non-solvent contact area (NSCA-explained schematically in Fig. 3)-based selection (Fig. 3).

This is based on a "sasa measure" VMD command, and the residue is selected if the area that is in contact with another structure is larger than 40 Å^2 (more detail in supporting information). The final results are visualized in the VMD graphics window, using the same color scheme of the summary and the detailed tables in the GUI.

The CompASM CORE is an independent set of procedures that load the structure and perform the algorithm proposed by Moreira et al. [3]. This CORE handles the files organization, performs the MMPBSA calculations, and returns an output file (ASM.out) with all the information necessary to evaluate the final results. To improve the speed and make the best use of the computational resources, we parallelized the slower procedure, the MMPBSA calculation, running each independent calculation in different CPU cores. In the CORE's procedures, the molecular dynamics simulation is performed in sets of 1/10 of the total time requested by the user. In each set, the program checks if the coordinates from the structure (backbone only) are stabilized, evaluating several parameters of the linear regression of the root-mean-square deviation (RMSD) calculated by the ptraj AMBER tool. The simulation is considered stabilized if the slope of the straight, the standard deviation and the correlation factor are in agreement with the values set by the user. For more information, please see the supporting information of this document. This software was developed in TCL/TK as the programming language, and it is

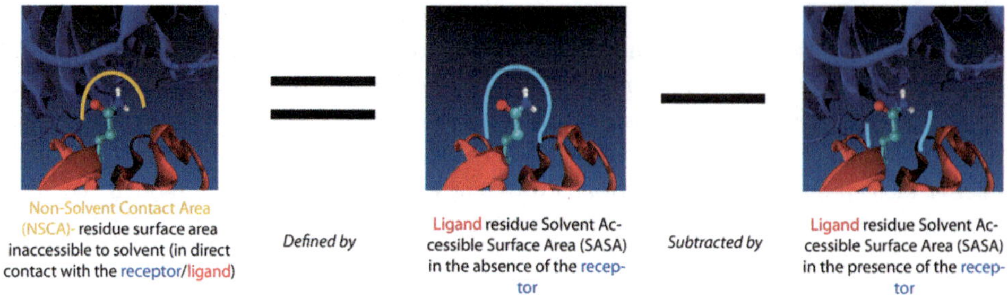

Fig. 3 Scheme representing the non-solvent contact area (NSCA). This area intends to represent by which amount each residue is buried in the protein surface. All surface areas are evaluated by the "sasa measure" command present in the visual molecular dynamics (VMD) software

available for the Amber 8 and 9 versions, requiring the DELPHI package. The application of this software using higher versions of Amber is not presently possible, only due to the inexistence of the DELPHI package in the MMPBSA protocol (i.e. not available in the Amber 10 package).

4 Results

The data resulting from CompASM are displayed in a simplified table and in a more detailed table are displayed all the values obtained from the MMPBSA calculation, including the NSCA values. All this data can be analyzed interactively in a VMD window, coloring the residue depending on the obtained score. Figure 4 shows the results obtained from CompASM concerning the protein–protein interface study of immunoglobulin complexed with an egg lysozyme, as an example.

In this section, we describe the validation process that was used to evaluate the performance of the CompASM software in the determination of the hot, warm, and null spots of three different protein–protein complexes. The results of this validation process are shown in Table 1. The structures analyzed are immunoglobulin complexed with an egg lysozyme (1VFB) [15]; complexes that mediate bacterial cell division (1F47) [16] and human immunoglobulin IgG complexed with the C2 fragment of streptococcal protein *G* (1FCC) [17]. Table 1 shows the values of NCSA, the experimental value of $\Delta\Delta G_{\text{binding}}$ of each mutation and the $\Delta\Delta G_{\text{binding}}$ evaluated by CompASM. All calculations involved molecular dynamics simulations using the default values of the CompASM GUI (Fig. 4) and modifying only the MMPBSA frequency to achieve low

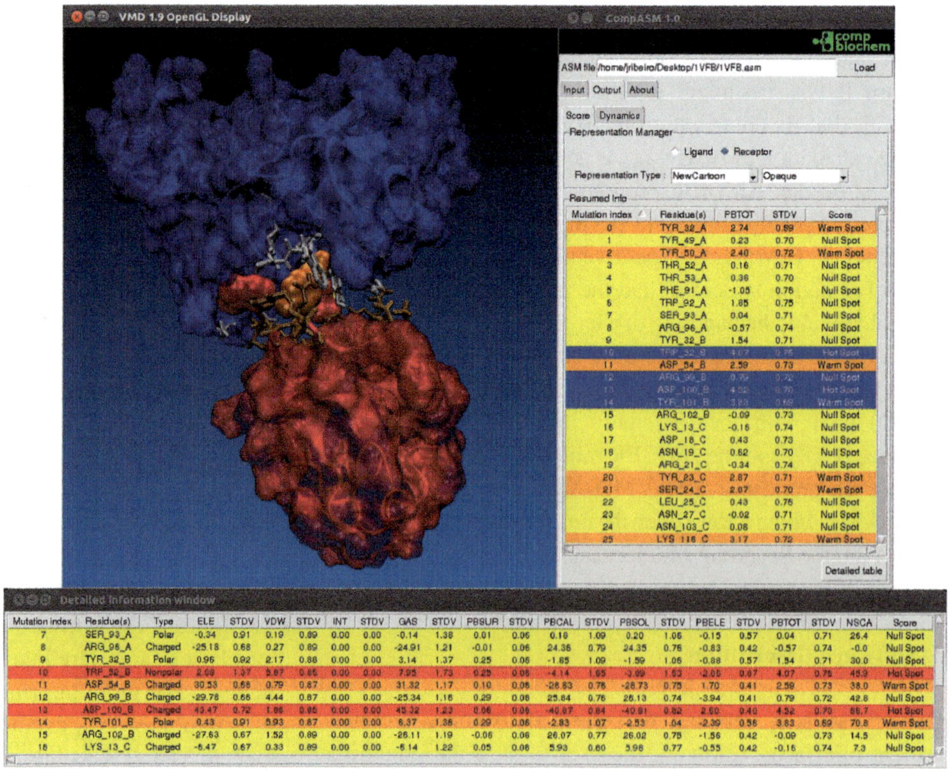

Fig. 4 Results of the protein–protein interface study of immunoglobulin complexed with an egg lysozyme (detailed results in supporting information)

Table 1 Results originated by CompASM

Protein	MMPBSA freq	Mutation	NSCA (Å^2)	Residue type	Experimental	ComPASM		Scores	
					$\Delta\Delta G_{exp}$ (kcal/mol)	$\Delta\Delta G_{ComASM}$ (kcal/mol)	STDV	Exp	CompASM
1VFB	67	TYR_32_A	65.00	Polar	1.30	3.20	0.89	Null spot	Warm spot
1VFB	67	TYR_49_A	20.60	Polar	0.80	0.39	0.89	Null spot	Null spot
1VFB	67	TYR_50_A	37.80	Polar	0.40	2.40	0.91	Null spot	Warm spot
1VFB	67	THR_53_A	15.00	Polar	−0.23	0.26	0.87	Null spot	Null spot
1VFB	67	TRP_92_A	48.30	Nonpolar	1.71	1.71	0.94	Null spot	Null spot
1VFB	67	SER_93_A	26.40	Polar	0.11	−0.10	0.90	Null spot	Null spot
1VFB	67	TYR_32_B	30.00	Polar	0.50	1.23	0.88	Null spot	Null spot
1VFB	67	TRP_52_B	45.90	Nonpolar	1.23	3.90	0.89	Null spot	Warm spot
1VFB	67	ASP_54_B	38.00	Charged	1.95	2.55	0.94	Null spot	Hot spot
1VFB	67	ARG_99_B	42.80	Charged	0.47	0.86	0.93	Null spot	Null spot
1VFB	67	ASP_100_B	66.70	Charged	3.10	4.47	0.92	Warm spot	Hot spot
1VFB	67	TYR_101_B	70.80	Polar	4.00	4.04	0.91	Hot spot	Hot spot
1VFB	67	ASP_18_C	27.90	Charged	0.30	0.51	0.97	Null spot	Null spot
1VFB	67	ASN_19_C	69.00	Polar	0.40	0.78	0.89	Null spot	Null spot
1VFB	67	TYR_23_C	12.10	Polar	0.80	2.84	0.91	Null spot	Warm spot
1VFB	67	SER_24_C	52.90	Polar	0.70	2.12	0.91	Null spot	Warm spot
1VFB	67	LYS_116_C	57.60	Charged	0.70	3.06	0.94	Null spot	Warm spot
1VFB	67	THR_118_C	42.70	Polar	0.80	0.19	0.91	Null spot	Null spot
1VFB	67	ASP_119_C	43.70	Charged	1.00	3.48	0.96	Null spot	Warm spot
1VFB	67	VAL_120_C	9.70	Nonpolar	0.90	−0.39	0.92	Null spot	Null spot
1VFB	67	GLN_121_C	108.10	Polar	2.90	2.53	0.91	Warm spot	Warm spot
1VFB	67	ILE_124_C	28.70	Nonpolar	1.20	0.09	0.92	Null spot	Null spot
1VFB	67	ARG_125_C	48.80	Charged	1.80	3.21	0.93	Null spot	Warm spot
1VFB	67	LEU_129_C	23.60	Nonpolar	0.20	−9.93	0.93	Null spot	Null spot
1F47	67	ASP_4_A	43.20	Charged	0.69	−1.60	0.8	Null spot	Null spot
1F47	67	TYR_5_A	63.30	Polar	0.86	4.21	0.74	Null spot	Hot spot
1F47	67	LEU_6_A	81.70	Nonpolar	0.92	1.59	0.78	Null spot	Null spot
1F47	67	ASP_7_A	22.30	Charged	1.73	−0.46	0.78	Null spot	Null spot
1F47	67	ILE_8_A	50.70	Nonpolar	2.50	3.12	0.77	Warm spot	Warm spot
1F47	67	PHE_11_A	77.00	Nonpolar	2.44	3.60	0.79	Warm spot	Warm spot
1F47	67	LEU_12_A	53.10	Nonpolar	2.29	2.21	0.79	Warm spot	Warm spot
1FCC	35	GLU_27_C	45.80	Charged	4.90	10.91	1.17	Hot spot	Hot spot
1FCC	35	LYS_28_C	122.30	Charged	1.30	2.40	1.22	Null spot	Warm spot
1FCC	35	LYS_31_C	56.10	Charged	3.50	4.83	1.20	Warm spot	Hot spot
1FCC	35	ASN_35_C	61.80	Polar	2.40	0.95	1.12	Warm spot	Null spot
1FCC	35	ASP_40_C	44.20	Charged	0.30	0.49	1.23	Null spot	Null spot
1FCC	35	GLU_42_C	26.00	Charged	0.40	0.34	1.23	Null spot	Null spot
1FCC	35	TRP_43_C	37.80	Nonpolar	3.80	−0.13	1.09	Warm spot	Null spot

The non-solvent contacting area (NSCA) is calculated by the CompASM program

standard deviation values. In the MMPBSA calculations of the 1VFB and 1F47 proteins, 50 structures were used, from the fourth and fifth nanosecond, respectively, of the simulation, and in the case of the 1FCC corresponding protein, 100 structures were used from the forth nanosecond. These same structures were also analyzed using well-known web servers available, and the positive predictive value or specificity P (Eq. 4), true positive rate or sensitivity R (Eq. 5), and the F measure test accuracy $F1$ (Eq. 6) values were calculated and shown in Table 2.

Table 2 Comparison of the values obtained using CompASM with those that result from the software/servers available

Abbrev.	ISIS	Promate	ROBETTA	KFC2-A	KFC2-B	HotPoint	CompASM
TP	1	0	7	5	6	6	8
TN	26	25	21	25	23	20	17
FP	2	3	7	3	5	8	11
FN	9	10	3	5	4	4	2
R	10	0	70	50	60	60	80
P	33	0	50	63	55	43	42
F1	15	0	58	56	57	50	55

TP true positive, *TF* true negative, *FN* false negative, *FP* false negative, *R* positive predictive, *P* negative predictive, *F1 F* measure

$$P = \frac{\mathrm{TP}}{\mathrm{TP} + \mathrm{FP}} \tag{4}$$

$$R = \frac{\mathrm{TP}}{\mathrm{TP} + \mathrm{FN}} \tag{5}$$

$$F1 = \frac{2\mathrm{PR}}{P + R} \tag{6}$$

In Eqs. (4), (5), and (6), TP corresponds to the correct computational prediction of the number of hot spots (TP—true positive), that is, when these residues are experimentally classified as hot spots; FP corresponds to the number of computationally predicted hot spots when these residues are experimentally classified as null spots (FP—false positive); and finally, FN corresponds to the number of computationally predicted null spots when these residues are experimentally classified as hot spots (FN—false negative). Therefore, the closer the values are to 100 the better they are, meaning in this case that the software is capable to predict correctly the numbers of hot spots and null spots.

To compare our software to those from web servers already available, we tested the same mutations in the proteins analyzed above using the following softwares: ISIS [18]; Promate [19]; Robetta [20]; K-FC2A and B [21]; and HotPoint [22].

Tables 1 and 2 demonstrate that the values resulting from CompASM are in the same range of those provided by experimental means, that is, the CompASM software was able to detect all the hot spots. This software was also sensitive enough to detect and to distinguish most warm- and null spots. The results are very close to those obtained by Moreira et al. [3] in which the methodology of CompASM was based on. The small differences result from the differences in the structures retrieved from the MD simulations and submitted to the MMPBSA calculations. CompASM has its own algorithm, which is used to control the molecular dynamics simulations. Once all the criteria are achieved, which the user can either set him/herself or accept the default, the MD simulation is stopped and the MMPBSA protocol starts. This algorithm is very useful because it makes the process straightforward and minimizes the computation time. Simultaneously, it allows the process to be reproducible. This makes it very handy not only for non-expert users, but also for more advanced users that can modify the criteria used in the CompASM GUI.

Table 2 shows the number of TP, TN, FP, and FN values, as well as the accuracy of the different tested softwares.

Looking at the data resulting from CompASM, we can notice only 2 false negative values. There are, however, 11 false positives. In the majority of the complexes studied, an analysis of the dominant interactions suggests that van der Waals interactions and hydrophobic effects provide a reasonable basis for understanding binding affinities. In fact, the breakdown of Eq. (2) for all the cases studied with CompASM shows that the $\Delta\Delta E_{\mathrm{vdw}}$ values are almost all positive indicating that the van der Waals interactions are favorable to complex binding and that alanine mutation of the residues diminishes the vdW contacts at the interfaces. This is explained by the hydrophobic character of the interfaces.

To further examine the reliability and usefulness of CompASM, we have compared our predictions to the predictions of some of the more popular softwares that are available nowadays. The results are presented in Table 2 pointing to the fact that the methods Robetta, KFC2-A, -B, HotPoint, and CompASM have similar performances, with the latter giving slightly better values. However, in a subsequent work [23], we provide a full study documenting the results obtained with CompASM based on a large number of structures, which points out to its good performance generating the better values among other methods.

We notice that only two of the above-mentioned approaches, ours and Robetta's, provide quantitative values, which can be fully compared to the experimental ones, as opposed to qualitative values that classify the mutations between hot, warm, and null spots only. However, in order to compare the results obtained with CompASM to the values returned by other softwares, we translated our values to the binominal qualitative classification (null- and hot spot) adopted by the tested softwares, which classify the residues in null spots if the $\Delta\Delta G$ values are below 2 kcal/mol and hot spots if the $\Delta\Delta G$ values are higher than 2 kcal/mol.

CompASM returns a value close to 100 (80) for the values of R (hot spots prediction over false negatives). However, as CompASM is a software with which we can obtain quantitative values, we know directly how close is the residue to be a null- or a hot spot by comparing the obtained $\Delta\Delta G$ with the reference barriers of 2 and 4 kcal/mol, respectively.

5 Conclusions

Computational alanine scanning mutagenesis [3] has proven to be an accurate means of detecting the residues that play an important role in protein–protein interfaces (hot spots). Here, we present a VMD plug-in, CompASM, which facilitates the application of this approach thus simplifying the study of protein interfaces. CompASM guides the user through all ASM steps, from the ligand/receptor selection to the molecular dynamics simulation, and provides the visualization of the final results in a VMD window. This program can run either in local machines or in a cluster (multicomputer system). The GUI package is multiplatform, and the CORE package works in a UNIX system (Mac OS and Linux).

The CompASM package is distributed as an independent platform, with script code under the GNU Public License from http://compbiochem.org/Software/compasm/Home.html.

Acknowledgments The authors would like to thank the FCT (Fundação para a Ciência e Tecnologia) for financial support (Grants SFRH/BD/61324/2009 and PTDC/QUI-QUI/103118/2008).

References

1. Bordner AJ, Abagyan R (2005) Statistical analysis and prediction of protein–protein interfaces. Proteins 60(3):353–366. doi:10.1002/Prot.20433
2. Thorn KS, Bogan AA (2001) ASEdb: a database of alanine mutations and their effects on the free energy of binding in protein interactions. Bioinformatics 17(3):284–285
3. Moreira IS, Fernandes PA, Ramos MJ (2007) Computational alanine scanning mutagenesis—an improved methodological approach. J Comput Chem 28(3):644–654. doi:10.1002/jcc.20566
4. Ramos MJ, Moreira IS, Fernandes PA (2007) Hot spots—a review of the protein–protein interface determinant amino-acid residues. Proteins 68(4):803–812. doi:10.1002/Prot.21396
5. Moreira IS, Fernandes PA, Ramos MJ (2007) Hot spot occlusion from bulk water: a comprehensive study of the complex between the lysozyme HEL and the antibody FVD1.3. J Phys Chem B 111(10):2697–2706. doi:10.1021/Jp067096p
6. Kollman PA, Massova I, Reyes C, Kuhn B, Huo SH, Chong L, Lee M, Lee T, Duan Y, Wang W, Donini O, Cieplak P, Srinivasan J, Case DA, Cheatham TE (2000) Calculating structures and free energies of complex molecules: combining molecular mechanics and continuum models. Acc Chem Res 33(12):889–897. doi:10.1021/Ar000033j
7. Fernandez-Recio J (2011) Prediction of protein binding sites and hot spots. Wires Comput Mol Sci 1(5):680–698. doi:10.1002/Wcms.45
8. Cornell WD, Cieplak P, Bayly CI, Gould IR, Merz KM, Ferguson DM, Spellmeyer DC, Fox T, Caldwell JW, Kollman PA (1996) A second generation force field for the simulation of proteins, nucleic acids, and organic molecules (vol 117, pg 5179, 1995). J Am Chem Soc 118(9):2309–2309
9. Rocchia W, Sridharan S, Nicholls A, Alexov E, Chiabrera A, Honig B (2002) Rapid grid-based construction of the molecular surface and the use of induced surface charge to calculate reaction field energies: applications to the molecular systems and geometric objects. J Comput Chem 23(1):128–137
10. Rocchia W, Alexov E, Honig B (2001) Extending the applicability of the nonlinear Poisson–Boltzmann equation: multiple dielectric constants and multivalent ions. J Phys Chem B 105(28):6507–6514
11. Massova I, Kollman PA (1999) Computational alanine scanning to probe protein–protein interactions: a novel approach to evaluate binding free energies. J Am Chem Soc 121(36):8133–8143
12. Huo S, Massova I, Kollman PA (2002) Computational alanine scanning of the 1:1 human growth hormone–receptor complex. J Comput Chem 23(1):15–27
13. Case DA, Darden TA, Cheatham, Simmerling CL, Wang J, Duke RE, Luo R, Merz KM, Pearlman DA, Crowley M, Walker RC, Zhang W, Wang B, Hayik S, Roitberg A, Seabra G, Wong KF, Paesani F, Wu X, Brozell S, Tsui V, Gohlke H, Yang L, Tan C, Mongan J, Hornak V, Cui G, Beroza P, Mathews DH, Schafmeister C, Ross WS, Kollman PA (2006) Amber 9
14. Humphrey W, Dalke A, Schulten K (1996) VMD: visual molecular dynamics. J Mol Graph 14(1):33–38
15. Bhat TN, Bentley GA, Boulot G, Greene MI, Tello D, Dallacqua W, Souchon H, Schwarz FP, Mariuzza RA, Poljak RJ (1994) Bound Water-molecules and conformational stabilization help mediate an antigen-antibody association. Proc Natl Acad Sci USA 91(3):1089–1093
16. Somers WS, Mosyak L, Zhang Y, Glasfeld E, Haney S, Stahl M, Seehra J (2000) The bacterial cell-division protein ZipA and its interaction with an FtsZ fragment revealed by X-ray crystallography. EMBO J 19(13):3179–3191
17. Sauereriksson AE, Kleywegt GJ, Uhl M, Jones TA (1995) Crystal-structure of the C2 fragment of streptococcal protein-G in complex with the Fc domain of human-Igg. Structure 3(3):265–278
18. Ofran Y, Rost B (2007) ISIS: interaction sites identified from sequence. Bioinformatics 23(2):E13–E16. doi:10.1093/Bioinformatics/Btl303
19. Neuvirth H, Raz R, Schreiber G (2004) ProMate: a structure based prediction program to identify the location of protein–protein binding sites. J Mol Biol 338(1):181–199. doi:10.1016/J.Jmb.2004.02.040
20. Kortemme T, Baker D (2002) A simple physical model for binding energy hot spots in protein–protein complexes. Proc Natl Acad Sci USA 99(22):14116–14121
21. Darnell SJ, Page D, Mitchell JC (2007) An automated decision-tree approach to predicting protein interaction hot spots. Proteins 68(4):813–823. doi:10.1002/Prot.21474
22. Tuncbag N, Keskin O, Gursoy A (2010) HotPoint: hot spot prediction server for protein interfaces. Nucleic Acids Res 38:W402–W406. doi:10.1093/Nar/Gkq323
23. Moreira IS, Martins JM, Ramos RR, Fernandes PA, Ramos MJ (2012) Computational alanine scanning mutagenesis: a new method for hot-spot detection (Results to be published)

Theor Chem Acc (2012) 131:1237
DOI 10.1007/s00214-012-1237-4

REGULAR ARTICLE

Spectrum simulation and decomposition with nuclear ensemble: formal derivation and application to benzene, furan and 2-phenylfuran

Rachel Crespo-Otero · Mario Barbatti

Received: 12 February 2012 / Accepted: 14 May 2012 / Published online: 9 June 2012

Abstract A formal derivation of the nuclear-ensemble method for absorption and emission spectrum simulations is presented. It includes discussions of the main approximations employed in the method and derivations of new features aiming at further developments. Additionally, a method for spectrum decomposition is proposed and implemented. The method is designed to provide absolute contributions of different classes of states (localized, diffuse, charge-transfer, delocalized) to each spectral band. The methods for spectrum simulation and decomposition are applied to the investigation of UV absorption of benzene, furan, and 2-phenylfuran, and of fluorescence of 2-phenylfuran.

Keywords Electronic spectrum · Absorption · Fluorescence · Spectrum simulation · Excimer · 2-Phenylfuran

Dedicated to Professor Marco Antonio Chaer Nascimento and published as part of the special collection of articles celebrating his 65th birthday.

Electronic supplementary material The online version of this article (doi:10.1007/s00214-012-1237-4) contains supplementary material, which is available to authorized users.

R. Crespo-Otero (✉) · M. Barbatti (✉)
Max-Planck-Institut für Kohlenforschung,
Kaiser-Wilhelm-Platz 1, 45470 Mülheim an der Ruhr, Germany
e-mail: crespo@kofo.mpg.de

M. Barbatti
e-mail: barbatti@kofo.mpg.de

1 Introduction

Visible and ultraviolet spectroscopy is a central technique for general material characterization. Its fundamental role is still enhanced when dealing with photo-active molecules employed in photo-receptors, light emitters and photovoltaics. From the theoretical standpoint, the characterization of the electronic spectrum of molecules is often restricted to characterization of vertical transitions and their energies and transition dipole moments. This procedure greatly aids the assignment of measured spectra, but is far from representing its complexity. Full spectrum simulation, beyond simple vertical-lines computation, is a much more involved task, demanding non-routine and computationally costly procedures, such as propagation of excited-state nuclear wave packets or determination of Franck–Condon factors [1, 2]. In general, such simulations are strongly limited in terms of nuclear degrees of freedom or number of electronic states considered. They are also non-black-box procedures, which require high degree of specialization to be conducted, precluding them of being adopted as routine procedures for researchers out of the computational-chemistry field.

Recently, there have been efforts to develop general methods for spectrum simulation more accessible to the quantum-chemical community and aiming at large molecules. Examples of such developments are the implementation of the Tannor–Heller method [3, 4] in the ORCA program, which allows computing vibrational progressions, and of the quadratic-coupling expansion in Gaussian, which allows computing vibrational progressions and dark vibronic bands [5]. Moreover, spectrum simulations based on excitation of an ensemble of nuclear geometries have become popular [6–8]. This kind of approach is based on the hypothesis that the spectral band shape is determined by the ground-state nuclear-geometry distribution. When this hypothesis is approximately fulfilled, the nuclear-ensemble approach can provide good estimates for several spectral features, including vibrational broadening of short lived states, dark vibronic bands, and absolute spectral intensities. Additionally, the nuclear-ensemble approach includes all degrees of freedom and can be extended to large number of excited states. The conceptual simplicity and relative low computational cost makes the nuclear-ensemble approach ideal for routine simulations. The nuclear-ensemble approach, however, has a number of handicaps, namely (1) it depends on arbitrary parameters; (2) it cannot properly predict vibrational broadening of long-lived states; (3) it cannot predict vibrational progressions; and (4) it does not properly take into account vibronic features of the energy gap of the spectral lines.

The origin of each element composing the nuclear-ensemble approach can be traced back to decades ago, first with the works of Heller, Wilson and others in the 1980s, where absorption bands were computed based on molecular dynamics [9]. It is also influenced by the works of Skinner [10], which provided a useful link between Kubo's stochastic theory of the line shape [11] and molecular dynamics, and by the reflection principle [12], which approaches bound to continuum transitions from the nuclear-ensemble perspective. The intuitive character of the nuclear-ensemble approach has created a situation where although the method is frequently employed, there is no clear derivation of its formalism. This information gap makes difficult to understand the reasons for its limitations and to propose ways to improve the method. In this contribution, we derive equations for absorption cross sections and radiative decay rates based on the nuclear-ensemble method. The main approximations are made explicit, and improvements on the method are proposed, in particular ways to get rid of arbitrary parameters.

One of the advantages of the nuclear-ensemble approach is that it provides straightforward ways to analyze different contributions summing up to the full spectrum. For instance, in Ref. [7], the UV absorption spectra of nucleobases have been decomposed in terms of their several diabatic contributions. In Ref. [13], the absorption spectrum of $Cr(CO)_6$ has been decomposed in terms of symmetry contributions. In Ref. [14], the spectrum of urocanic acid has been decomposed in terms of isomeric contributions. In all these cases, the spectral decomposition helped to reveal the character of states forming the several bands. In the present work, we develop a new way to perform spectral decomposition analysis, now, based on the electronic-density distribution of the states. This approach allows characterizing the contributions of localized states, charge-transfer states, diffuse states and excimer states to each band.

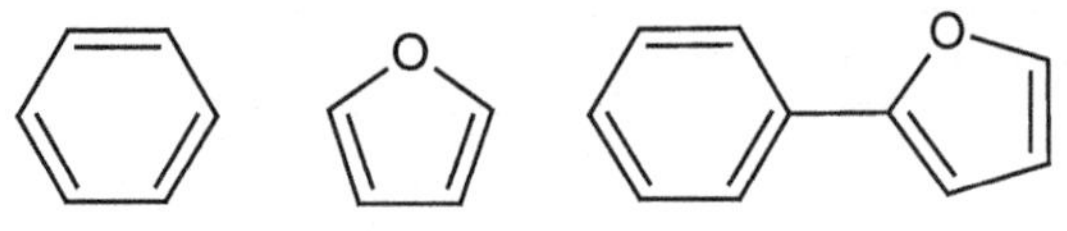

Fig. 1 Illustration of benzene, furan and 2-phenylfuran

Another advantage of the nuclear-ensemble approach is that it is naturally a post-Condon approximation. Because the transition moments are evaluated for geometries displaced from equilibrium position, vibronic contributions to the spectrum are computed without need of Herzberg–Teller type of expansions [5]. Thus, even dark vibronic bands are described by the simulations [15].

Often in the literature, computed vertical excitations are compared to experimental band maxima. Although this is a valid procedure to provide assignments and to check the overall quality of theoretical method, it can also be misleading, when one does not take into account that the band maximum is normally red-shifted in relation to the vertical excitation. The spectrum simulation allows computing the shift between these two quantities. This shift is used to estimate the experimental value of the vertical excitation, which is a more appropriate quantity to have the theoretical data compared with.

To illustrate several of these methodological features, we have investigated the absorption and emission spectra of 2-phenylfuran and of its constituent monomers, benzene and furan (Fig. 1). A series of features lead us to this particular choice of molecules for our benchmark investigations. First, a good deal of experimental and theoretical data is available for all three molecules. Second, all three molecules are small enough to have their spectra simulated with even more accurate (and costly) methods than the nuclear-ensemble approach, allowing methodological comparisons. Third, there is an interesting physical chemistry associated with these molecules: While benzene and furan are non-fluorescent species, 2-phenylfuran is fluorescent [16].

2 Derivation of the nuclear-ensemble method

2.1 Absorption cross section

With the formalism of the time-dependent perturbation theory for interaction between an electron and the classical radiation field, within the electric dipole and Born–Oppenheimer approximations, the isotropic absorption cross section is given by [17]

$$\sigma(E) = \frac{\pi}{3\hbar c\varepsilon_0 n_r E}\sum_{n,k}\int \left|\Delta E_{00,nk}(\mathbf{R})\chi^*_{00}(\mathbf{R})\langle\phi_0|\mu(\mathbf{r},\mathbf{R})|\phi_n\rangle_{\mathbf{r}}\chi_{nk}(\mathbf{R})\right|^2\delta(\Delta E_{00,nk}(\mathbf{R}) - E)\mathrm{d}\mathbf{R}, \quad (1)$$

where E is the photon energy of the probe radiation, ε_0 is the vacuum permittivity, c is the speed of light, n_r is the refractive index of the medium, and μ is the dipole operator. The sum runs over the final electronic states n and vibrational states k. The molecule is supposed to be initially in the electronic and vibrational ground state. The electronic and nuclear wavefunctions are represented by ϕ and χ, respectively. The δ function selects the resonant frequency between the initial and the final states. The ket indexes indicate the integration coordinates, $\mathbf{r}$ for electronic and $\mathbf{R}$ for nuclear. The resonance occurs for

$$\Delta E_{00,nk} = E_{nk} - E_{00} + \Delta E_{0,n}, \quad (2)$$

where $\Delta E_{0,n}$ is the vertical excitation energy and E_{00} and E_{nk} are the vibrational energies of the ground and excited states.

The cross section can be recast in the time domain as [3]

$$\sigma(E) = \frac{1}{6\hbar^2 c\varepsilon_0 n_r E}\sum_n \mathrm{Re}\int \Delta E^2_{0,n}(\mathbf{R})M^2_{0n}(\mathbf{R}) \times \left[\int \chi^*_{00}(\mathbf{R})\chi_n(\mathbf{R},t)\mathrm{e}^{i(E+E_{00})t/\hbar}\mathrm{d}t\right]\mathrm{d}\mathbf{R}, \quad (3)$$

where M_{0n} is the electronic transition dipole moment given by

$$M_{0n}(\mathbf{R}) = \langle\phi_0|\mu_e(\mathbf{r};\mathbf{R})|\phi_n\rangle_{\mathbf{r}} \quad (4)$$

and where the following approximation was employed

$$\Delta E_{0,n} \approx \langle\Delta E_{00,nk}\rangle_k. \quad (5)$$

The term between brackets in Eq. (3) is the inverse Fourier transform of the overlap between the ground-state nuclear wavefunction and the wave packet $|\chi_n(t)\rangle$ given by the operation of the excited-state Hamiltonian H_n on the ground-state wavefunction:

$$|\chi_n(t)\rangle = \mathrm{e}^{-iH_n t/\hbar}|\chi_{00}\rangle. \quad (6)$$

The extinction coefficient ε for photoabsorption (in L mol^{-1} cm^{-1}) can be written in terms of the absorption cross section σ (in cm^2) as

$$\varepsilon = \frac{10^{-3}N_A}{\ln(10)}\sigma, \quad (7)$$

where N_A is the Avogadro constant.

The core of the method is to compute the overlap functions

$$\varphi(\mathbf{R},t) = \chi^*_{00}(\mathbf{R})\chi_n(\mathbf{R},t), \quad (8)$$

which are needed to integrate Eq. (3). In the Tannor–Heller approach [3], for instance, the computation of the overlap function is done within the Condon approximation employing harmonic oscillator wavefunctions. For the nuclear-ensemble approaches explored in the present work, which goes beyond the Condon approximation, we proceed without explicit computation of overlap functions. In the next sections, two phenomenological models for the overlap function are discussed. These models, however, depend on arbitrary parameters. In Sect. 2.3, we discuss how to get rid of these parameters. Although the results discussed here are restricted to these two models, the derivation itself is rather general and can be adapted to other choices of overlap functions, including explicitly computed overlap functions obtained from wave packet propagation.

2.2 Overlap function: approximation 1

To get the result discussed in Ref. [7], the overlap function is approximated by

$$\chi^*_{00}(\mathbf{R})\chi_n(\mathbf{R},t) = |\chi_{00}(\mathbf{R})|^2\exp\left[-\frac{\delta_n}{2\hbar}|t| - \frac{i}{\hbar}\Delta E_{0,n}(\mathbf{R})t - \frac{i}{\hbar}E_{00}(\mathbf{R})t\right], \quad (9)$$

where δ_n is a parameter associated with the lifetime of state n. Inserting Eq. (9) in Eq. (3) and performing the integration over t result in

$$\sigma(E) = \frac{\pi\hbar e^2}{2mc\varepsilon_0 n_r E}\sum_n\int |\chi_{00}(\mathbf{R})|^2\Delta E_{0,n}(\mathbf{R})f_{0n}(\mathbf{R}) \times g_{\mathrm{Lorentz}}(E - \Delta E_{0,n}(\mathbf{R}), \delta_n)\mathrm{d}\mathbf{R}. \quad (10)$$

where e and m are the electron charge and mass. For convenience, the transition dipole moment has been expressed in terms of oscillator strengths [18]:

$$M^2_{0n} = \frac{3\hbar^2 e^2}{2m\Delta E_{0,n}}f_{0n}. \quad (11)$$

In Eq. (10), g_{Lorentz} is a normalized Lorentzian line shape given by

$$g_{\mathrm{Lorentz}}(E - \Delta E_{0,n}, \delta_n) = \frac{1}{\pi}\frac{\delta_n/2}{(E - \Delta E_{0,n}(\mathbf{R}))^2 + (\delta_n/2)^2}. \quad (12)$$

2.3 Overlap function: approximation 2

In the harmonic approximation, neglecting Duschinksky effects and frequency shifts between the ground and excited states, the nuclear wavefunction overlap (integrated over $\mathbf{R}$) can be written as [3]

$$\langle \chi_{00} \mid \chi_n(t) \rangle = \exp\left[\sum_j \left(\frac{-\Delta_{nj}^2}{2}\left(1 - e^{-i\omega_j t}\right) - \frac{i\omega_j t}{2} \right) + \frac{i}{\hbar}\Delta E_{0,n}^a t\right], \tag{13}$$

where sum runs over all normal modes j with frequency ω_j. Δ_{nj} (dimensionless) is the component along normal mode j of the displacement vector between the ground- and excited-state equilibrium positions. $\Delta E_{0,n}^a$ is the energy difference between the minima of the excited state n and of the ground state. By expanding the exponential term in the exponential argument in a power series of t, we see that the first non-imaginary term in the equation above is proportional to t^2. The equation including all terms until the second order becomes

$$\langle \chi_{00} \mid \chi_n(t) \rangle \approx \exp\left[-i\left(\frac{1}{2}\sum_j \left(\Delta_{nj}^2 + 1\right)\omega_j + \omega_{0n}^a \right) t - \frac{1}{2}\left(\sum_j \Delta_{nj}^2 \omega_j^2 \right) t^2 \right]. \tag{14}$$

Since the non-imaginary terms are actually those responsible for the line width, Eq. (14) motivates a new functional form for the overlap function:

$$\chi_{00}^* \chi_n(\mathbf{R}, t) = |\chi_{00}(\mathbf{R})|^2 \exp\left[-\frac{i}{\hbar}\Delta E_{0,n}(\mathbf{R})t - \frac{i}{\hbar}\varepsilon_n t - \frac{i}{\hbar}E_{00}t - \frac{1}{8\hbar^2}\delta_n^2 t^2 \right]. \tag{15}$$

As we shall see, Eq. (14) is obtained from Eq. (15) under the condition that $\Delta E_{0,n}(\mathbf{R})$ is approximately constant over $\mathbf{R}$.

Inserting Eq. (15) in Eq. (3) and integrating over t leads to

$$\sigma(E) = \frac{\pi\hbar e^2}{2mc\varepsilon_0 n_r E}\sum_n \int |\chi_{00}(\mathbf{R})|^2 \Delta E_{0,n}(\mathbf{R}) f_{0n}(\mathbf{R}) \times g_{\text{Gauss}}\left(E - \Delta E_{0,n}(\mathbf{R}) - \varepsilon_n(\mathbf{R}), \delta_n\right) d\mathbf{R}, \tag{16}$$

which, different from the previous result, Eq. (10), has normalized Gaussian line shapes

$$g_{\text{Gauss}}\left(E - \Delta E_{0,n}(\mathbf{R}) + \varepsilon_n(\mathbf{R}), \delta_n\right) = \frac{1}{\left(2\pi(\delta_n/2)^2\right)^{1/2}} \exp\left(\frac{-\left(E - \Delta E_{0,n}(\mathbf{R}) - \varepsilon_n\right)^2}{2(\delta_n/2)^2} \right) \tag{17}$$

instead of Lorentzian line shapes and includes vibronic shifts through the ε_n term.

Another difference regarding the comparison between Eqs. (10) and (16) is that this second approximation for the overlap function gives us a direct way to determine the parameters δ_n and ε_n. Although we will not investigate state-specific line widths and vibronic shifts in the simulations discussed in this work, for completeness of the derivation, it is worth showing how these parameters can be computed. If $\Delta E_{0,n}(\mathbf{R})$ is approximately constant over $\mathbf{R}$, then δ_n and ε_n in Eq. (15) can be obtained by comparison with Eq. (14) and be defined in terms of and of ω_j, Δ_{nj}, and $\delta E_{0,n} = \Delta E_{0,n} - \Delta E_{0,n}^a$ as:

$$\varepsilon_n = \frac{\hbar}{2}\sum_j \Delta_{nj}^2 \omega_j - \delta E_{0,n}, \tag{18}$$

$$\delta_n = 2\hbar\left(\sum_j \Delta_{nj}^2 \omega_j^2 \right)^{1/2}. \tag{19}$$

Still in the harmonic approximation without Duschinsky rotation or frequency shifts, these quantities are simply (see details in the Online Resource 1)

$$\varepsilon_n = -\frac{1}{2}\sum_j \frac{G_{nj}^2}{\mu_j \omega_j^2} = -\frac{1}{2}\delta E_{0,n}, \tag{20}$$

$$\delta_n = 2\left(\hbar \sum_j \frac{G_{nj}^2}{\mu_j \omega_j} \right)^{1/2}, \tag{21}$$

where

$$G_{nj} = \left.\frac{dE_n}{dq_j}\right|_{q_j=0} \tag{22}$$

is the excited-state potential energy gradient evaluated at the ground-state minimum geometry with respect to the ground-state normal coordinates $\mathbf{q}$.

2.4 Nuclear-ensemble approximation

The integrals on $\mathbf{R}$ in Eqs. (10) and (16) can be computed employing a Monte Carlo procedure sampling N_p random $\mathbf{R}_l$values according to the distribution $|\chi_{00}(\mathbf{R}_l)|^2$ [19]:

$$\sigma(E) = \frac{\pi e^2 \hbar}{2mc\varepsilon_0 n_r E}\sum_n^{N_{fs}} \frac{1}{N_p}\sum_l^{N_p} \Delta E_{0,n}(\mathbf{R}_l) f_{0n}(\mathbf{R}_l) \times g(E - \Delta E(\mathbf{R}_l), \delta_n). \tag{23}$$

In this equation, g is a normalized Lorentzian line shape centered at $\Delta E = \Delta E_{0,n}$ for the integration of Eq. (10) and a normalized Gaussian line shape centered at $\Delta E = \Delta E_{0,n} + \varepsilon_n$ for integration of Eq. (16). Note that the sum over the states was restricted to a number N_{fs} of final states.

A convenient way of sampling the ground-state density at time zero is to suppose that the harmonic approximation is valid and to employ a Wigner distribution

$$|\chi_{00}(\mathbf{q})|^2 = \prod_{j=1}^{3N-6} \left(\frac{\mu_j \omega_j}{\pi \hbar}\right)^{1/2} \exp\left(-\mu_j \omega_j q_j^2/\hbar\right). \tag{24}$$

Using a stochastic algorithm, an ensemble of N_p normal coordinates $\{\mathbf{q}_l\}$ ($l = 1 - N_p$) can be generated according to the Wigner distribution and converted into an ensemble of Cartesian geometries $\{\mathbf{R}_l\}$. Alternatively, a ground-state ensemble $\{\mathbf{R}_l\}$ can also be generated by trajectories simulations in the ground state. In this case, however, one should care that the kinetic energy in the internal vibrational degrees is high enough as to provide the zero-point energy for each degree of freedom. As discussed in Ref. [20], when the vibrational temperature is properly taken into account, Wigner and trajectory samplings tend to produce equivalent results. Either way, after having an ensemble of geometries distributed according to $|\chi_{00}(\mathbf{R})|^2$, and $\Delta E_{0,n}$ and f_{0n} computed for each geometry in the ensemble, $\sigma(E)$ can be evaluated using Eq. (23).

The error in the cross section due to the statistical sampling can be estimated by

$$\delta\sigma(E) \simeq \frac{\pi e^2 \hbar}{2mc\varepsilon_0 n_r E} \sum_n^{N_{fs}} \frac{1}{N_p^{1/2}(N_p - 1)^{1/2}} \times \left[\sum_l^{N_p} \left(\Delta E_{0n}(\mathbf{R}_l) f_{0n}(\mathbf{R}_l) g\left(E - \Delta E_{0,n}(\mathbf{R}_l), \delta_n\right) - \langle s_n \rangle\right)^2\right]^{1/2}, \tag{25}$$

where

$$\langle s_n \rangle = \frac{1}{N_p} \sum_{l'}^{N_p} \Delta E_{0n}(\mathbf{R}_{l'}) f_{0n}(\mathbf{R}_{l'}) g\left(E - \Delta E_{0,n}(\mathbf{R}_{l'}), \delta_n\right). \tag{26}$$

2.5 Emission spectrum

The differential rate for radiative emission (dimensionless) is given by [2, 21]

$$\Gamma_{\text{rad}}(E) = \frac{n_r^3}{3\pi \hbar^3 c^3 \varepsilon_0} \times \sum_{k,m} \rho_k^T \int \Delta E_{1k,0m}(\mathbf{R})^3 |\chi_{0m}^*(\mathbf{R}) \times \langle \phi_0 | \mu(\mathbf{r}, \mathbf{R}) | \phi_1 \rangle_{\mathbf{r}} \chi_{1k}(\mathbf{R})|^2 \delta(E - \Delta E_{1k,0m}(\mathbf{R})) d\mathbf{R}, \tag{27}$$

where the emission is supposed to occur from the kth vibrational state of the first-excited electronic state, into the mth vibrational state of the ground electronic state. At a certain temperature T, state k is populated according to the Boltzmann distribution

$$\rho_k^T = \frac{e^{-(E_{1k} - E_{10})/k_B T}}{\sum_{k'} e^{-(E_{1k'} - E_{10})/k_B T}}. \tag{28}$$

Following Ref. [21], we have adopted an empirical n_r^3 dependence of the differential emission rate on the refractive index. Since the final emission rate should also take into account the dependence on the absorption intensity, which depends itself on n_r^{-1}, the final emission rate should bear an n_r^2 dependence, as experimentally observed [22].

In the time domain, the differential emission rate becomes

$$\Gamma_{\text{rad}}(E) = \frac{n_r^3}{6\pi^2 \hbar^3 c^3 \varepsilon_0} \text{Re} \int \Delta E_{1,0}(\mathbf{R})^3 M_{01}^2(\mathbf{R}) \times \left[\sum_k \rho_k^T \int \chi_{1k}^*(\mathbf{R}) \chi_0(\mathbf{R}, t) e^{i(E_{1k} - E)t/\hbar} dt\right] d\mathbf{R}, \tag{29}$$

where the following approximation was employed

$$\Delta E_{1,0} \approx \langle \Delta E_{1k,0m} \rangle_{k,m}. \tag{30}$$

Now, the wave packet is given by operating the ground-state Hamiltonian H_0 on the excited-state wavefunction:

$$|\chi_0(t)\rangle = e^{-iH_0 t/\hbar} |\chi_{1k}\rangle. \tag{31}$$

With an overlap-function approximation similar to the first approximation discussed above for absorption,

$$\chi_{1k}^*(\mathbf{R}) \chi_0(\mathbf{R}, t) = |\chi_{1k}(\mathbf{R}_l)|^2 \times \exp\left[-\alpha_n(\mathbf{R})|t| + i\omega_{01}(\mathbf{R})t - i\omega_{1k}(\mathbf{R})t\right], \tag{32}$$

we obtain

$$\Gamma_{\text{rad}}(E) = \frac{e^2 n_r^3}{2\pi \hbar m c^3 \varepsilon_0} \int \left[\sum_k \rho_k^T |\chi_{1k}(\mathbf{R})|^2\right] \times \Delta E_{1,0}(\mathbf{R})^2 |f_{10}(\mathbf{R})| g_{\text{Lorentz}}\left(E - \Delta E_{1,0}(\mathbf{R}), \delta\right) d\mathbf{R}. \tag{33}$$

This last expression can be integrated with a Monte Carlo procedure:

$$\Gamma_{\text{rad}}(E) = \frac{e^2 n_r^3}{2\pi \hbar m c^3 \varepsilon_0} \frac{1}{N_p} \sum_l^{N_p} \Delta E_{1,0}(\mathbf{R}_l)^2 |f_{10}(\mathbf{R}_l)| \times g_{\text{Lorentz}}\left(E - \Delta E_{1,0}(\mathbf{R}_l), \delta\right), \tag{34}$$

where N_p random $\mathbf{R}_l$ points are sampled according to the $\sum_k \rho_k^T |\chi_{1k}(\mathbf{R}_l)|^2$ distribution. Within the harmonic approximation, the density of the nuclear ensemble is simply given by the Wigner distribution for a specific temperature T [23]:

$$\sum_k \rho_k^T |\chi_{1k}(\mathbf{q})|^2 = |\chi_{10}^T(\mathbf{q})|^2$$
$$= \prod_{j=1}^{3N-6} \left(\frac{\mu_j^1 \omega_j^1}{2\pi\hbar \sinh\left(\hbar\omega_j^1/k_B T\right)} \right)^{1/2}$$
$$\times \exp\left(-\frac{\mu_j^1 \omega_j^1}{\hbar} q_j^2 \tanh\left(\frac{\hbar\omega_j^1}{2k_B T}\right)\right), \quad (35)$$

where μ_j^1 and ω_j^1 refer to reduced masses and angular frequencies of the electronic excited state.

The quantity Γ_{rad} expresses the rate of spontaneous emission per molecule per unit of angular frequency between $E/\hbar$ and $(E + dE)/\hbar$ [2]. For a single fluorescence band, in the absence of non-radiative processes, the radiative decay rate (κ_{rad}) and the maximum lifetime (τ_0) are given by

$$\kappa_{rad} = 1/\tau_0 = \frac{1}{\hbar}\int \Gamma_{rad}(E)dE. \quad (36)$$

2.6 Comments on the line shapes

Usually, the discussion on the line shape of the broadening of spectroscopic lines is made in terms of homogeneous and inhomogeneous broadening [2, 10]. Lorentzian line shapes are associated with homogeneous broadening caused by the natural lifetime of the lines, while Gaussian line shapes are associated with inhomogeneous broadening caused by thermodynamics events like collisions. In the derivation of the nuclear-ensemble method, we have seen that the Lorentzian and Gaussian line shapes were directly connected to the time dependence of the vibrational overlap function. While the exponential decay (Eq. 9) resulted in a Lorentzian line shape, a Gaussian decay (Eq. 15) resulted in a Gaussian line shape.

We can also relate these two approximations through the stochastic theory of the line shape developed by Kubo [11] and applied to molecular line shapes by Saven and Skinner [10]. As shown by Kubo, the overlap function given by Eq. (13) is a general result for a Gaussian-distributed random variable in a Markovian process [11]. In the limit of a very slow decay of the time-correlation function of this random variable, the overlap function reduces to Eq. (9) and the line has a Lorentzian shape. In the limit of a very fast decay of the time-correlation function, the overlap function reduces to Eq. (15) and the line has a Gaussian shape. Employing molecular dynamics simulations of chromophores within non-polar fluids, Saven and Skinner [10] showed that Kubo's model works well in these two limits, but cannot produce accurate line shapes in intermediary regimes. They also found out that inhomogeneous broadening (Gaussian lines) dominates the line shapes, unless the spectra are measured at very low temperature.

Although in practical terms, the difference between spectra computed with Eq. (23) employing Lorentzian or Gaussian line shapes is minimal, Gaussian line shapes lend a more physical interpretation to the simulations. When the molecule absorbs a photon, the excited wave packet should accelerate out of the Franck–Condon region [24]. Thus, the overlap function should change slowly initially, as expressed by a Gaussian overlap function. Moreover, as we have seen, the line width is proportional to the energy gradient at the Franck–Condon region (Eq. 21). When the gradient is small, the excited wave packet remains in the Franck–Condon region for a long time, resulting in a narrow spectral line, like those often observed for Rydberg states. When the energy gradient is large, the wave packet moves quickly out of the Franck–Condon region, creating broad bands, like those typically observed in $\pi\pi^*$ excitations.

In the case of molecules holding conical intersections between excited states, the wave packet will quickly populate different states due to the non-adiabatic transitions [25]. Part of this effect is captured by the nuclear-ensemble approach because it populates different states according to their diabatic features (which are determined by the transition dipole moments). But the contribution of the non-adiabatic phenomenon itself for the population of different states is not accounted at all with the overlap functions proposed in the previous sections.

3 Spectrum decomposition

In a system composed of various molecular units, it is of general interest to know how, for a specific electronic state, the electronic density is distributed among the units. Here, we will considerer systems composed of two molecular units, but the generalization for a larger number of units is straightforward. The aim is to implement an automatic analysis algorithm, which can be applied for every electronic state, for each point in the nuclear ensemble, and that returns the state classification in terms of pre-established classes. These classes are (1) local excitation within unit A; (2) local excitation within unit B; (3) delocalized excitation (excimer) involving A and B; (4) diffuse excitation (Rydberg) involving A and B; (5) charge transfer from A to B; and (6) charge transfer from B to A.

We start by expressing the excited-state electronic wavefunction Ψ_I for state I as a singly excited configuration interaction wavefunction (CIS)

$$\Psi_I = \sum_{i\to j} C^I_{i\to j}\Phi_{i\to j}, \tag{37}$$

where $\Phi_{i\to j}$ is a Slater determinant for single excitation from molecular orbital ψ_i into molecular orbital ψ_j. In the case of electronic states computed with TD-DFT approach, the CI coefficients $C^I_{i\to j}$ are given by the time-dependent amplitudes, while the molecular orbitals are given by Kohn–Sham orbitals.

If the ground state can be represented by a restricted closed shell determinant, the density difference between state I and the ground state is given by (after integrating over the electron coordinates)

$$\begin{aligned}\Delta P_{I0} \equiv P_I - P_0 &= \sum_{i\to j}\left(C^I_{i\to j}\right)^2 \sum_{\mu\nu}\left(c_{\mu j}c_{\nu j} - c_{\mu i}c_{\nu i}\right)S_{\mu\nu} \\ &\equiv \sum_{i\to j}\left(C^I_{i\to j}\right)^2\left(\rho^j - \rho^i\right),\end{aligned} \tag{38}$$

where $S_{\mu\nu} \equiv \langle\phi_\mu|\phi_\nu\rangle$ is the overlap integral elements between basis functions ϕ_μ, and ϕ_ν and $c_{\mu a}$ are the molecular-orbital coefficients for orbital a and basis function. (Detailed derivation of Eq. (38) is given in the Online Resource 1.)

If the molecular system is split in two units A and B, ΔP_{I0} can be partitioned between basis functions centered in atoms belonging to unit A and atoms belonging to unit B:

$$\begin{aligned}\Delta P_{I0} &= \sum_{i\to j}\left(C^I_{i\to j}\right)^2\Bigg[\sum_{\mu\in A,\nu\in A}\left(c_{\mu j}c_{\nu j} - c_{\mu i}c_{\nu i}\right)S_{\mu\nu} \\ &\quad + \sum_{\mu\in B,\nu\in B}\left(c_{\mu j}c_{\nu j} - c_{\mu i}c_{\nu i}\right)S_{\mu\nu} \\ &\quad + 2\sum_{\mu\in A,\nu\in B}\left(c_{\mu j}c_{\nu j} - c_{\mu i}c_{\nu i}\right)S_{\mu\nu}\Bigg] \\ &\equiv \sum_{i\to j}\left(C^I_{i\to j}\right)^2\Big[\left(\rho^j_{AA} - \rho^i_{AA}\right) + \left(\rho^j_{BB} - \rho^i_{BB}\right) \\ &\quad + 2\left(\rho^j_{AB} - \rho^i_{AB}\right)\Big].\end{aligned} \tag{39}$$

The distribution of the ρ^k_{AB} electronic density between the molecular units can be done employing standard schemes for the calculation of atomic charges [26]. Here, we employ the simplest approach, the Mulliken partition, where ρ^k_{AB} is distributed equally between A and B. In spite of the well-known handicaps of this partition scheme, it serves well our purposes of qualitatively assigning a large number of electronic states. Employing the Mulliken partition, Eq. (39) is recast as

$$\begin{aligned}\Delta P_{I0} &= \sum_{i\to j}\left(C^I_{i\to j}\right)^2\Big[\left(\rho^j_{AA} + \rho^j_{AB} - \rho^i_{AA} - \rho^i_{AB}\right) \\ &\quad + \left(\rho^j_{BB} + \rho^j_{AB} - \rho^i_{BB} - \rho^i_{AB}\right)\Big] \\ &= \sum_{i\to j}\left(C^I_{i\to j}\right)^2\Big[\left(\rho^j_A - \rho^i_A\right) + \left(\rho^j_B - \rho^i_B\right)\Big],\end{aligned} \tag{40}$$

where

$$\begin{aligned}\rho^k_A &\equiv \rho^k_{AA} + \rho^k_{AB}, \\ \rho^k_B &\equiv \rho^k_{BB} + \rho^k_{AB}\end{aligned} \tag{41}$$

with $k = i, j$.

In the case of Rydberg orbitals, this partition of the electronic density may not work well because of their very diffuse character. Therefore, the first step in the classification of the electronic states is to detect transitions into Rydberg orbitals. In some particular cases, the detection of these orbitals can be simply done by monitoring the quantity $2\rho^j_{AB}/\left(\rho^j_{AA} + \rho^j_{BB}\right)$. (See Online Resource 1 for a discussion on this point.) After detecting the Rydberg orbitals, their contribution to each state I is computed by $cR = \sum_m \left(C^I_m\right)^2$, with m running over all transition where j is a Rydberg orbital. If cR is equal or larger than a certain threshold t_R, then state I is classified as a diffuse (Rydberg) state.

If $cR < t_R$, the charge transfer is calculated. Because $\Delta P_{I0} = 0$, all change of density in unit A in Eq. (40) must correspond to a complementary change in unit B. Therefore, to obtain the amount of charge transfer between the two units, it is enough to compute the quantity:

$$\mathrm{TCT}^I_A = \sum_{i\to j}\left(C^I_{i\to j}\right)^2\left(\rho^j_A - \rho^i_A\right), \tag{42}$$

where transitions into Rydberg states are excluded from the summation. If $\mathrm{TCT}^I_A > t_{\mathrm{CT}}$, state I is classified as a charge transfer from B to A. If $\mathrm{TCT}^I_A < -t_{\mathrm{CT}}$, the state is classified as charge transfer from A to B. In both cases, t_{CT} is a positive threshold value.

If $\left|\mathrm{TCT}^I_A\right| \le t_{\mathrm{CT}}$, the electronic state can still be either a localized state or a delocalized state (excimer). To distinguish between them, a third threshold value t_L is used. For each transition within state I, if $\rho^i_A/\rho^i_B \ge t_L$, then the transition belongs to the sub-set m_A. If $\rho^i_B/\rho^i_A \ge t_L$, then the transition belongs to a sub-set m_B. The total localization of state I within units A and B is evaluated as $cA = \sum_{m_A}\left(C^I_{m_A}\right)^2$ and $cB = \sum_{m_B}\left(C^I_{m_B}\right)^2$, respectively. A last threshold, t_D, is employed: if $cA > t_D$, state I is localized in unit A; if $cB > t_D$, it is localized in unit B. In all other cases, state I is classified as delocalized. The values of the thresholds t_R, t_{CT}, t_L, and t_D are discussed below.

The method for spectrum decomposition proposed above is certainly not unique, and other criteria and threshold values could be invoked. Besides that, it depends on the approximate validity of a few hypotheses. First, we assume the adequacy of CIS wavefunctions to describe the electronic density. Second, we also assume that the usage of time-dependent DFT amplitudes together with Kohn–Sham orbitals results in an acceptable representation of the CIS wavefunction. Third, we assume that the density partition among the units is uniquely defined, even though the molecular orbitals (Kohn–Sham or Hartree–Fock) are not unique and the Mulliken partition employed is somewhat arbitrary. Due to all these factors, we should take the decomposition as a qualitative analysis of the several contributions to each band, rather than an exact numerical analysis.

4 Computational details

Minimum-energy geometries of ground and excited states were optimized at density functional theory (DFT) and time-dependent functional theory (TDDFT) [27] levels. The long-range corrected CAM-B3LYP functional [28] was employed for most of calculations, since it provides a good description of both localized and delocalized excitations [29]. The aug-cc-pVDZ, aug-cc-pVTZ [30], def2-TZVPP [31], and the TZVP + mod basis sets were used. The latter is the standard def2-TZVP basis set [31] with an extra set of diffuse s and p functions on the heavy atoms. The exponents of these extra Gaussian functions were obtained as 1/3 of the most diffuse exponent of each kind. Complementary calculations were performed with the resolution-of-identity coupled-cluster to the second-order method (RI-CC2) [32–34]. Cartesian coordinates are given in the Online Resource 1.

Time-dependent functional theory electronic structure calculations were performed with Gaussian 09 [35]. RI-CC2 calculations were performed with Turbomole [36]. Spectra were simulated with the Newton-X program [37, 38] interfaced to Gaussian 09.

5 Vertical spectra

5.1 Vertical excitation of furan

Due to several reasons, including its small size, complex spectrum mixing valence and Rydberg states, relation to other pentacyclic molecules, and availability of gas-phase experimental data, furan has been adopted as a benchmark molecule for most of quantum-chemical methods. Good comparative investigations involving diverse theoretical methods are reported in Refs. [39, 40].

In Table 1, we report results for the vertical spectrum of furan at TDDFT and RI-CC2. The multireference general-model-space coupled-cluster (GMS CCSD) results from Ref. [40] are given as well. The low-energy region of furan vertical spectrum is dominated by the $^1B_2$ $\pi\pi^*$ transition, whose experimental maximum is at 6.04 eV [41]. In Sect. 6.1, we will see that the vertical excitation is placed by 0.15 eV higher than the band maximum. Applying this same shift to the experimental result implies that the "experimental vertical excitation" should lie at 6.19 eV. The TD-CAM-B3LYP results, red-shifted by only 0.06 eV in relation to this estimate, are in very good agreement with the experimental result. Both coupled-cluster results are blue-shifted by more than 0.2 eV. This shift is due to the intrinsic difficulty of Hartree–Fock-based methods (multireference or not) to describe the ionic V state [42].

5.2 Vertical excitation of benzene

The vertical spectrum of benzene is reported in Table 2. Additionally to TDDFT and RI-CC2 results computed in this work, the results of the symmetry adapted cluster configuration interaction (SAC-CI) method from Ref. [43] are reported for comparison as well.

The first state is a dark $^1B_{2u}$ $\pi\pi^*$ transition. Its experimental value is 4.9 eV [44] at the band maximum. We discuss later (Sect. 6.2) that the "experimental vertical excitation" is at 5.07 eV. TD-CAM-B3LYP substantially overestimates this transition, which lies at 5.50 eV for both basis sets investigated. RI-CC2 result (5.25 eV) and, especially, SAC-CI (5.06 eV) are in much better agreement with the experiment. The next state, also dark, is a $^1B_{1u}$ $\pi\pi^*$ transition. The "experimental vertical excitation" is at 6.53 eV in this case. RI-CC2 has the smallest deviation to this value. SAC-CI and TD-CAM-B3LYP underestimate the energy of this transition by more than 0.3 eV. The experimental bright $\pi\pi^*$ transition ($^1E_{1u}$) is centered at 6.94 eV, with vertical excitation at 7.07 eV. TD-CAM-B3LYP and RI-CC2 results are in very good agreement with this value. SAC-CI result is blue-shifted by 0.4 eV. All three Rydberg states for which experimental information is available in Table 2 are well described by all methods.

5.3 Vertical excitation and emission of 2-phenylfuran

The ground-state geometry of 2-phenylfuran was optimized at B3LYP/TZVPP and CC2/TZVPP levels. Geometries

Table 1 Vertical excitations of furan with different methods

	TD-CAM-B3LYP				RI-CC2		GMS CCSD[a]	Expt.[b]	Assignment
	aug-cc-pVTZ		TZVP + mod		aug-cc-pVTZ				
	ΔE (eV)	f	ΔE (eV)	f	ΔE (eV)	f	ΔE (eV)	ΔE (eV)	
A_2	5.93	0.000	6.02	0.000	6.02	0.000	5.94	5.91	π–Ryd(s)
B_2	**6.13**	**0.167**	**6.13**	**0.164**	**6.41**	**0.186**	**6.51**	**6.04 (6.19)**[c]	**π–π***
B_1	**6.44**	**0.037**	**6.56**	**0.041**	**6.55**	**0.038**	**6.46**	**6.47**	**π–Ryd(p_{yz})**
A_2	6.63	0.000	6.81	0.000	6.71	0.000	6.61	6.61	π–Ryd(p_{yz})
A_1	6.97	0.000	6.97	0.000	6.75	0.000	6.89		π–π*
B_1	7.12	0.001	7.32	0.002	7.23	0.003	7.14		π–Ryd(d_{yz})
A_2	7.10	0.000	7.09	0.000	7.20	0.000	7.00		π–Ryd (p_{yz})
B_2	7.11	0.008	7.07	0.005	7.21	0.004	6.87	6.75	π–π*

TDDFT results with the B3LYP/TVPP ground-state geometry. RICC2 results with CC2/TZVPP ground-state geometry. Bright states are shown in bold face

[a] Ref. [40]

[b] Data reported in Ref. [39]

[c] "Experimental vertical excitation." See text

Table 2 Vertical excitations of benzene with different methods

	TD-CAM-B3LYP				RI-CC2		SAC-CI[a]	Expt.[b]	Assignment
	TZVP + mod		aug-cc-pVTZ		aug-cc-pVTZ				
	ΔE (eV)	f	ΔE (eV)	f	ΔE (eV)	f	ΔE (eV)	ΔE (eV)	
B_{2u}	5.50	0.000	5.50	0.000	5.25	0.000	5.06	4.90 (5.07)[c]	π–π*
B_{1u}	6.16	0.000	6.16	0.000	6.47	0.000	6.22	6.20 (6.53)	π–π*
E_{1g}	6.55	0.000	6.45	0.000	6.47	0.000	6.42	6.33	π–Ryd(s)
A_{2u}	**7.03**	**0.068**	**6.95**	**0.068**	**7.00**	**0.064**	**7.06**	**6.93**	**π–Ryd(p_{yz})**
E_{1u}	**7.05**	**0.607**	**7.05**	**0.614**	**7.15**	**0.665**	**7.48**	**6.94 (7.07)**	**π–π***
E_{2u}	7.14	0.000	7.05	0.000	7.06	0.000	7.12	6.95	π–Ryd(p_{yz})
A_{1u}	7.28	0.000	7.17	0.000	7.14	0.000	7.19	–	π–Ryd(p_{yz})

TDDFT results with the B3LYP/TVPP ground-state geometry. RICC2 results with CC2/TZVPP ground-state geometry. Bright states are shown in bold face

[a] Ref. [43]

[b] Band maxima as surveyed in Ref. [43]

[c] The values in parenthesis are the "experimental vertical excitations." See text

calculated at both levels are similar. In particular, the phenyl and furans rings are twisted with a dihedral angle around 14° for both levels of theory.

The bright states of the vertical excitation spectrum of 2-phenylfuran in gas phase are reported in Table 3. The full set of vertical transitions up to 7.4 eV is reported in Table S1 of the Online Resource 1. The first two transitions are into a bright and a dark excimer states at 4.66 and 4.98 eV, respectively. The experimental band maximum in this region is at 4.55 eV in hexanes [16] and at 4.44 eV in methanol [45]. After a series of dark transitions into Rydberg states, a second band is characterized by a local ππ* excitation within the benzene ring (5.98 eV) and a ππ* excitation delocalized over the whole molecule. A very bright benzene-localized transition appears high in the spectrum at 7.02 eV. A less intense transition at 7.38 eV marks the fourth band.

The dependence of the vertical spectrum on the theoretical method can be accessed with the data of Table 3. TD-CAM-B3LYP energy of the bright states computed with aug-cc-pVDZ and aug-cc-pVTZ basis sets differs by less than 0.03 eV, with systemically lower values when using the triple-ζ basis set. RI-CC2 and TD-CAM-B3LYP predict very consistent results for the spectrum as well. The largest difference is observed in the lowest state, for which CC2 result is 0.19 eV higher than that at TDDFT level. For the other states, the energy difference is always smaller than 0.08 eV.

Table 3 Bright vertical excitations and emission of 2-phenylfuran with different methods

Assignment	TD-CAM-B3LYP				RI-CC2		Expt.
	aug-cc-pVDZ		aug-cc-pVTZ				
	ΔE (eV)	f	ΔE (eV)	f	ΔE (eV)	f	ΔE (eV)
Absorption							
π(PF)–π*(PF)	4.66	0.530	4.65	0.529	4.84	0.546	4.44[a] (4.54)[c] 4.55[b] (4.64)[c]
π(P)–π*(P)	5.98	0.092	5.96	0.089	5.96	0.085	
π(PF)–π*(PF)	6.19	0.107	6.16	0.116	6.08	0.080	
π(P)–π*(P)	7.02	0.399	6.99	0.335	7.07	0.387	
Emission							
π(PF)–π*(PF)	3.94	−0.562	3.99	−0.559	4.28	−0.575	3.97[b]

TDDFT using B3LYP/TZVPP ground-state geometry. RI-CC2 absorption: aug-cc-pVTZ with CC2/TZVPP ground-state geometry. Emission: S_1-minimum geometry optimized at the same level as the reported single point

[a] Band maximum in methanol. Ref. [45]

[b] Band maximum in hexanes. Ref. [16]

[c] "Experimental vertical excitation." See text

The geometry of the S_1 state of 2-phenylfuran was optimized at RI-CC2/TZVPP, TD-CAM-B3LYP/aug-cc-pVDZ and TD-CAMB3LYP/aug-cc-pVTZ levels of theory. The emission energy is given in Table 3 for these levels. TDDFT geometries are planar, while the CC2 geometry holds a small degree of puckering at the bridge carbon atom of the phenyl ring (CCCO dihedral = 11°). The CC bridge is shorten from 1.45 Å in the ground state to 1.40 Å in the excited state.

For all cases, the S_1 state corresponds to a delocalized $\pi\pi^*$ transition with large absolute value of oscillator strength. With the triple-ζ basis set, TDDFT vertical emission energy is 3.99 eV, which is 0.05 eV higher than the double-ζ result. Both results are in good agreement with the band maximum, located at 3.97 eV [16]. The RI-CC2 result, 4.28 eV, is blue-shifted in comparison to the experiment.

6 Spectrum simulations

6.1 Absorption spectrum of furan

Furan absorption cross section is shown in Fig. 2. Absorption was computed with Eq. (23), and the parameters employed in the simulations are given in Table 4. Vibronic shifts were set to zero, and all lines were assumed to have the same width δ_n. Eight excited states for each one of the 350 ensemble points were computed. The gray area indicates the error in the numerical integration computed with Eq. (25).

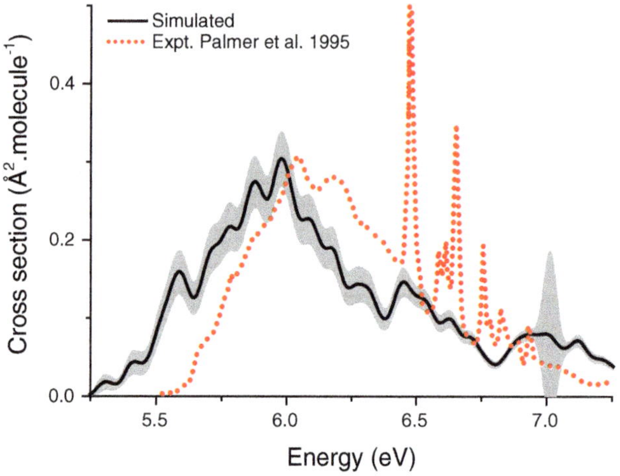

Fig. 2 Simulated (TD-CAM-B3LYP/TZVP-mod) and experimental absorption cross section of furan in gas phase. Experimental data from Ref. [41]

The experimental data, also in gas phase, is from Ref. [41]. In the region below 7 eV, furan shows a series of Rydberg states over-imposed to a broad band. The nuclear-ensemble method provides a good qualitative prediction of the spectrum. The intensity and the shape of the broad band are in very good agreement with the experiment. The energy shift is caused by the electronic structure method (see Sect. 5.1), rather than by the spectrum simulation method itself.

The broad band is due to the bright $\pi\pi^*$ state. The maximum of this band in the simulated spectrum is at 5.98 eV, slightly red-shifted in comparison with the experimental result, 6.04 eV [41]. The corresponding vertical excitation is the 1^1B_2 transition (see Table 1), which

Table 4 Parameters employed for spectrum simulations in this work

		δ_n (eV)	N_p	N_{fs}	n_r
Absorption					
Furan	Fig. 2	0.05	350	8	1
Benzene	Fig. 3-left	0.02	10,000	1	1
Benzene	Fig. 3-center	0.02	10,000	4	1
Benzene	Fig. 3-right	0.05	500	10	1
2-Phenylfuran	Fig. 4-top	0.05	850	23	1.375
2-Phenylfuran	Fig. 4-bottom	0.05	850	23	1.375
Emission ($T = 0$ K)					
2-Phenylfuran	Fig. 5	0.05	850	1	1

Absorption: Eq. (23) with Gaussian line shapes and $\varepsilon_n = 0.0$ eV. Emission: Eq. (34) with Lorentzian line shapes

lies at 6.13 eV, therefore, 0.15 eV higher than the band maximum. The sum of this shift to the experimental band maximum defines the "experimental vertical excitation" shown in Table 1.

The two main limitations of the current implementation of the nuclear-ensemble approach are clear in the simulations for furan: First, the lack of vibrational resolution: while the experimental results show a vibrational structure near the maximum, the simulations predict only the envelope of the band (the apparent oscillations are numerical noise). Second, the wrong band width for long-lived states: long-lived states give rise to very thin peaks in the spectrum, which are not correctly described in the simulations. Both limitations are caused by the overlap-function approximation (Eq. 15), which neglects the excited-state wave packet evolution (see discussion in Sect. 2.6).

6.2 Absorption spectrum of benzene

The simulated and experimental results for the three bands below 7 eV are shown in Fig. 3. The simulated absorption spectrum was computed with Eq. (23) with numerical-integration error (gray area) given by Eq. (25). Parameters are given in Table 4. The vibronic shift ε_n was assumed to be zero. An estimate for this shift is discussed below. Specific line widths were not computed, and all lines were assumed to have the same width δ_n given in Table 4. The number of excited states (up to 10) and the number of points in the ensemble (up to 10,000) were set differently for each region of the spectrum.

In the region below 7 eV, benzene absorbs in three distinct bands. The first is a dark band around 5 eV; the second is low-intensity band around 6.2 eV; and the third is a bright band around 7 eV. Very high-resolution measurements of the dark band along with a good review of previous spectroscopic data for benzene can be found in Ref. [46]. The UV absorption spectra of benzene vapor over a large wavelength domain are available in Refs. [44, 47, 48]. For comparison with the present simulations, we have taken the results from Ref. [44].

There is an overall good qualitative agreement between the simulations and the experimental results (Fig. 3). Absolute intensities are very well reproduced, especially if we take into account that they span three orders of magnitude over the three bands. Once more, vibrational resolution and long-lived states (narrow peaks) are not described. Besides that, the main deviations between theory and experiment are the energy shifts observed in the dark and in the intermediary bands. These shifts, however, are not caused by the spectrum simulation itself, but they are caused by the uncertainties of the electronic structure method and level, in this case, the TD-CAM-B3LYP/TZVP-mod, as we have discussed in Sect. 5.2.

Other source of band shift is vibronic corrections, but they are much smaller than the shift caused by the electronic structure method. In the case of the first energy band of benzene (Fig. 3-left), we can employ Eq. (20) to estimate the vibronic shift. At TD-CAM-B3LYP/TZVP-mod level, $\Delta E_{0,1} = 5.548$ eV and $\Delta E^a_{0,1} = 5.421$ eV, and therefore $\varepsilon_1 = -0.06$ eV, which is much smaller than the shift observed between theory and experiment in Fig. 3-left, about -0.6 eV from maximum to maximum.

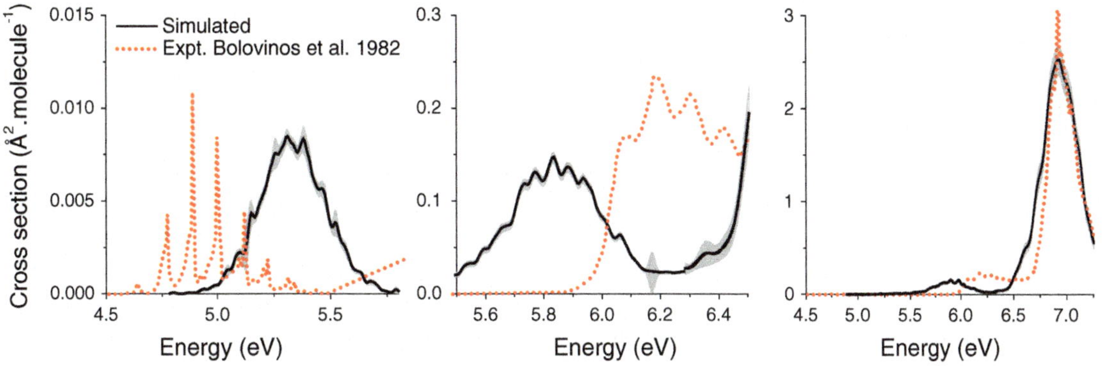

Fig. 3 Simulated (TD-CAM-B3LYP/TZVP-mod) and experimental absorption cross section of benzene vapor. Experimental data from Ref. [44]

For each of the three bands, the comparison between the simulated band maximum and the corresponding vertical excitation is 0.17 eV for the first band (B_{2u}), 0.33 eV for the second band (B_{1u}), and 0.13 eV for the third band (E_{1u}). These deviations were summed to the experimental band centers to give the "experimental vertical excitations" shown in Table 2.

6.3 Absorption spectrum of 2-phenylfuran

The absorption cross section of 2-phenylfuran is shown in Fig. 4. As for the previous cases, the simulated absorption was computed with Eq. (23) with numerical-integration error (gray area) given by Eq. (25). Parameters are given in Table 4. Vibronic shifts were set to zero, and all lines were assumed to have the same width δ_n. To cover the excitation-energy domain under 7 eV, 23 excited states were computed for each one of the 850 ensemble points. Note that the number of points in the ensembles for furan (350), benzene (500) and 2-phenylfuran (850) was chosen to be proportional to the number of degrees of freedom of each one of these molecules. The number of excited states was chose in each case as the minimum necessary do describe the spectrum up to 7 eV.

Experimental data for absorption of 2-phenylfuran were reported in Ref. [16] in hexanes (also shown in Fig. 4-top) and in Ref. [45] in methanol. Different from benzene and furan, 2-phenylfuran absorbs with appreciable intensity below 5 eV, with a band centered at 4.56 eV (expt.: 4.55 eV [16]). The simulations predict other two band of similar intensity at 6.00 eV and 6.86 eV, and a less intense band at 7.29 eV. The experimental result also indicates the raise of a second band around 6 eV, but the information is limited in this region. Overall, the simulated spectrum is in good agreement with the experiment. Intensities are very well reproduced, and the band shift is much smaller than in the simulations of furan and benzene, probably by error compensation due to the comparison between gas-phase simulations and solvated experiments.

Another noticeable difference between the spectrum of 2-phenylfuran and that of the isolated monomers is the intensity around 7 eV. While the absorption cross section of benzene is very high in this region (2.5 $Å^2$ $molecule^{-1}$), the absorption cross section of 2-phenylfuran is only about 1/3 of this value. This reduction in the absorption intensity is a direct effect of the delocalization of the orbitals over the whole molecule, reducing the transition moments.

Figure 4—bottom—shows the decomposition of the spectrum of 2-phenylfuran in terms of the several classes of states. The spectrum-decomposition method explained in Sect. 3 was applied for each of the $N_p \times N_{fs} = 19{,}550$ lines composing the spectrum. The threshold values are $t_R = 0.9$, $t_{CT} = 0.6$, $t_L = 0.8$, and $t_D = 0.35$. These values

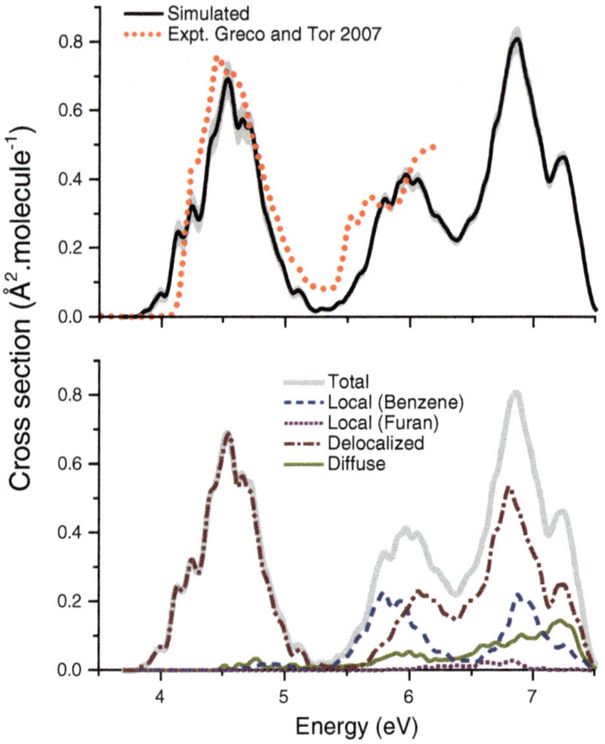

Fig. 4 *Top* simulated (TD-CAM-B3LYP/aug-cc-pVDZ) and experimental absorption cross section of 2-phenylfuran. Experimental data in hexanes from Ref. [16]. *Bottom* Spectrum decomposition. CT contributions are negligible and are not shown

were chosen after making a visual assignment of the states at the ground-state geometry. Having the decomposition, the spectrum for each class was recomputed with Eq. (23) including only the transitions belonging to the respective class, but still keeping the total N_p value.

As expected, the lowest-energy band (4.56 eV) is due to delocalized excitations involving both phenyl and furan rings. Delocalized excitations are also the major contributions for the other two bands. In the case of the band at 6.00 eV, localized excitations within benzene ring are as important as the delocalized states. In the third band (6.86 eV), benzene-localized states and diffuse states give minor contributions to the band. The fourth band (7.29 eV) is mainly due to delocalized excitations with large contributions of diffuse states. For the whole spectrum, local excitations within the furan ring are negligible, while charge-transfer states do not register in the scale of this graph.

It is instructive to compare the vertical excitations provided in Table 3 and the bands show in Fig. 4. The bright vertical excitation in the lowest band (S_1: 4.66 eV) is 0.1 eV higher than the band maximum. This shift is summed to the experimental band maximum to provide the "experimental vertical excitations" in Table 3 and in Table S1 of the Online Resource 1. The delocalized $\pi\pi^*$ character of this vertical excitations is the same as the character of the whole band. The second band, whose corresponding vertical excitations are S_6 and S_7, is centered exactly in between the energies of these two states. While S_6 is a benzene-localized state, S_7 is a delocalized state, which corresponds to the two sub-bands revealed by the spectrum decomposition shown in Fig. 4-bottom. The third band at 6.86 eV is related to the vertical excitation at 7.02. The assignment in this case is not straightforward as for the other two bands. The vertical excitation S_{17} is a benzene-localized $\pi\pi^*$ excitation. The spectrum decomposition, however, shows that the major contribution to this band comes from delocalized excitations, with secondary contributions from benzene-localized excitations. This difference occurs because the degree of localization depends on the nuclear geometry, and for this specific state, this quantity is very near the adopted threshold. For most of points in the nuclear-ensemble, the degree of localization was below the threshold $t_D = 0.35$. Specifically for the ground state minimum, this quantity was 0.36 (see *cA* in Table S2 of the Online Resource 1), just enough to make the transition to be classified as localized.

6.4 Emission spectrum of 2-phenylfuran

The simulated differential emission rate for 2-phenylfuran computed with Eq. (34) is shown in Fig. 5, together with experimental data from Ref. [16]. The simulated emission was computed with Eq. (34). The numerical-integration error (gray area) was computed with an expression similar to Eq. (25), but for the differential emission rate. Parameters are given in Table 4. All lines were assumed to have the same width δ_n. An ensemble of 850 points was built around the minimum of the S_1 state assuming $T = 0$ K.

Because the experimental data are given in arbitrary units, in Fig. 5, it has been normalized to match the maximum intensity of the simulated spectrum. The fluorescence of 2-phenylfuran shows a single band. The maximum of the simulated data occurs at 3.85 eV, while for the experimental data, it is at 3.97 eV. In spite of the good agreement between the theoretical and experimental results, one can observe that the asymmetry of the experimental band is not fully reproduced in the simulations.

Using Eq. (36), the maximum radiative lifetime is $\tau_0 = 3.1 \pm 0.2$ ns. With the normalization adopted in Fig. 5, the experimental value is $\tau_0 = 3.2$ ns. This same quantity is usually estimated based on the values of the energy gap and oscillator strength obtained at the excited-state minimum geometry ($\mathbf{R}_{S_1\,\mathrm{min}}$), where it is given by

$$\frac{1}{\tau_0} = \frac{e^2 n_r^3}{2\pi\hbar^2 mc^3 \varepsilon_0} \Delta E_{1,0}(\mathbf{R}_{S_1\,\mathrm{min}})^2 |f_{01}(\mathbf{R}_{S_1\,\mathrm{min}})|. \tag{43}$$

Employing this last equation, which assumes the validity of the Condon approximation, the lifetime is

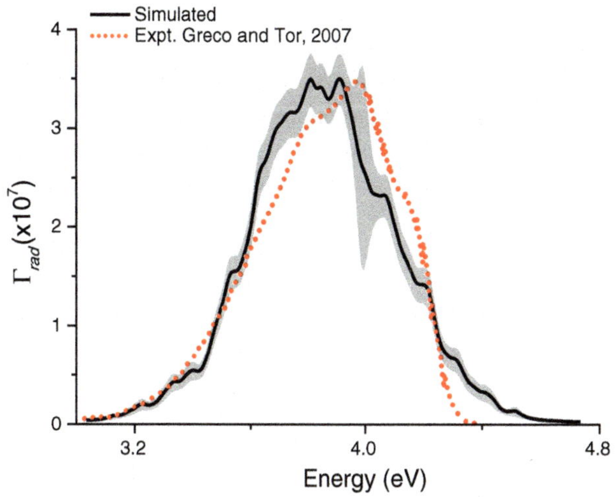

Fig. 5 Simulated (TD-CAM-B3LYP/aug-cc-pVDZ) and experimental emission spectrum of 2-phenylfuran. Experimental data of Ref. [16] normalized by the maximum of the simulated data

2.6 ns ($n_r = 1$), somewhat shorter than the value obtained from the simulated spectrum.

7 Conclusions

Simulations of electronic spectra based on nuclear-ensemble are an efficient approach for theoretical investigations of large molecules, especially when post-Condon features play a major role. The formal derivation of the method presented in this work enhances the approximations employed in the method and allows proposing new developments that will be explored in further works.

The nuclear-ensemble approach is especially well tailored to investigate how different properties contribute to each band in the spectrum. In particular, we have proposed and implemented a method for spectrum decomposition in terms of the contributions from different classes of states (localized, delocalized, diffuse and charge-transfer).

Using the nuclear-ensemble approach, we have simulated the absorption spectra of benzene, furan and 2-phenylfuran in gas phase. Based on these simulations, "experimental vertical excitations" were estimated. For the three molecules, the main bands were assigned and the spectrum-decomposition method was applied to 2-phenylfuran.

The absorption spectrum of 2-phenylfuran is composed of four bands between 4.5 and 7.5 eV. The lowest band is almost purely due to excimer states. The other bands are still dominated by delocalized states, but with relevant contributions from benzene-localized states and diffuse states. Furan-localized and charge-transfer states play a minor role in the absorption of 2-phenylfuran.

The emission spectrum of 2-phenylfuran is characterized by a single band centered at 3.9 eV and with a radiative lifetime of 3.2 ns. This value, which is obtained by integration of the full emission spectrum, is larger than the 2.6 ns predicted by a usual Condon approximation.

Spectrum simulations were based on TDDFT with the CAM-B3LYP functional. Methodological comparisons with RI-CC2 and other methods showed that TD-CAM-B3LYP have a very good performance for furan, 2-phenylfuran and the bright band of benzene. Substantial energy shifts, however, were observed for the dark bands of benzene.

Acknowledgments The authors acknowledge the fruitful discussions with Dr. Jan Goetze. This work is a contribution to the Festschrift issue in honor of Prof. Chaer Nascimento, who was co-advisor of the doctoral work of one of the authors (MB). Nascimento's insightful analysis of fundamental issues in quantum chemistry [49] has inspired a whole generation of computational theoretical chemists.

References

1. Koppel H, Domcke W, Cederbaum LS (1984) Adv Chem Phys 57:59
2. Niu Y, Peng Q, Deng C, Gao X, Shuai Z (2010) J Phys Chem A 114:7817
3. Tannor DJ, Heller EJ (1982) J Chem Phys 77:202
4. Petrenko T, Neese F (2007) J Chem Phys 127:164319
5. Barone V, Bloino J, Biczysko M, Santoro F (2009) J Chem Theory Comput 5:540
6. Improta R, Barone V, Santoro F (2007) Angew Chem 119:409
7. Barbatti M, Aquino AJA, Lischka H (2010) Phys Chem Chem Phys 12:4959
8. Svoboda O, Oncak M, Slavícek P (2011) J Chem Phys 135:154301
9. Bergsma JP, Berens PH, Wilson KR, Fredkin DR, Heller EJ (1984) J Phys Chem 88:612
10. Saven JG, Skinner JL (1993) J Chem Phys 99:4391
11. Kubo R (1969) Adv Chem Phys 15:101
12. Schinke R (1995) Photodissociation dynamics: spectroscopy and fragmentation of small polyatomic molecules. Cambridge University Press, Cambridge
13. Crespo-Otero R, Barbatti M (2011) J Chem Phys 134:164305
14. Barbatti M (2011) Phys Chem Chem Phys 13:4686
15. Szalay PG, Aquino AJA, Barbatti M, Lischka H (2011) Chem Phys 380:9
16. Greco NJ, Tor Y (2007) Tetrahedron 63:3515
17. Sakurai JJ (1994) Modern quantum mechanics. Addison-Wesley, Massachusetts
18. Hilborn RC (1982) Am J Phys 50:982
19. Lepage GP (1978) J Comput Phys 27:192
20. Lukes V, Solc R, Barbatti M, Lischka H, Kauffmann HF (2010) J Theor Comput Chem 9:249
21. Rikken GLJA (1995) Physica B 204:353
22. Lampert RA, Meech SR, Metcalfe J, Phillips D, Schaap AP (1983) Chem Phys Lett 94:137
23. Feynman RP (1982) Statistical mechanics: a set of lectures. The Benjamin/Cummings Publishing Company, London
24. Heller EJ (1981) Acc Chem Res 14:368
25. Worth GA, Cederbaum LS (2004) Annu Rev Phys Chem 55:127
26. Meister J, Schwarz WHE (1994) J Phys Chem 98:8245
27. Casida M (1995) Time-dependent density functional response theory for molecules. In: Chong D (ed) Recent advances in density functional methods, part I. World Scientific, Singapore, p 155
28. Yanai T, Tew DP, Handy NC (2004) Chem Phys Lett 393:51
29. Peach MJG, Benfield P, Helgaker T, Tozer DJ (2008) J Chem Phys 128:044118
30. Dunning TH (1989) J Chem Phys 90:1007
31. Weigend F, Häser M, Patzelt H, Ahlrichs R (1998) Chem Phys Lett 294:143
32. Christiansen O, Koch H, Jorgensen P (1995) Chem Phys Lett 243:409
33. Hättig C, Weigend F (2000) J Chem Phys 113:5154
34. Hättig C, Köhn A (2002) J Chem Phys 117:6939
35. Frisch MJ, Trucks GW, Schlegel HB, Scuseria GE, Robb MA, Cheeseman JR, Scalmani G, Barone V, Mennucci B, Petersson GA, Nakatsuji H, Caricato M, Li X, Hratchian HP, Izmaylov AF, Bloino J, Zheng G, Sonnenberg JL, Hada M, Ehara M, Toyota K, Fukuda R, Hasegawa J, Ishida M, Nakajima T, Honda Y, Kitao O, Nakai H, Vreven T, Montgomery J, J. A., Peralta JE, Ogliaro F, Bearpark M, Heyd JJ, Brothers E, Kudin KN, Staroverov VN, Kobayashi R, Normand J, Raghavachari K, Rendell A, Burant JC, Iyengar SS, Tomasi J, Cossi M, Rega N, Millam NJ, Klene M, Knox JE, Cross JB, Bakken V, Adamo C, Jaramillo J, Gomperts R, Stratmann RE, Yazyev O, Austin AJ, Cammi R, Pomelli C, Ochterski JW, Martin RL, Morokuma K, Zakrzewski VG, Voth GA, Salvador P, Dannenberg JJ, Dapprich S, Daniels AD, Farkas Ö, Foresman JB, Ortiz JV, Cioslowski J, Fox DJ (2009). Gaussian 09, revision A02. Gaussian, Inc., Wallingford
36. Ahlrichs R, Bar M, Haser M, Horn H, Kolmel C (1989) Chem Phys Lett 162:165
37. Barbatti M, Granucci G, Ruckenbauer M, Plasser F, Pittner J, Persico M, Lischka H (2011). NEWTON-X: a package for Newtonian dynamics close to the crossing seam. www.newtonx.org
38. Barbatti M, Granucci G, Persico M, Ruckenbauer M, Vazdar M, Eckert-Maksic M, Lischka H (2007) J Photochem Photobiol A 190:228
39. Wan J, Meller J, Hada M, Ehara M, Nakatsuji H (2000) J Chem Phys 113:7853
40. Li X, Paldus J (2010) J Phys Chem A 114:8591
41. Palmer MH, Walker IC, Ballard CC, Guest MF (1995) Chem Phys 192:111
42. Angeli C (2009) J Comput Chem 30:1319
43. Li Y, Wan J, Xu X (2007) J Comput Chem 28:1658
44. Bolovinos A, Philis J, Pantos E, Tsekeris P, Andritsopoulos G (1982) J Mol Spectrosc 94:55
45. Abu-Eittah R, Hilal R, Hamed MM (1981) Int J Quantum Chem 19:383
46. Fally S, Carleer M, Vandaele AC (2009) J Quant Spectrosc Radiat Transf 110:766
47. Pantos E, Philis J, Bolovinos A (1978) J Mol Spectrosc 72:36
48. Suto M, Wang X, Shan J, Lee LC (1992) J Quant Spectrosc Radiat Transf 48:79
49. Nascimento M, Barbosa A (2003) Quantum mechanics of many-electrons systems and the theories of chemical bond. In: Brändas EJ, Kryachko E (eds) Fundamental world of quantum chemistry, vol 1. Kluwer, Dordrecht, p 371

Theor Chem Acc (2012) 131:1222
DOI 10.1007/s00214-012-1222-y

REGULAR ARTICLE

Methods of continuous translation of the origin of the current density revisited

P. Lazzeretti

Please note an erratum was published subsequent to publication of the original article. For reasons of consistency with the electronic record, the article and the corresponding erratum are reproduced here as published in the journal.

Received: 9 March 2012 / Accepted: 8 April 2012 / Published online: 29 April 2012

Abstract Approaches to the calculation of magnetizability and nuclear magnetic shieldings in a molecule, based on continuous translation of the origin of the magnetic field-induced electronic current density, are reviewed. The connections among apparently unrelated philosophies (Geertsen propagator methods, Keith-Bader continuous set of gauge transformations, and analytical reformulation by Lazzeretti, Malagoli, and Zanasi) are emphasized, and a unitary theoretical scheme is given.

Keywords Molecular magnetic response · Magnetic field-induced current density · Continuous translation of the origin of the current density

1 Introduction

The van Vleck theory of magnetic susceptibilities [1] and the Ramsey approach to nuclear magnetic shielding [2–4] are based on the quantum mechanical Rayleigh–Schrödinger perturbation theory (RSPT). [5] The van Vleck–Ramsey (VR) method [1–4] can advantageously be reformulated via an equivalent approach, in terms of induced electronic current densities, allowing for "phenomenological" relationships that actually restate the validity of the laws of classical electrodynamics [6] in a quantum mechanical context. Magnetizability [1] $\xi_{\alpha\beta}$ and nuclear magnetic shielding [2–4] $\sigma^I_{\alpha\beta}$ can therefore be determined via the quantum mechanical electronic current densities $\mathbf{J}^{\mathbf{B}}$ and $\mathbf{J}^{\mathbf{m}_I}$, induced by an applied magnetic field with flux density $\mathbf{B}$ and by an intramolecular magnetic dipole moment $\mathbf{m}_I$ on nucleus I.

The theories of magnetic response [1–3] are gauge invariant in the limit of exact eigenfunctions to a model Hamiltonian. The conditions for gauge invariance of magnetic response properties calculated via optimal variational wavefunctions have been investigated by Epstein [7, 8] and Sambe [9] in connection with charge conservation [10]. Arrighini, Maestro, and Moccia (AMM) reported quantum mechanical sum rules for invariance in the gauge transformation that amounts to a translation of the origin of the coordinate system [11–13]. The AMM relationships, which also restate charge-current conservation [10] in integral form, as shown later on [14, 15], are exactly satisfied only in the ideal cases investigated by Epstein [7, 8] and Sambe [9]. In actual calculations employing a common gauge origin (CGO) and ordinary (gaugeless) basis sets, they are only approximately fulfilled. Numerical experience and simple theoretical considerations show that the quality of CGO calculations increases by improving basis set size and quality. Only in the limit of complete basis set would the AMM sum rules be numerically verified and calculated magnetic properties origin independent.

This state of affairs prompted the implementation of computational methods using gauge-including atomic orbitals (GIAO) [16][1], first introduced by London [17]. Computer packages available nowadays [18–20] adopt London orbitals for accurate calculation of magnetic properties including electronic correlation [21].

Dedicated to Professor Marco Antonio Chaer Nascimento and published as part of the special collection of articles celebrating his 65th birthday.

P. Lazzeretti (✉)
Dipartimento di Chimica dell'Università degli Studi di Modena e Reggio Emilia, Via Campi 183, 41124 Modena, Italy
e-mail: lazzeret@unimore.it

[1] This is the first reference in which the reinterpretation of the GIAO acronym for gauge-including-atomic-orbitals has been proposed, see footnote 6, p. 5047

A different solution of the gauge-origin problem for magnetizability and nuclear magnetic shielding was proposed by Geertsen [22–24]. In his treatment, the diamagnetic contributions to magnetic properties, evaluated as expectation values within the conventional RSPT approaches [2–4], are rewritten as polarization propagators. However, the Geertsen claim was later reconsidered, showing that his method provides only origin-independent average nuclear magnetic shieldings [25]. A computational scheme *à la* Geertseen has been employed by Smith et al. [26].

An approach referred to as CSGT (for continuous set of gauge transformations), which, at first sight, appears not linked to that of Geertsen, was put forward by Keith and Bader (KB) [27–29]. Within the computational scheme implemented by these authors, the diamagnetic contribution to the $\mathbf{J}^{\mathbf{B}}(\mathbf{r})$ current density vector field is formally annihilated at each point $\mathbf{r}$, by choosing that point as origin of the coordinate system. The procedure was reformulated in analytical form afterward and denominated "continuous transformation of the origin of the current density" (CTOCD), whereby the diamagnetic term is set to zero (DZ), in previous papers [25, 30], in which the reasons why the acronym CTOCD seems preferable to CSGT have been discussed. The denomination *ipsocentric* was suggested by Steiner and Fowler [31]. A large body of work is based on use of their methods for provision of interpretation of molecular magnetic response via orbital-contribution analysis [32–35], which can therefore be applied in a uniform way to both localized and delocalized description of the electronic structure. Thus, the advantages of the ipsocentric approach carry over into the description of integrated magnetic properties by a resolution into orbital components.

Analogous procedures, based on formal annihilation of the paramagnetic contribution to the current density, and indicated by CTOCD-PZ, have been proposed [36–38].

The central aim of this article is to give an updated, compact and self-contained, theoretical outline of CTOCD methods, discussing results scattered in different journals and illustrating in detail the connections with previous formulations, that of Geertsen in particular [22–24], which, as it were, has been reformulated allowing for Hermitian operators. The $\xi_{\alpha\beta}$ CTOCD-DZ and CTOCD-PZ magnetizabilities have been re-defined as tensors symmetrical in $\alpha \leftrightarrow \beta$, as physically required, and the relationships describing the change of $\xi_{\alpha\beta}$ in a translation of coordinate system have accordingly been modified. The paper is organized as follows. The conventional Rayleigh–Schrödinger approach to magnetic properties is summarized in Sect. 2. In Sect. 3, magnetizability and NMR shieldings are represented by second-rank tensors in terms of electron current densities induced by a static magnetic field and by a nuclear magnetic dipole. A comprehensive discussion of CTOCD is given in Sect. 4.

2 The Rayleigh–Schrödinger approach to magnetic response properties

Standard tensor formalism is employed throughout this paper, for example, the Einstein convention of implicit summation over two repeated Greek subscripts is in force. The notation adopted in previous references [15, 39, 40] is used. The SI system of units has been adopted.

Within the Born-Oppenheimer (BO) approximation, for a molecule with n electrons and N clamped nuclei, charge, mass, position, canonical, and angular momentum of the i-th electron are indicated, in the configuration space, by $-e, m_e, \mathbf{r}_i, \hat{\mathbf{p}}_i, \hat{\mathbf{l}}_i = \mathbf{r}_i \times \hat{\mathbf{p}}_i, \quad i = 1, 2 \ldots n$. Analogous quantities for nucleus I are $Z_I e$, M_I, $\mathbf{R}_I$, etc., for $I = 1, 2 \ldots N$. Capital letters denote total electronic operators, for example, $\hat{\mathbf{R}} = \sum_{i=1}^{n} \mathbf{r}_i, \hat{\mathbf{P}} = \sum_{i=1}^{n} \hat{\mathbf{p}}_i, \hat{\mathbf{L}} = \sum_{i=1}^{n} \hat{\mathbf{l}}_i$, etc. N' nuclei are endowed with an intrinsic magnetic dipole $\mathbf{m}_I = \gamma_I \hbar \mathbf{I}_I$, expressed via the magnetogyric ratio γ_I and spin $\hbar \mathbf{I}_I$ of nucleus I. Within the framework of the BO approximation, $\mathbf{m}_I$ is regarded as a mere phenomenological parameter.

The magnetic Hamiltonians describe the interaction of electrons with the intramolecular perturbation, that is, the intrinsic magnetic dipoles $\mathbf{m}_I$, via the vector potential $\sum_{I=1}^{N'} \mathbf{A}^{\mathbf{m}_I}$, and with an external, spatially uniform and time-independent magnetic field $\mathbf{B} = \nabla \times \mathbf{A}^{\mathbf{B}}$,

$$\mathbf{A} = \mathbf{A}^{\mathbf{B}} + \sum_{I=1}^{N'} \mathbf{A}^{\mathbf{m}_I}, \tag{1}$$

$$\mathbf{A}^{\mathbf{B}} = \frac{1}{2} \mathbf{B} \times \mathbf{r}, \tag{2}$$

$$\mathbf{A}^{\mathbf{m}_I} = \frac{\mu_0}{4\pi} \frac{\mathbf{m}_I \times (\mathbf{r} - \mathbf{R}_I)}{|\mathbf{r} - \mathbf{R}_I|^3}. \tag{3}$$

within the constraint of Coulomb gauge

$$\nabla \cdot \mathbf{A} = 0, \tag{4}$$

explicitly satisfied by the vector potentials (1).

The operator for the orbital magnetic dipole moment is related to the angular momentum operator with respect to the origin of the coordinate system, that is,

$$\hat{\mathbf{m}}_i = -\frac{e}{2m_e} \hat{\mathbf{l}}_i, \quad \hat{\mathbf{m}} = \sum_{i=1}^{n} \hat{\mathbf{m}}_i. \tag{5}$$

The operator for the magnetic field exerted by the electrons on nucleus I is

$$\hat{\mathbf{B}}_I^n = \sum_{i=1}^{n} \hat{\mathbf{B}}_I^i, \quad \hat{\mathbf{B}}_I^i = -\frac{e}{m_e} \hat{\mathbf{M}}_I^i, \tag{6}$$

introducing the operator

$$\hat{\mathbf{M}}_I^i = \frac{\mu_0}{4\pi}\frac{\mathbf{r}_i - \mathbf{R}_I}{|\mathbf{r} - \mathbf{R}_I|^3} \times \hat{\mathbf{p}}_i = \frac{\mu_0}{4\pi}\frac{\hat{\mathbf{l}}_i(\mathbf{R}_I)}{|\mathbf{r} - \mathbf{R}_I|^3}, \quad \hat{\mathbf{M}}_I^n = \sum_{i=1}^{n} \hat{\mathbf{M}}_I^i, \tag{7}$$

where $\hat{\mathbf{l}}_i(\mathbf{R}_I)$ is the angular momentum operator for electron i with respect to the origin at nucleus I. We also define the operators

$$\hat{\xi}^{\mathrm{d}}_{\alpha\beta} = -\frac{e^2}{4m_e}\sum_{i=1}^{n}\left(r_\gamma^2\delta_{\alpha\beta} - r_\alpha r_\beta\right)_i, \tag{8}$$

$$\hat{\sigma}^{\mathrm{d}I}_{\alpha\beta} = \frac{e}{2m_e c^2}\sum_{i=1}^{n}\left(r_{i\gamma}\hat{E}^i_{I_\gamma}\delta_{\alpha\beta} - r_{i\alpha}\hat{E}^i_{I_\beta}\right), \tag{9}$$

$$r_\gamma^2 \equiv r_\gamma r_\gamma \tag{10}$$

where $\hat{\mathbf{E}}_I^i$ denotes the multiplicative operator for the electric field of electron i on nucleus I,

$$\hat{\mathbf{E}}_I^i = \frac{1}{4\pi\epsilon_0} e \frac{\mathbf{r}_i - \mathbf{R}_I}{|\mathbf{r}_i - \mathbf{R}_I|^3}, \tag{11}$$

and

$$\mu_0\epsilon_0 c^2 = 1. \tag{12}$$

In this notation the VR Hamiltonians [1–3] are cast in the form

$$\hat{H}^{\mathbf{B}} = \frac{e}{m_e}\sum_{i=1}^{n}\mathbf{A}_i^{\mathbf{B}} \cdot \hat{\mathbf{p}}_i = -\hat{m}_\alpha B_\alpha, \tag{13}$$

$$\hat{H}^{\mathbf{m}_I} = \frac{e}{m_e}\sum_{i=1}^{n}\mathbf{A}_i^{\mathbf{m}_I} \cdot \hat{\mathbf{p}}_i = -\hat{B}^n_{I\alpha} m_{I_\alpha}, \tag{14}$$

$$\hat{H}^{\mathbf{BB}} = \frac{e^2}{2m_e}\sum_{i=1}^{n}\mathbf{A}_i^{\mathbf{B}} \cdot \mathbf{A}_i^{\mathbf{B}} = -\frac{1}{2}\hat{\xi}^{\mathrm{d}}_{\alpha\beta}B_\alpha B_\beta, \tag{15}$$

$$\hat{H}^{\mathbf{m}_I\mathbf{B}} = \frac{e^2}{m_e}\sum_{i=1}^{n}\mathbf{A}_i^{\mathbf{B}} \cdot \mathbf{A}_i^{\mathbf{m}_I} = \hat{\sigma}^{\mathrm{d}I}_{\alpha\beta} m_{I_\alpha} B_\beta. \tag{16}$$

The total electronic energy of a molecule, in the presence of external magnetic field $\mathbf{B}$ and intramolecular magnetic dipoles $\mathbf{m}_I$, is

$$W = W^{(0)} - \frac{1}{2}\xi_{\alpha\beta}B_\alpha B_\beta + \sigma^I_{\alpha\beta} m_{I_\alpha} B_\beta + \cdots, \tag{17}$$

where

$$\xi_{\alpha\beta} = -\left.\frac{\partial^2 W}{\partial B_\alpha \partial B_\beta}\right|_{\mathbf{B}\to\mathbf{0}} \tag{18}$$

is the magnetizability, and

$$\sigma^I_{\alpha\beta} = \left.\frac{\partial^2 W}{\partial m_{I_\alpha} \partial B_\beta}\right|_{\mathbf{m}_I,\mathbf{B}\to\mathbf{0}} \tag{19}$$

is the magnetic shielding at nucleus I. Within the Rayleigh–Schrödinger perturbation theory, the expressions for first- and second-order contributions to the electronic energy of a molecule in the reference electronic state $|a\rangle \equiv |\Psi_a^{(0)}\rangle$ are

$$W_a^{(1)} = \left\langle a\middle|\hat{H}^{(1)}\middle|a\right\rangle, \tag{20}$$

$$W_a^{(2)} = \left\langle a\middle|\hat{H}^{(2)}\middle|a\right\rangle - \frac{1}{\hbar}\sum_{j\neq a}\omega_{ja}^{-1}\left\langle a\middle|\hat{H}^{(1)}\middle|j\right\rangle\left\langle j\middle|\hat{H}^{(1)}\middle|a\right\rangle, \tag{21}$$

where $\hat{H}^{(1)} = \hat{H}^{\mathbf{B}} + \hat{H}^{\mathbf{m}_I}$ and the second-order Hamiltonians are specified by Eqs. (15) and (16). $\hat{H}^{(0)}$ is the BO unperturbed electronic Hamiltonian of a molecule, so that

$$\hat{H}^{(0)}|j\rangle = W_j^{(0)}|j\rangle,$$

and

$$\omega_{ja} = (W_j^{(0)} - W_a^{(0)})/\hbar$$

is a natural transition frequency for the excited state $|j\rangle \equiv |\Psi_j^{(0)}\rangle$.

Allowing for relationships (17), (20) and (21), for the Hellmann-Feynman theorem, and for the expression

$$\{\hat{A},\hat{B}\}_{-1} = \frac{2}{\hbar}\sum_{i\neq a}\omega_{ia}^{-1}\Re\{\langle a|\hat{A}|i\rangle\langle i|\hat{B}|a\rangle\} \equiv -\langle\langle A;B\rangle\rangle_{\omega=0}, \tag{22}$$

for the polarization propagator [41, 42], the magnetizability is evaluated as

$$\xi_{\alpha\beta} = \xi^{\mathrm{d}}_{\alpha\beta} + \xi^{\mathrm{p}}_{\alpha\beta}, \tag{23}$$

$$\xi^{\mathrm{d}}_{\alpha\beta} = \left\langle a\middle|\hat{\xi}^{\mathrm{d}}_{\alpha\beta}\middle|a\right\rangle, \tag{24}$$

$$\xi^{\mathrm{p}}_{\alpha\beta} = \{\hat{m}_\alpha, \hat{m}_\beta\}_{-1}, \tag{25}$$

and the magnetic shielding at nucleus I is

$$\sigma^I_{\alpha\beta} = \sigma^{\mathrm{d}I}_{\alpha\beta} + \sigma^{\mathrm{p}I}_{\alpha\beta}, \tag{26}$$

$$\sigma^{\mathrm{d}I}_{\alpha\beta} = \left\langle a\middle|\hat{\sigma}^{\mathrm{d}I}_{\alpha\beta}\middle|a\right\rangle, \tag{27}$$

$$\sigma^{\mathrm{p}I}_{\alpha\beta} = -\left\{\hat{B}^n_{I_\alpha}, \hat{m}_\beta\right\}_{-1}. \tag{28}$$

3 The electronic current density induced by a static magnetic field

Adopting the McWeeny normalization and allowing for his notation [43–45], the probability density matrix for an n-electron quantum state $|\Psi\rangle$ is defined by the relationship

$$\gamma(\mathbf{x};\mathbf{x}') = n\int \Psi(\mathbf{x},\mathbf{x}_2,\ldots\mathbf{x}_n)\Psi^*(\mathbf{x}',\mathbf{x}_2,\ldots\mathbf{x}_n)\,\mathrm{d}\mathbf{x}_2\ldots\mathrm{d}\mathbf{x}_n, \tag{29}$$

where $\mathbf{x}_i \equiv \mathbf{r}_i \otimes \eta_i$ is a space-spin coordinate, and $\mathrm{d}\mathbf{x}_i \equiv \mathrm{d}^3r_i \otimes \mathrm{d}\eta_i$. By integrating over the spin variable η, a spatial density matrix is obtained,

$$\gamma(\mathbf{r};\mathbf{r}') = \int_{\eta'=\eta} \gamma(\mathbf{x};\mathbf{x}')\mathrm{d}\eta. \tag{30}$$

The diagonal elements of the density matrix, Eq. (30),

$$\gamma(\mathbf{r}) = \gamma(\mathbf{r};\mathbf{r}), \tag{31}$$

give the electronic charge distribution of the state

$$\rho(\mathbf{r}) = -e\gamma(\mathbf{r}). \tag{32}$$

For the unperturbed reference state $\Psi_a^{(0)}$, the probability and charge density, Eqs. (31) and (32), become

$$\gamma^{(0)}(\mathbf{r}) = n\int \mathrm{d}\mathbf{x}_2\ldots\mathrm{d}\mathbf{x}_n\,\Psi_a^{(0)}(\mathbf{r},\mathbf{x}_2,\ldots\mathbf{x}_n)\Psi_a^{(0)*}(\mathbf{r},\mathbf{x}_2,\ldots\mathbf{x}_n), \tag{33}$$

$$\rho^{(0)}(\mathbf{r}) = -e\gamma^{(0)}(\mathbf{r}). \tag{34}$$

The probability current density is obtained from Eq. (30) for the density matrix,

$$\mathbf{j}(\mathbf{r}) = \frac{1}{m_e}\Re[\hat{\boldsymbol{\pi}}\gamma(\mathbf{r};\mathbf{r}')]_{\mathbf{r}'=\mathbf{r}}, \tag{35}$$

where $\Re$ takes the real part of the content within brackets. The operator for the electronic mechanical momentum in the presence of external magnetic field and intramolecular magnetic dipoles $\mathbf{m}_I$ at the nuclei is defined by

$$\hat{\boldsymbol{\pi}} = \hat{\mathbf{p}} + e\left(\mathbf{A}^{\mathbf{B}} + \sum_{I=1}^{N'}\mathbf{A}^{\mathbf{m}_I}\right) \tag{36}$$

within the Gell-Mann minimal coupling principle [46]. The electronic current density is given by

$$\mathbf{J}(\mathbf{r}) = -e\mathbf{j}(\mathbf{r}). \tag{37}$$

The zero-order current density, see Eqs. (35) and (37), vanishes identically for a reality condition, if the reference state is an electronic singlet, which can always be expressed in real form.

If the analytic expressions for the spatial density functions, Eqs. (32) and (37), were known, the calculation of the electronic properties (23)–(25) and (26)–(28) of a molecule would be straightforward. The response tensors might be evaluated via phenomenological equations, formally identical to those applied in classical theory, see Eqs. (46)–(47) hereafter. In practice, however, the density functions, Eqs. (32) and (37), are evaluated from an electronic wavefunction $|\Psi\rangle$, which is usually given as a perturbation series in powers of some physically meaningful parameter. For instance, for a molecule in the presence of an external spatially uniform, time-independent magnetic field $\mathbf{B}$ and an intramolecular magnetic dipole $\mathbf{m}_I$ on nucleus I, the electronic wave function for the a reference state will be expressed in the general form

$$\Psi_a = \Psi_a^{(0)} + \Psi_a^{\mathbf{B}}\cdot\mathbf{B} + \Psi_a^{\mathbf{m}_I}\cdot\mathbf{m}_I + \cdots, \tag{38}$$

where $\Psi_a^{\mathbf{B}}$ and $\Psi_a^{\mathbf{m}_I}$ are axial vectors with three independent components, see Eqs. (39) and (40) hereafter.

Allowing for the Rayleigh–Schrödinger perturbation theory, see Sect. 2, we assume that a complete set $|\Psi_j^{(0)}\rangle \equiv |j\rangle$ of eigenstates to the unperturbed Hamiltonian $H^{(0)}$ is available, together with first-order perturbed functions,

$$|\Psi_a^{B_\alpha}\rangle = \frac{1}{\hbar}\sum_{j\neq a}\omega_{ja}^{-1}|j\rangle\langle j|\hat{m}_\alpha|a\rangle, \tag{39}$$

and

$$|\Psi_a^{m_{I_\alpha}}\rangle = \frac{1}{\hbar}\sum_{j\neq a}\omega_{ja}^{-1}|j\rangle\left\langle j|\hat{B}_{I_\alpha}^{n}|a\right\rangle. \tag{40}$$

Employing Eqs. (35)–(37), the first-order electronic current density induced by the external magnetic field can be written as a sum of paramagnetic and diamagnetic contributions,

$$\mathbf{J}^{\mathbf{B}} = \mathbf{J}_p^{\mathbf{B}} + \mathbf{J}_d^{\mathbf{B}}, \tag{41}$$

where

$$\mathbf{J}_d^{\mathbf{B}}(\mathbf{r}) = -\frac{e^2}{2m_e}\mathbf{B}\times\mathbf{r}\gamma^{(0)}(\mathbf{r}), \tag{42}$$

is related to the probability density of the unperturbed molecule, see Eq. (34). Using Eq. (39), $\hat{p}_\alpha^* = -\hat{p}_\alpha$ and $\Psi_a^{\mathbf{B}*} = -\Psi_a^{\mathbf{B}}$, the paramagnetic contribution is given by

$$\begin{aligned}\mathbf{J}_p^{\mathbf{B}}(\mathbf{r}) = -\frac{ne}{m_e}\int \mathrm{d}\mathbf{x}_2\ldots\mathrm{d}\mathbf{x}_n \\ \times\Big[\mathbf{B}\cdot\Psi_a^{\mathbf{B}*}(\mathbf{r},\mathbf{x}_2,\ldots\mathbf{x}_n)\hat{\mathbf{p}}\Psi_a^{(0)}(\mathbf{r},\mathbf{x}_2,\ldots\mathbf{x}_n) \\ +\Psi_a^{(0)*}(\mathbf{r},\mathbf{x}_2,\ldots\mathbf{x}_n)\hat{\mathbf{p}}\mathbf{B}\cdot\Psi_a^{\mathbf{B}}(\mathbf{r},\mathbf{x}_2,\ldots\mathbf{x}_n)\Big].\end{aligned} \tag{43}$$

The terms of the vector potential, Eq. (36), involving nuclear magnetic dipoles give rise to the classical Larmor contributions, as discussed previously [15],

$$\mathbf{J}_d^{\mathbf{m}_I}(\mathbf{r}) = -\frac{e^2}{m_e}\mathbf{A}^{\mathbf{m}_I}\gamma^{(0)}(\mathbf{r}), \tag{44}$$

due to diamagnetic nuclear spin/electron orbit interaction discussed by Ramsey [47].

The paramagnetic spin-orbital contributions to the electronic current density induced by the nuclear magnetic dipole are obtained from Eqs. (35) and (36). Since $\Psi_a^{\mathbf{m}_I *} = -\Psi_a^{\mathbf{m}_I}$, the current density expression

$$\mathbf{J}_p^{\mathbf{m}_I}(\mathbf{r}) = -\frac{en}{m_e}\int \mathrm{d}\mathbf{x}_2\ldots\mathrm{d}\mathbf{x}_n \times \Big[\mathbf{m}_I \cdot \Psi_a^{\mathbf{m}_I *}(\mathbf{r},\mathbf{x}_2,\ldots\mathbf{x}_n)\,\hat{\mathbf{p}}\,\Psi_a^{(0)}(\mathbf{r},\mathbf{x}_2,\ldots\mathbf{x}_n) + \Psi_a^{(0)*}(\mathbf{r},\mathbf{x}_2,\ldots\mathbf{x}_n)\,\hat{\mathbf{p}}\,\Psi_a^{\mathbf{m}_I}(\mathbf{r},\mathbf{x}_2,\ldots\mathbf{x}_n)\cdot\mathbf{m}_I\Big] \tag{45}$$

is found for the contribution due to paramagnetic spin-orbit interaction. The three-dimensional vector fields defined via Eqs. (43) and (45) have the direction of the $\hat{\mathbf{p}}$ vector.

The response properties introduced via Eqs. (17)–(19) are conveniently re-expressed via the induced current densities allowing for the relationships of classical electrodynamics [6]

$$W^{\mathbf{BB}} = -\frac{1}{2}\int \mathbf{A}^{\mathbf{B}}\cdot\mathbf{J}^{\mathbf{B}}\mathrm{d}^3r, \tag{46}$$

$$W^{\mathbf{m}_I\mathbf{B}} = -\int \mathbf{A}^{\mathbf{m}_I}\cdot\mathbf{J}^{\mathbf{B}}\mathrm{d}^3r = -\int \mathbf{A}^{\mathbf{B}}\cdot\mathbf{J}^{\mathbf{m}_I}\mathrm{d}^3r, \tag{47}$$

which define the second-order contribution to the Rayleigh–Schrödinger electronic energy, Eq. (21).

The explicit expressions for the response tensors, obtained by differentiating the perturbed second-order energies (46), (47) as in Eqs. (18), (19), are identical to the relationships, Eqs. (23)–(26), from the Rayleigh–Schrödinger perturbation theory. Therefore, Eqs. (46) and (47) can be used as computational recipes alternative to Eqs. (23) and (26). Formulae rewritten in terms of current density tensors are obtained in the following.

3.1 Magnetizability and nuclear magnetic shielding tensors from electronic current density

The orbital magnetic dipole moment induced in the n electrons of a molecule by an external magnetic field with flux density $\mathbf{B}$ is evaluated assuming linear response,

$$\Delta\langle\hat{m}_\alpha\rangle = \xi_{\alpha\beta}B_\beta = -\frac{1}{2}\epsilon_{\alpha\beta\gamma}\int J_\beta^{\mathbf{B}}(\mathbf{r})r_\gamma\,\mathrm{d}^3r. \tag{48}$$

The magnetic field induced at an observation point $\mathbf{R}$ is determined by the Biot-Savart law [6],

$$\Delta\langle B_\alpha^n(\mathbf{R})\rangle = -\sigma_{\alpha\beta}(\mathbf{R})B_\beta = \frac{\mu_0}{4\pi}\epsilon_{\alpha\beta\gamma}\int J_\beta^{\mathbf{B}}(\mathbf{r})\frac{R_\gamma - r_\gamma}{|\mathbf{R}-\mathbf{r}|^3}\,\mathrm{d}^3r. \tag{49}$$

Introducing the current density tensor [25] via the derivative

$$\mathcal{J}_\alpha^{B_\beta}(\mathbf{r}) = \frac{\partial}{\partial B_\beta}J_\alpha^{\mathbf{B}}(\mathbf{r}), \tag{50}$$

with paramagnetic components

$$\mathcal{J}_{p_\alpha}^{B_\beta}(\mathbf{r}) = -\frac{ne}{m_e}\int \mathrm{d}\mathbf{x}_2\ldots\mathrm{d}\mathbf{x}_n\Big[\Psi_a^{B_\beta *}(\mathbf{r},\mathbf{x}_2,\ldots\mathbf{x}_n)\hat{p}_\alpha\Psi_a^{(0)}(\mathbf{r},\mathbf{x}_2,\ldots\mathbf{x}_n) + \Psi_a^{(0)*}(\mathbf{r},\mathbf{x}_2,\ldots\mathbf{x}_n)\hat{p}_\alpha\Psi_a^{B_\beta}(\mathbf{r},\mathbf{x}_2,\ldots\mathbf{x}_n)\Big], \tag{51}$$

and diamagnetic components

$$\mathcal{J}_{d_\alpha}^{B_\beta}(\mathbf{r}) = -\frac{e^2}{2m_e}\epsilon_{\alpha\beta\gamma}r_\gamma\gamma^{(0)}(\mathbf{r}), \tag{52}$$

the magnetizability tensor is evaluated by (18) and (46)

$$\xi_{\alpha\beta} = \frac{1}{4}\int r_\gamma\Big(\epsilon_{\alpha\gamma\delta}\mathcal{J}_\delta^{B_\beta} + \epsilon_{\beta\gamma\delta}\mathcal{J}_\delta^{B_\alpha}\Big)\mathrm{d}^3r, \tag{53}$$

and the shielding tensor at $\mathbf{R}$ is obtained as

$$\sigma_{\alpha\delta}(\mathbf{R}) = -\frac{\mu_0}{4\pi}\epsilon_{\alpha\beta\gamma}\int \frac{r_\beta - R_\beta}{|\mathbf{r}-\mathbf{R}|^3}\mathcal{J}_\gamma^{B_\delta}(\mathbf{r})\,\mathrm{d}^3r. \tag{54}$$

If $\mathbf{R}$ coincides with the position $\mathbf{R}_I$ of the I-th nucleus, carrying an intrinsic magnetic dipole m_{I_α}, the quantity $\sigma_{\alpha\beta}(\mathbf{R}_I) \equiv \sigma_{\alpha\beta}^I$ defines the magnetic shielding tensor of that nucleus.

3.2 Invariance of magnetizability, nuclear magnetic shielding and electronic current density in a gauge translation

In a gauge transformation of the vector potential (2),

$$\mathbf{A}^{\mathbf{B}} \rightarrow \mathbf{A}^{\mathbf{B}} + \nabla f, \tag{55}$$

induced by an arbitrary generating function $f(\mathbf{r})$ well-behaved for $\mathbf{r}\rightarrow\infty$, the interaction energy and the molecular properties (magnetizability, nuclear shieldings, and the induced current densities) are invariant for exact [12, 13, 15, 25, 40, 48–51] and optimal variational eigenfunctions [7]. Gauge invariance is related to charge-current conservation [7, 8, 10], as can immediately be seen. For instance, in the change of gauge (55) considered above, one gets an additional term on the r.h.s. of the second identity, Eq. (47), which is required to identically vanish for the energy to stay the same, that is,

$$\int \mathbf{J}^{\mathbf{m}_I}\cdot\nabla f\mathrm{d}^3r = \int \nabla\cdot(f\mathbf{J}^{\mathbf{m}_I}) - \int f\nabla\cdot\mathbf{J}^{\mathbf{m}_I}. \tag{56}$$

In fact, owing to the Gauss theorem, the first volume integral on the r.h.s. of Eq. (56) is converted into a surface integral, vanishing due to the boundary conditions $\gamma^{(0)}(\mathbf{r}), \mathrm{J}_\alpha^{\mathbf{m}_I}(\mathbf{r}) \rightarrow 0$ for $\mathbf{r}\rightarrow\infty$. Therefore, the integral on the l.h.s. of Eq. (56), arising in the gauge transformation induced by the generating function f, vanishes, and the interaction energy (47) is invariant, if the continuity equation for the stationary state $\nabla \cdot \mathbf{J}^{\mathbf{m}_I} = 0$ is satisfied. As f is fully arbitrary, one finds in particular, for $f \equiv x, y, z$, the integral conservation condition

$$\int J_\alpha^{\mathbf{m}_I} d^3 r = \int d^3 r \left(J_{p\alpha}^{\mathbf{m}_I} + J_{d\alpha}^{\mathbf{m}_I} \right) = \frac{e^2}{m_e^2} \left(\left\{ \hat{P}_\alpha, \hat{M}_{I_\beta}^n \right\}_{-1} - \frac{m_e}{ec^2} \epsilon_{\alpha\beta\gamma} \left\langle a \left| \hat{E}_{I_\gamma}^n \right| a \right\rangle \right) m_{I_\beta} = 0 \tag{57}$$

for the electronic current density induced by a nuclear magnetic dipole. The condition for charge-current conservation is expressed via the AMM sum rule [12, 13]

$$\left\{ \hat{P}_\alpha, \hat{M}_{I_\beta}^n \right\}_{-1} = \frac{m_e}{ec^2} \epsilon_{\alpha\beta\gamma} \left\langle a \left| \hat{E}_{I_\gamma}^n \right| a \right\rangle, \tag{58}$$

which is also a constraint for translational invariance of calculated magnetic shieldings, see Eqs. (76)–(77) hereafter.

The analogous integral constraint for conservation of the $\mathbf{J}^{\mathbf{B}}$ current density, Eq. (41), is obtained from Eqs. (42) and (43), relying on an equation similar to (56),

$$\int J_\alpha^{\mathbf{B}} d^3 r = \int d^3 r \left(J_{p\alpha}^{\mathbf{B}} + J_{d\alpha}^{\mathbf{B}} \right) = \frac{e^2}{m_e^2} \left(\{ \hat{P}_\alpha, \hat{L}_\beta \}_{-1} - m_e \epsilon_{\alpha\beta\gamma} \langle a | \hat{R}_\gamma | a \rangle \right) B_\beta = 0. \tag{59}$$

This is satisfied by the AMM sum rule [11]

$$\{ \hat{P}_\alpha, \hat{L}_\beta \}_{-1} = m_e \epsilon_{\alpha\beta\gamma} \langle a | \hat{R}_\gamma | a \rangle, \tag{60}$$

which is also a condition for origin independence of calculated magnetizabilities [11, 15, 25, 40, 48–51], see Eqs. (74)–(75) hereafter. The Condon sum rule for rotational strengths within the dipole velocity formalism [52] is obtained from the more general Eq. (60) by putting $\alpha = \beta$.

In a change

$$\mathbf{r}' \to \mathbf{r}'' = \mathbf{r}' + \mathbf{d} \tag{61}$$

of the origin of coordinate system, which can be associated with the change of gauge

$$\mathbf{A}^{\mathbf{B}}(\mathbf{r} - \mathbf{r}') \to \mathbf{A}^{\mathbf{B}}(\mathbf{r} - \mathbf{r}'') = \mathbf{A}^{\mathbf{B}}(\mathbf{r} - \mathbf{r}') + \nabla[\mathbf{A}^{\mathbf{B}}(\mathbf{r} - \mathbf{r}') \cdot \mathbf{d}], \tag{62}$$

the diamagnetic contribution (42) to the current density induced by the external magnetic field changes:

$$\mathbf{J}_d^{\mathbf{B}}(\mathbf{r} - \mathbf{r}') \to \mathbf{J}_d^{\mathbf{B}}(\mathbf{r} - \mathbf{r}'') = \mathbf{J}_d^{\mathbf{B}}(\mathbf{r} - \mathbf{r}') + \mathbf{J}_d^{\mathbf{d}\times\mathbf{B}}(\mathbf{r}), \tag{63}$$

where

$$\mathbf{J}_d^{\mathbf{d}\times\mathbf{B}}(\mathbf{r}) = -\frac{e^2}{2m_e} \mathbf{d} \times \mathbf{B} \gamma^{(0)}(\mathbf{r}). \tag{64}$$

The wavefunction $\mathbf{\Psi}_a^{\mathbf{B}}$ changes according to

$$\mathbf{B} \cdot \mathbf{\Psi}_a^{\mathbf{B}} \to \mathbf{B} \cdot \mathbf{\Psi}_a^{\mathbf{B}} + \mathbf{d} \times \mathbf{B} \cdot \mathbf{\Psi}_a^{\mathbf{d}\times\mathbf{B}}, \tag{65}$$

where

$$\left| \mathbf{\Psi}_a^{\mathbf{d}\times\mathbf{B}} \right\rangle = -\frac{e}{2m_e \hbar} \sum_{j \neq a} \omega_{ja}^{-1} |j\rangle \langle j | \hat{\mathbf{P}} | a \rangle, \tag{66}$$

therefore, the paramagnetic contribution to the current density induced by the external magnetic field, Eq. (43), transforms

$$\mathbf{J}_p^{\mathbf{B}}(\mathbf{r} - \mathbf{r}') \to \mathbf{J}_p^{\mathbf{B}}(\mathbf{r} - \mathbf{r}'') = \mathbf{J}_p^{\mathbf{B}}(\mathbf{r} - \mathbf{r}') + \mathbf{J}_p^{\mathbf{d}\times\mathbf{B}}(\mathbf{r}), \tag{67}$$

where

$$\mathbf{J}_p^{\mathbf{d}\times\mathbf{B}}(\mathbf{r}) = -\frac{ne}{m_e} \int d\mathbf{x}_2 \ldots d\mathbf{x}_n \times \left[\mathbf{d} \times \mathbf{B} \cdot \mathbf{\Psi}_a^{\mathbf{d}\times\mathbf{B}*}(\mathbf{r}, \mathbf{x}_2, \ldots \mathbf{x}_n) \hat{\mathbf{p}} \Psi_a^{(0)}(\mathbf{r}, \mathbf{x}_2, \ldots \mathbf{x}_n) + \Psi_a^{(0)*}(\mathbf{r}, \mathbf{x}_2 \ldots \mathbf{x}_n) \hat{\mathbf{p}} \mathbf{B} \cdot \mathbf{\Psi}_a^{\mathbf{d}\times\mathbf{B}}(\mathbf{r}, \mathbf{x}_2, \ldots \mathbf{x}_n) \right]. \tag{68}$$

Allowing for the Ehrenfest off-diagonal hypervirial relationship [7]

$$-\frac{i}{m_e} \omega_{ja}^{-1} \langle j | \hat{\mathbf{P}} | a \rangle = \langle j | \hat{\mathbf{R}} | a \rangle \tag{69}$$

one finds

$$\mathbf{\Psi}_a^{\mathbf{d}\times\mathbf{B}} = -\frac{ie}{2\hbar} \hat{\overline{\mathbf{R}}} \Psi_a^{(0)}, \quad \hat{\overline{\mathbf{R}}} = \hat{\mathbf{R}} - \langle a | \hat{\mathbf{R}} | a \rangle, \tag{70}$$

then

$$\mathbf{J}_p^{\mathbf{d}\times\mathbf{B}}(\mathbf{r}) = -\mathbf{J}_d^{\mathbf{d}\times\mathbf{B}}(\mathbf{r}). \tag{71}$$

The conservation condition for $\mathbf{J}^{\mathbf{d}\times\mathbf{B}}(\mathbf{r})$, is obtained from Eqs. (64) and (68),

$$\int J_\alpha^{\mathbf{d}\times\mathbf{B}} d^3 r = \int d^3 r \left(J_{p\alpha}^{\mathbf{d}\times\mathbf{B}} + J_{d\alpha}^{\mathbf{d}\times\mathbf{B}} \right) = -\frac{e^2}{2m_e^2} \epsilon_{\beta\gamma\delta} d_\gamma B_\delta \left(\{ \hat{P}_\alpha, \hat{P}_\beta \}_{-1} - m_e n \delta_{\alpha\beta} \right) = 0. \tag{72}$$

This is satisfied by the AMM sum rule [11]

$$\{ \hat{P}_\alpha, \hat{P}_\beta \}_{-1} = m_e n \delta_{\alpha\beta}, \tag{73}$$

which is the Thomas–Reiche–Kuhn (TRK) sum rule for oscillator strengths within the dipole velocity gauge, also providing a condition for origin independence of calculated magnetizabilities [11, 15, 25, 40, 48–51]. We emphasize that the Ehrenfest relationship (69) yields the basic condition for the existence of sum rules (58), (60), (73) and identities (70), (71).

In the gauge translation (62), the components of the magnetizability tensor change according to the relationships [11, 15, 40, 53]

$$\xi^{d}_{\alpha\beta}(\mathbf{r}'') = \xi^{d}_{\alpha\beta}(\mathbf{r}') + \frac{e^2}{4m_e} \times \left[2\langle a|\hat{R}_\gamma(\mathbf{r}')|a\rangle\delta_{\alpha\beta}d_\gamma - d_\beta\langle a|\hat{R}_\alpha(\mathbf{r}')|a\rangle - d_\alpha\langle a|\hat{R}_\beta(\mathbf{r}')|a\rangle - n(d_\gamma d_\gamma\delta_{\alpha\beta} - d_\alpha d_\beta)\right], \quad (74)$$

$$\xi^{p}_{\alpha\beta}(\mathbf{r}'') = \xi^{p}_{\alpha\beta}(\mathbf{r}') + \frac{e^2}{4m_e^2} \times \left[d_\delta\left(\epsilon_{\alpha\gamma\delta}\left\{\hat{P}_\gamma, \hat{L}_\beta(\mathbf{r}')\right\}_{-1} + \epsilon_{\beta\gamma\delta}\left\{\hat{P}_\gamma, \hat{L}_\alpha(\mathbf{r}')\right\}_{-1}\right) + d_\delta d_\mu \epsilon_{\alpha\gamma\delta}\epsilon_{\beta\lambda\mu}\left\{\hat{P}_\gamma, \hat{P}_\lambda\right\}_{-1}\right], \quad (75)$$

and the analogous change for the magnetic shielding tensor of nucleus I is given by [12, 13, 15, 25, 40, 48–51]

$$\sigma^{dI}_{\alpha\beta}(\mathbf{r}'') = \sigma^{dI}_{\alpha\beta}(\mathbf{r}') - \frac{e}{2m_e c^2}\left(d_\gamma\left\langle a\left|\hat{E}^n_{I_\gamma}\right|a\right\rangle\delta_{\alpha\beta} - d_\alpha\left\langle a\left|\hat{E}^n_{I_\beta}\right|a\right\rangle\right), \quad (76)$$

$$\sigma^{pI}_{\alpha\beta}(\mathbf{r}'') = \sigma^{pI}_{\alpha\beta}(\mathbf{r}') - \frac{e^2}{2m_e^2} d_\delta\epsilon_{\beta\gamma\delta}\left\{\hat{M}^n_{I_\alpha}, \hat{P}_\gamma\right\}_{-1}. \quad (77)$$

The total magnetizability, Eq. (23), is origin independent if the sum rules (60) and (73) are satisfied, and the condition for origin independence of nuclear magnetic shielding, Eq. (26), is given by the sum rule (58).

4 Methods of continuous translation of the origin of the current density

Geertsen proposed to rewrite the diamagnetic contributions to total magnetic properties of the conventional theory [1, 2, 4] in propagator form [22–24]. The nice feature characterizing the Geertsen approach is that the calculated average magnetic shielding at a nucleus is origin independent, irrespective of size and quality of the gaugeless basis set retained within the algebraic approximation. Correlated and gauge-origin-independent Geertsen-type calculations of magnetic properties of triply bonded molecules [54, 55] and simple singly bonded molecules [56, 57] at the second-order polarization propagator approximation (SOPPA) [58] level have been reported.

Geertsen developed his approach via the commutators

$$A^{\mathbf{B}}_\alpha A^{\mathbf{B}}_\alpha = \frac{i}{4\hbar}B_\gamma B_\delta\epsilon_{\alpha\beta\gamma}[r_\alpha, r_\beta\hat{l}_\delta],$$
$$A^{\mathbf{B}}_\alpha A^{\mathbf{m}_I}_\alpha = \frac{i}{2\hbar}B_\gamma m_{I_\delta}\epsilon_{\alpha\beta\gamma}[r_\alpha, r_\beta\hat{M}_{I_\delta}]. \quad (78)$$

In the spirit of the Geertsen method, Smith et al. [26] proposed a sum-over-state expression for the diamagnetic contribution to the nuclear magnetic shielding tensor. They used the commutator

$$r_{0_\beta}\left[r_\alpha\nabla_\beta - r_\beta\nabla_\alpha, r_\gamma\right] = r_\alpha r_{0_\gamma} - r_\beta r_{0_\beta}\delta_{\alpha\gamma}, \quad (79)$$

see their Eqs. (23)–(25) for a gauge origin $\mathbf{r}_0$.

Slightly modifying the Geertsen formulation [22–24], it is expedient to introduce the Hermitian one-electron operators

$$\hat{u}_{\alpha\beta} = \frac{1}{2}(r_\alpha\hat{l}_\beta + \hat{l}_\beta r_\alpha), \quad (80)$$

$$\hat{t}_{I_{\alpha\beta}} = \frac{1}{2}(r_\alpha\hat{M}_{I_\beta} + \hat{M}_{I_\beta} r_\alpha), \quad (81)$$

so that

$$A^{\mathbf{B}}_\alpha A^{\mathbf{B}}_\alpha = \frac{i}{4\hbar}B_\gamma B_\delta\epsilon_{\alpha\beta\gamma}\left[r_\alpha, \hat{u}_{\beta\delta}\right], \quad (82)$$

$$A^{\mathbf{B}}_\alpha A^{\mathbf{m}_I}_\alpha = \frac{i}{2\hbar}B_\gamma m_{I_\delta}\epsilon_{\alpha\beta\gamma}\left[r_\alpha, \hat{t}_{I_{\beta\delta}}\right]. \quad (83)$$

Therefore, allowing for relationships (2), (3), (15), (16), (82) and (83), the n-electron operators for the diamagnetic contributions can be rewritten in the form

$$\hat{\xi}^{\Delta}_{\alpha\beta} = \frac{ie^2}{8m_e\hbar}\sum_{i=1}^{n}\left\{\epsilon_{\beta\gamma\delta}\left[r_\gamma, \hat{u}_{\delta\alpha}\right]_i + \epsilon_{\alpha\gamma\delta}\left[r_\gamma, \hat{u}_{\delta\beta}\right]_i\right\} = \frac{ie^2}{8m_e\hbar}\left\{\epsilon_{\beta\gamma\delta}\left[\hat{R}_\gamma, \hat{U}_{\delta\alpha}\right] + \epsilon_{\alpha\gamma\delta}\left[\hat{R}_\gamma, \hat{U}_{\delta\beta}\right]_i\right\}, \quad (84)$$

$$\hat{\sigma}^{\Delta I}_{\alpha\beta} = \frac{ie^2}{2m_e\hbar}\epsilon_{\beta\gamma\delta}\sum_{i=1}^{n}\left[r_\gamma, \hat{t}_{I_{\delta\alpha}}\right]_i = \frac{ie^2}{2m_e\hbar}\epsilon_{\beta\gamma\delta}\sum_{i=1}^{n}\left[\hat{R}_\gamma, \hat{T}_{I_{\delta\alpha}}\right], \quad (85)$$

where

$$\hat{U}_{\alpha\beta}(\mathbf{r}') = \sum_{i=1}^{n}\hat{u}_{i\alpha\beta}(\mathbf{r}') = \frac{1}{2}\sum_{i=1}^{n}\left[(r_{i\alpha} - r'_\alpha)\hat{l}_{i\beta}(\mathbf{r}') + \hat{l}_{i\beta}(\mathbf{r}')(r_{i\alpha} - r'_\alpha)\right], \quad (86)$$

is related to the Hermitian magnetic quadrupole operator [59]

$$\hat{m}_{\beta\alpha} = -\frac{e}{3m_e}\hat{U}_{\alpha\beta}, \quad (87)$$

and

$$\hat{T}^n_{I_{\alpha\beta}}(\mathbf{r}') = \sum_{i=1}^{n}\hat{t}_{I_{\alpha\beta}}(\mathbf{r}') = \frac{1}{2}\sum_{i=1}^{n}\left[(r_{i\alpha} - r'_\alpha)\hat{M}^i_{I_\beta} + \hat{M}^i_{I_\beta}(r_{i\alpha} - r'_\alpha)\right]. \quad (88)$$

Now, using the resolution of the identity

$$\sum_j |j\rangle\langle j| = I,$$

the hypervirial relationship (69), with $\langle a|\hat{P}_\alpha|a\rangle = 0$, and the expression (22) for the propagator, Δ contributions to magnetizability and nuclear magnetic shielding are obtained,

$$\xi^{\Delta}_{\alpha\beta} = \frac{e^2}{8m_e^2}\left(\epsilon_{\alpha\gamma\delta}\{\hat{P}_\gamma, \hat{U}_{\delta\beta}\}_{-1} + \epsilon_{\beta\gamma\delta}\{\hat{P}_\gamma, \hat{U}_{\delta\alpha}\}_{-1}\right), \tag{89}$$

$$\sigma^{\Delta I}_{\alpha\beta} = -\frac{e^2}{2m_e^2}\epsilon_{\beta\gamma\delta}\left\{\hat{P}_\gamma, \hat{T}^n_{I_{\delta\alpha}}\right\}_{-1}, \tag{90}$$

which reduce to the conventional diamagnetic terms, Eqs. (24) and (27), in the limit of exact eigenstates and optimal variational wavefunctions.

Within the computational procedure developed by KB [27, 28, 29], the current density induced in the electrons of a molecule by a spatially uniform static magnetic field is evaluated at every point of space assuming that the same point is also the origin of the coordinate system. Magnetic properties are then calculated by differentiating the relationships of classical electrodynamics, Eqs. (46) and (47), involving the current density, according to the definitions (18) and (19). The KB approach has been implemented developing a pointwise procedure [27–29].

We will now show that the procedure of Keith and Bader [27–29] is computationally equivalent to that of Geertsen [22–24] reformulated in Eqs. (80)–(90).

The total current density vector field induced in a molecule by an external magnetic field, Eq. (41), is a function of position $\mathbf{J}^{\mathbf{B}} = \mathbf{J}^{\mathbf{B}}(\mathbf{r})$, whose origin can arbitrarily be chosen in the case of exact and optimal variational wavefunctions [7]. In a change of coordinate system, Eq. (61), diamagnetic and paramagnetic contributions change according to Eqs. (63) and (67), but the total function should remain the same [14, 15], that is,

$$\begin{aligned}\mathbf{J}^{\mathbf{B}}(\mathbf{r}-\mathbf{r}'') &= \mathbf{J}^{\mathbf{B}}(\mathbf{r}-\mathbf{r}') + \mathbf{J}_d^{(\mathbf{r}''-\mathbf{r}')\times\mathbf{B}}(\mathbf{r}) + \mathbf{J}_p^{(\mathbf{r}''-\mathbf{r}')\times\mathbf{B}}(\mathbf{r}) \\ &= \mathbf{J}^{\mathbf{B}}(\mathbf{r}-\mathbf{r}') \equiv \mathbf{J}^{\mathbf{B}}(\mathbf{r}).\end{aligned} \tag{91}$$

This notation implies that diamagnetic and paramagnetic components, which depend on the coordinate system, are evaluated corresponding to different origins.

Let us now consider continuous coordinate translations whereby either the diamagnetic or paramagnetic contributions to the total current density are systematically annihilated at every point $\mathbf{r}$, all over the domain of a molecule. These procedures were referred to as CTOCD-DZ and CTOCD-PZ, setting to zero either the diamagnetic or paramagnetic terms of the $\mathbf{J}^{\mathbf{B}}$ field.

The n-electron (diamagnetic) Larmor current density, $\mathbf{J}_d^{\mathbf{B}}(\mathbf{r}-\mathbf{r}'')$ in Eq. (63), is formally killed within the CTOCD-DZ scheme for every point $\mathbf{r}$ by choosing that point as origin of the coordinate system [27–29], so that

$$\mathbf{J}_d^{\mathbf{B}}(\mathbf{r}-\mathbf{r}') = -\mathbf{J}_d^{(\mathbf{r}-\mathbf{r}')\times\mathbf{B}}(\mathbf{r}). \tag{92}$$

Such a procedure amounts to killing everywhere the vector potential (2), appearing in the definition of the diamagnetic current density (42). However, it is impractical to regard this procedure as a continuous gauge transformation, as this would imply that also the second-order energies, Eq. (46) and second identity of Eq. (47), vanish. That's the reason why the denomination CTOCD seems preferable to CSGT.

The expression for the total CTOCD-DZ current density contains two non-Larmor terms, both referred to the original coordinate system, that is,

$$\mathbf{J}^{\mathbf{B}}(\mathbf{r}) = \mathbf{J}_p^{\mathbf{B}}(\mathbf{r}-\mathbf{r}') + \mathbf{J}_p^{(\mathbf{r}-\mathbf{r}')\times\mathbf{B}}(\mathbf{r}-\mathbf{r}'), \tag{93}$$

where, allowing for the CTOCD prescription $\mathbf{r}'' \equiv \mathbf{r}$ in Eq. (68), the second term on the r.h.s. is given by

$$\begin{aligned}\mathbf{J}_p^{(\mathbf{r}-\mathbf{r}')\times\mathbf{B}}(\mathbf{r}-\mathbf{r}') = -\frac{ne}{m_e}\int d\mathbf{x}_2\ldots d\mathbf{x}_n \\ \times\Big[(\mathbf{r}''-\mathbf{r}')\times\mathbf{B}\cdot\Psi_a^{\mathbf{d}\times\mathbf{B}*}\hat{\mathbf{p}}\Psi_a^{(0)} \\ + \Psi_a^{(0)*}\hat{\mathbf{p}}(\mathbf{r}''-\mathbf{r}')\times\mathbf{B}\cdot\Psi_a^{\mathbf{d}\times\mathbf{B}}\Big]_{\mathbf{r}''=\mathbf{r}}.\end{aligned} \tag{94}$$

This notation means that $\mathbf{r}''$ is put equal to $\mathbf{r}$ after operating with $\hat{\mathbf{p}}$. It is convenient to recast relationship (94) in tensor notation,

$$\begin{aligned}J_{p\alpha}^{(\mathbf{r}-\mathbf{r}')\times\mathbf{B}}(\mathbf{r}-\mathbf{r}') = -\frac{ne}{m_e}\epsilon_{\beta\gamma\delta}(r_\beta - r'_\beta)B_\gamma\int d\mathbf{x}_2\ldots d\mathbf{x}_n \\ \times\left[\Psi_a^{(\mathbf{d}\times\mathbf{B})_\delta *}\hat{p}_\alpha\Psi_a^{(0)} + \Psi_a^{(0)*}\hat{p}_\alpha\Psi_a^{(\mathbf{d}\times\mathbf{B})_\delta}\right].\end{aligned} \tag{95}$$

The total quantum mechanical current density is an invariant quantity, mapped onto itself in a gauge transformation, in the ideal case of electronic wavefunctions satisfying hypervirial theorems, for example, optimal variational eigenfunctions [7]. In particular, it remains the same in a change of coordinate system, as recalled above. Then comparison between Eqs. (41) and (93) necessarily implies that the new, formally paramagnetic, term should be equivalent to the diamagnetic contribution in the conventional formulation, that is,

$$\mathbf{J}_p^{(\mathbf{r}-\mathbf{r}')\times\mathbf{B}}(\mathbf{r}) = \mathbf{J}_d^{\mathbf{B}}(\mathbf{r}-\mathbf{r}'), \tag{96}$$

for every $\mathbf{r}$. This relationship can be directly proven via Eqs. (42) and (94), using off-diagonal hypervirial relationships [15], for every plane perpendicular to $\mathbf{B}$, where the original diamagnetic flow takes place. However, it should be recalled that the formal replacement, according to Eq. (96) of the diamagnetic term described by Eq. (42) with a paramagnetic one in Eq. (93), introduces a spurious paramagnetic component along the inducing magnetic field in Eq. (94). Nonetheless, this quantity does not contribute to the diagonal components of response properties [15].

In a change of coordinate system (61), the transformation laws for (43) and (95) are, respectively,

$$J^{\mathbf{B}}_{p\alpha}(\mathbf{r}-\mathbf{r}'') = J^{\mathbf{B}}_{p\alpha}(\mathbf{r}-\mathbf{r}') - \frac{ne}{m_e}\epsilon_{\beta\gamma\delta}(r''_\beta - r'_\beta)B_\gamma \int \mathrm{d}\mathbf{x}_2\ldots\mathrm{d}\mathbf{x}_n \times \left[\Psi_a^{(\mathbf{d}\times\mathbf{B})_\delta *}\hat{p}_\alpha\Psi_a^{(0)} + \Psi_a^{(0)*}\hat{p}_\alpha\Psi_a^{(\mathbf{d}\times\mathbf{B})_\delta}\right], \tag{97}$$

and

$$J^{(\mathbf{r}-\mathbf{r}')\times\mathbf{B}}_{p\alpha}(\mathbf{r}-\mathbf{r}'') = J^{(\mathbf{r}-\mathbf{r}')\times\mathbf{B}}_{p\alpha}(\mathbf{r}-\mathbf{r}') + \frac{ne}{m_e}\epsilon_{\beta\gamma\delta}(r''_\beta - r'_\beta)B_\gamma \int \mathrm{d}\mathbf{x}_2\ldots\mathrm{d}\mathbf{x}_n \times \left[\Psi_a^{(\mathbf{d}\times\mathbf{B})_\delta *}\hat{p}_\alpha\Psi_a^{(0)} + \Psi_a^{(0)*}\hat{p}_\alpha\Psi_a^{(\mathbf{d}\times\mathbf{B})_\delta}\right]. \tag{98}$$

Two equal terms with opposite sign appear on the r.h.s of these equations, and then, the total CTOCD-DZ current, obtained by the algebraic sum of (97) and (98), is origin independent. Its origin independence is guaranteed also for approximate wavefunctions, whereas the property (41) expressed as a sum of conventional diamagnetic (42) and paramagnetic (43) contributions is invariant only in the ideal cases recalled above.

Within the analytical CTOCD-DZ procedure [25, 60–62], expressions for $\boldsymbol{\xi}$ and $\boldsymbol{\sigma}^I$ tensors are obtained from the second-order energies (46) and (47), substituting the current density, Eq. (93), and differentiating according to the definitions, Eqs. (18) and (19). The "diamagnetic" Δ-contributions are given by exactly the same relationships, Eqs. (89) and (90), arrived at via a modified Geertsen method. Therefore, the analytical CTOCD-DZ method [25, 30, 60–62] is fully equivalent to that of Geertsen [22–24] and to that implemented by Keith and Bader [27, 28], using the same philosophy to annihilate the diamagnetic current density term and numerical integration.

In a translation of origin $\mathbf{r}' \rightarrow \mathbf{r}'' = \mathbf{r}' + \mathbf{d}$,

$$\hat{U}_{\alpha\beta}(\mathbf{r}'') = \hat{U}_{\alpha\beta}(\mathbf{r}') - d_\gamma\epsilon_{\beta\gamma\delta}\hat{V}_{\alpha\delta}(\mathbf{r}') - d_\alpha\hat{L}_\beta(\mathbf{r}') + d_\alpha d_\gamma\epsilon_{\beta\gamma\delta}\hat{P}_\delta, \tag{99}$$

$$\hat{T}^n_{I_{\alpha\beta}}(\mathbf{r}'') = \hat{T}^n_{I_{\alpha\beta}}(\mathbf{r}') - d_\alpha\hat{M}^n_{I_\beta}, \tag{100}$$

where the Hermitian virial tensor operator [7] $\hat{V}_{\alpha\beta}$ appearing in the transformation law, Eq. (99), for $\hat{U}_{\alpha\beta}$, see Eq. (86), is defined

$$\hat{V}_{\alpha\beta}(\mathbf{r}') = \frac{1}{2}\sum_{i=1}^{n}\left[(r_{i\alpha} - r'_\alpha)\hat{p}_{i\beta} + \hat{p}_{i\beta}(r_{i\alpha} - r'_\alpha)\right]. \tag{101}$$

Therefore, the changes of the diamagnetic CTOCD-DZ contributions are evaluated from

$$\begin{aligned}\xi^\Delta_{\alpha\beta}(\mathbf{r}'') = \xi^\Delta_{\alpha\beta}(\mathbf{r}') - \frac{e^2}{8m_e^2} &\times \Big[d_\delta\Big(\epsilon_{\alpha\gamma\delta}\Big\{\hat{L}_\beta(\mathbf{r}'), \hat{P}_\gamma\Big\}_{-1} + \epsilon_{\beta\gamma\delta}\Big\{\hat{L}_\alpha(\mathbf{r}'), \hat{P}_\gamma\Big\}_{-1}\Big) \\ &+ d_\lambda\big(\epsilon_{\alpha\gamma\delta}\epsilon_{\beta\lambda\mu} + \epsilon_{\beta\gamma\delta}\epsilon_{\alpha\lambda\mu}\big)\Big\{\hat{P}_\gamma, \hat{V}_{\delta\mu}(\mathbf{r}')\Big\}_{-1} \\ &- d_\delta d_\lambda\big(\epsilon_{\alpha\gamma\delta}\epsilon_{\beta\lambda\mu} + \epsilon_{\beta\gamma\delta}\epsilon_{\alpha\lambda\mu}\big)\big\{\hat{P}_\mu, \hat{P}_\gamma\big\}_{-1}\Big],\end{aligned} \tag{102}$$

$$\sigma^{\Delta I}_{\alpha\beta}(\mathbf{r}'') = \sigma^{\Delta I}_{\alpha\beta}(\mathbf{r}') + \frac{e^2}{2m_e^2}d_\delta\epsilon_{\beta\gamma\delta}\Big\{\hat{M}^n_{I_\alpha}, \hat{P}_\gamma\Big\}_{-1}. \tag{103}$$

On summing paramagnetic p and diamagnetic Δ contributions in Eqs. (77) and (103), full cancellation of equal terms linear in d_δ takes place on the r.h.s., then the total CTOCD-DZ shieldings $\sigma^I_{\alpha\beta} = \sigma^{pI}_{\alpha\beta} + \sigma^{\Delta I}_{\alpha\beta}$ are origin independent. Terms depending on the square of the $\mathbf{d}$ shift, Eq. (61), in Eqs. (75) and (102) cancel out on summing, then total CTOCD-DZ magnetizabilities are origin independent if the AMM sum rule (60) and the additional constraint

$$\big\{\hat{L}_\alpha, \hat{P}_\beta\big\}_{-1} = \epsilon_{\alpha\gamma\epsilon}\big\{\hat{P}_\epsilon, \hat{V}_{\gamma\beta}\big\}_{-1} \tag{104}$$

are fulfilled.

In Eq. (61), the shift of origin is represented by an arbitrary constant vector $\mathbf{d}$. Within the CTOCD-PZ approach, a general transformation function $\mathbf{d} = \mathbf{d}(\mathbf{r})$ is sought, specifying the origin of the coordinate system in which the paramagnetic contributions to the current density is formally annihilated. This function is evaluated pointwise via the condition determined by Eq. (67). The l.h.s. of this relationship vanishes for

$$\mathbf{J}^{\mathbf{B}}_p(\mathbf{r}-\mathbf{r}') = -\mathbf{J}^{(\mathbf{r}''-\mathbf{r}')\times\mathbf{B}}_p(\mathbf{r}), \tag{105}$$

which gives the 3 × 3 system of linear equations

$$\mathbf{Md} = \mathbf{T}, \tag{106}$$

where, allowing for Eqs. (66) and (67),

$$M_{\delta\beta} = \frac{ne}{m_e}\epsilon_{\alpha\beta\gamma}B_\gamma \int \mathrm{d}\mathbf{x}_2\ldots\mathrm{d}\mathbf{x}_n \times \left[\Psi_a^{(\mathbf{d}\times\mathbf{B})_\alpha *}\hat{p}_\delta\Psi_a^{(0)} + \Psi_a^{(0)*}\hat{p}_\delta\Psi_a^{(\mathbf{d}\times\mathbf{B})_\alpha}\right], \tag{107}$$

and

$$T_\delta = -\frac{ne}{m_e}B_\alpha \int \mathrm{d}\mathbf{x}_2\ldots\mathrm{d}\mathbf{x}_n\left[\Psi_a^{B_\alpha *}\hat{p}_\delta\Psi_a^{(0)} + \Psi_a^{(0)*}\hat{p}_\delta\Psi_a^{B_\alpha}\right]. \tag{108}$$

The 3 × 3 $\mathbf{M}$ matrix defined by Eq. (107) is singular, for example, for $\mathbf{B} = B\epsilon_3$, its last column vanishes.

$$\mathbf{M} = \begin{pmatrix} M_{xx} & M_{xy} & 0 \\ M_{yx} & M_{yy} & 0 \\ M_{zx} & M_{zy} & 0 \end{pmatrix}. \quad (109)$$

In physical terms, the quantum mechanical paramagnetic current flowing in the direction of $\mathbf{B}$ cannot be annihilated [15]. Therefore, a 2 × 2 subsystem of Eq. (106),

$$\begin{pmatrix} M_{xx} & M_{xy} \\ M_{yx} & M_{yy} \end{pmatrix}\begin{pmatrix} d_x \\ d_y \end{pmatrix} = \begin{pmatrix} T_x \\ T_y \end{pmatrix}, \quad (110)$$

is solved, over a grid of points in real space, to determine the components

$$d_x = \frac{T_x M_{yy} - T_y M_{xy}}{M_{xx}M_{yy} - M_{xy}M_{yx}}, \qquad d_y = \frac{T_y M_{xx} - T_x M_{yx}}{M_{xx}M_{yy} - M_{xy}M_{yx}} \quad (111)$$

of the shift vector function that annihilates the paramagnetic current over planes perpendicular to $\mathbf{B}$.

Thus, within the CTOCD-PZ scheme, the transverse current density contains only contributions that are formally diamagnetic,

$$\mathbf{J}^{\mathbf{B}}(\mathbf{r}) = \mathbf{J}_d^{\mathbf{B}}(\mathbf{r}-\mathbf{r}') + \mathbf{J}_d^{\mathbf{d}\times\mathbf{B}}(\mathbf{r}-\mathbf{r}') = -\frac{e^2}{2m_e c}\mathbf{B}\times[\mathbf{r}-\mathbf{d}(\mathbf{r})]\gamma^{(0)}(\mathbf{r}). \quad (112)$$

To show that the PZ current density, Eq. (112), is origin independent, it is sufficient to verify that the shift functions transform like a vector in a translation of coordinates, that is,

$$d_\alpha(\mathbf{r}-\mathbf{r}') - d_\alpha(\mathbf{r}-\mathbf{r}'') \equiv d'_\alpha - d''_\alpha = r''_\alpha - r'_\alpha \equiv s_\alpha, \quad (113)$$

see Fig. 1.

From the invariance constraint, Eqs. (91), and (105), the identity

$$\mathbf{J}_d^{\mathbf{d}\times\mathbf{B}}(\mathbf{r}-\mathbf{r}') = \mathbf{J}_p^{\mathbf{B}}(\mathbf{r}-\mathbf{r}') \quad (114)$$

is obtained. This relationship does not provide a recipe for calculating the shift functions in the approximate case but yields the definition of exact $\mathbf{d}(\mathbf{r})$,

$$d_x(\mathbf{r}) = \frac{2m_e}{e^2}\left[\gamma^{(0)}(\mathbf{r})\right]^{-1}\mathcal{J}_{p_y}^{B_z}(\mathbf{r}), \qquad d_y(\mathbf{r}) = -\frac{2m_e}{e^2}\left[\gamma^{(0)}(\mathbf{r})\right]^{-1}\mathcal{J}_{p_x}^{B_z}(\mathbf{r}) \quad (115)$$

using the paramagnetic contribution to the current density tensor, Eq. (51).

The Eqs. (109)–(111) are valid for cyclic permutations of x, y, z; therefore, the transverse PZ current density (112)

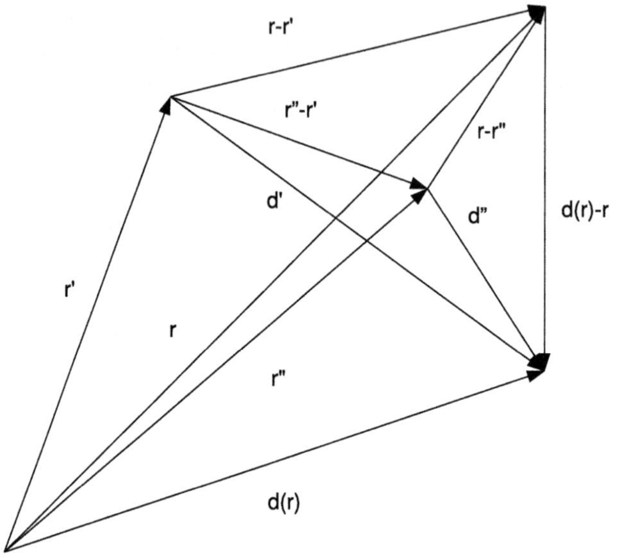

Fig. 1 Coordinate systems used in the CTOCD-PZ procedure. For every point $\mathbf{r}$, the origin is translated to a point $\mathbf{d}(\mathbf{r})$, so that the paramagnetic contribution $\mathbf{J}_p^{\mathbf{B}}$ to the current density, evaluated with respect to the new origin, vanishes

is explicitly origin independent also for approximate electronic wavefunctions, since it depends on the difference $\mathbf{r} - \mathbf{d}(\mathbf{r})$ of two vectors whose origin can arbitrarily be chosen.

The total CTOCD-PZ magnetizability is obtained by differentiating the second-order energy

$$\xi_{\alpha\beta} = \xi^{\mathrm{d}}_{\alpha\beta} + \xi^{\Pi}_{\alpha\beta} = -\frac{\partial^2}{\partial B_\alpha \partial B_\beta}\left(-\frac{1}{2}\int \mathbf{J}^{\mathbf{B}}\cdot\mathbf{A}^{\mathbf{B}}\mathrm{d}^3 r\right), \quad (116)$$

in which $\xi^{\mathrm{d}}_{\alpha\beta}$ is the conventional diamagnetic term (24) of the van Vleck theory [1], and the $\xi^{\Pi}_{\alpha\beta}$ term is obtained by numerical integration from the second addendum within brackets on the r.h.s. of Eq. (112). A formal expression is obtained,

$$\xi^{\Pi}_{\alpha\beta} = \frac{e^2}{4m_e}\int \gamma^{(0)}(\mathbf{r}) \times \left\{d_\gamma(\mathbf{r})r_\gamma\delta_{\alpha\beta} - \frac{1}{2}\left[d_\alpha(\mathbf{r})r_\beta + r_\alpha d_\beta(\mathbf{r})\right]\right\}\mathrm{d}^3 r, = \frac{e^2}{4m_e}\left\langle a\left|\sum_{i=1}^{n}\left\{d_\gamma(\mathbf{r})r_\gamma\delta_{\alpha\beta} - \frac{1}{2}\left[d_\alpha(\mathbf{r})r_\beta + r_\alpha d_\beta(\mathbf{r})\right]\right\}_i\right|a\right\rangle, \quad (117)$$

replacing $\xi^{\mathrm{p}}_{\alpha\beta}$ of the canonical theory [1].

It is expedient to define a multiplicative operator

$$\hat{\mathbf{D}}(\mathbf{r}) = \sum_{i=1}^{n}\mathbf{d}_i(\mathbf{r})$$

for n electrons, with expectation value

$$\langle a|\hat{D}_\alpha|a\rangle = \left\langle a\left|\sum_{i=1}^{n} d_{i\alpha}\right|a\right\rangle, \tag{118}$$

then, denoting the origin shift in Eq. (61), $\mathbf{r}'' - \mathbf{r}' = \mathbf{s} = \mathbf{d}' - \mathbf{r}''$, as in Eq. (113), the change of the formally paramagnetic contribution (117) to the magnetizability is written

$$\begin{aligned}\xi^{\Pi}_{\alpha\beta}(\mathbf{r}'') = \xi^{\Pi}_{\alpha\beta}(\mathbf{r}') - \frac{e^2}{4m_e}\{ & s_\gamma(\langle a|\hat{R}_\gamma(\mathbf{r}')|a\rangle + \langle a|\hat{D}_\gamma(\mathbf{r}')|a\rangle)\delta_{\alpha\beta} \\ & -\frac{1}{2}[s_\alpha(\langle a|\hat{R}_\beta(\mathbf{r}')|a\rangle + \langle a|\hat{D}_\beta(\mathbf{r}')|a\rangle) \\ & + s_\beta(\langle a|\hat{R}_\alpha(\mathbf{r}')|a\rangle + \langle a|\hat{D}_\alpha(\mathbf{r}')|a\rangle)] \\ & - n(s_\gamma s_\gamma \delta_{\alpha\beta} - s_\alpha s_\beta)\}.\end{aligned} \tag{119}$$

Therefore, the condition for invariance of total CTOCD-PZ magnetizability is obtained by comparison with Eq. (74),

$$\langle a|\hat{R}_\alpha|a\rangle = \langle a|\hat{D}_\alpha|a\rangle, \tag{120}$$

valid for any coordinate system, since the operator $\hat{D}_\alpha$ transforms like a vector, according to Eq. (113). In particular, $\langle a|\hat{D}_\alpha(\mathbf{R}_e)|a\rangle = 0$ in the limit of a complete basis set calculation, if the origin of the coordinate system lies at $\mathbf{R}_e$, the electronic centroid: the *allocentric* [63] PZ procedure scatters the origin of the current density in such a way that the statistical average of the $d_\alpha(\mathbf{r})$ functions vanishes.

The Π contribution to the magnetic shielding at nucleus I is obtained by differentiating the second-order energy (47), from the second addendum within brackets on the r.h.s. of Eq. (112). Using numerical integration, it becomes

$$\begin{aligned}\sigma^{\Pi I}_{\alpha\beta} &= -\frac{e}{2m_e c^2}\int \gamma^{(0)}(\mathbf{r})\left[d_\gamma(\mathbf{r})\hat{E}_{I_\gamma}(\mathbf{r})\delta_{\alpha\beta} - d_\alpha(\mathbf{r})\hat{E}_{I_\beta}(\mathbf{r})\right]\mathrm{d}^3r \\ &= -\frac{e}{2m_e c^2}\left\langle a\left|\sum_{i=1}^{n}\left(d_{i\gamma}\hat{E}^i_{I_\gamma}\delta_{\alpha\beta} - d_{i\alpha}\hat{E}^i_{I_\beta}\right)\right|a\right\rangle.\end{aligned} \tag{121}$$

The change in the Π contributions to the nuclear magnetic shieldings in the translation (61) of the origin of the coordinate system is

$$\begin{aligned}\sigma^{\Pi I}_{\alpha\beta}(\mathbf{r}'') &= \sigma^{\Pi I}_{\alpha\beta}(\mathbf{r}') \\ &+ \frac{e}{2m_e c^2}\left(s_\gamma\left\langle a\left|\hat{E}^n_{I_\gamma}\right|a\right\rangle\delta_{\alpha\beta} - s_\alpha\left\langle a\left|\hat{E}^n_{I_\beta}\right|a\right\rangle\right),\end{aligned} \tag{122}$$

then there is complete cancellation with the corresponding change in Eq. (76), and the total CTOCD-PZ shielding is origin independent irrespective of basis set size and quality.

Calculations of magnetizability and nuclear magnetic shielding in molecules employing the CTOCD-PZ approach have been reported [37, 38, 64–66].

5 Concluding remarks and outlook

A review and new perspectives are presented on the connections among various methods of calculation of molecular magnetic response properties, all constructed with the aim of finessing the troublesome gauge-origin problem that plagued calculations of these types in the latter half of the twentieth century. The analytical formulation of CTOCD-DZ procedures, based on the ipsocentric choice of origin that formally annihilates the diamagnetic contribution to magnetic field-induced quantum mechanical current density, provides a compact and unitary theoretical framework, showing the equivalence of apparently unrelated work of different authors. CTOCD-PZ methods, formally destroying the paramagnetic contribution to the electronic current density via a systematic allocentric choice of origin, have only been implemented at numerical level: an analytical formulation of the PZ philosophy has not so far been reported. Attempts at developing CTOCD-PZ algorithms via closed-form equations would seem theoretically interesting, as well as able to be used for practical purposes.

Whereas current computational techniques based on gauge-including atomic orbitals meet the requirement of translational invariance of calculated magnetic properties, they do not necessarily guarantee current/charge conservation. On the other hand, CTOCD schemes account for the fundamental identity between these constraints, which are expressed by the same quantum mechanical sum rules.

References

1. van Vleck JH (1932) The theory of electric and magnetic susceptibilities. Oxford University Press, Oxford
2. Ramsey NF (1950) Phys Rev 78:699
3. Ramsey NF (1951) Phys Rev 83:540
4. Ramsey NF (1952) Phys Rev 86:243
5. Hirschfelder JO, Byers-Brown W, Epstein ST (1964) Adv Quantum Chem 1:255 (references therein)
6. Jackson JD (1999) Classical electrodynamics. 3rd edn. Wiley, New York, pp 175–178
7. Epstein ST (1974) The variation method in quantum chemistry. Academic Press, New York
8. Epstein ST (1973) J Chem Phys 58:1592
9. Sambe H (1973) J Chem Phys 59:555
10. Landau LD, Lifshitz EM (1979) The classical theory of fields, 4th revised english edition. Pergamon Press, Oxford
11. Arrighini GP, Maestro M, Moccia R (1968) J Chem Phys 49:882
12. Arrighini GP, Maestro M, Moccia R (1970) J Chem Phys 52:6411

13. Arrighini G, Maestro M, Moccia R (1970) Chem Phys Lett 7:351
14. Lazzeretti P, Malagoli M, Zanasi R (1991) Chem Phys 150:173
15. Lazzeretti P (2003) Electric and magnetic properties of molecules. In: Handbook of molecular physics and quantum chemistry, vol 3, Part 1, Chapter 3. Wiley, Chichester, pp 53–145
16. Hansen AE, Bouman TD (1985) J Chem Phys 82:5035
17. London F (1937) J Phys Radium 8:397 (7ème Série)
18. DALTON (2008) An electronic structure program, release 2.0, 2005 (http://www.kjemi.uio.no/software/dalton/)
19. Frisch MJ, Trucks GW et al (2003) Gaussian 2003, revision B.05. Gaussian, Inc., Pittsburgh
20. Stanton JF, Gauss J, Harding ME, Szalay PG (2010) CFOUR, coupled-cluster techniques for computational chemistry (http://www.cfour.de)
21. Gauss J (2002) J Chem Phys 116:4773
22. Geertsen J (1989) J Chem Phys 90:4892
23. Geertsen J (1991) Chem Phys Lett 179:479
24. Geertsen J (1992) Chem Phys Lett 188:326
25. Lazzeretti P, Malagoli M, Zanasi R (1994) Chem Phys Lett 220:299
26. Smith CM, Amos RD, Handy NC (1992) Mol Phys 77:381
27. Keith TA, Bader RFW (1993) Chem Phys Lett 210:223
28. Keith TA, Bader RFW (1993) J Chem Phys 99:3669
29. Keith TA, Bader RFW (1996) Can J Chem 74:185
30. Coriani S, Lazzeretti P, Malagoli M, Zanasi R (1994) Theor Chim Acta 89:181
31. Steiner E, Fowler PW (2001) J Phys Chem A 105:9553
32. Steiner E, Fowler PW (2001) J Chem Soc Chem Comm 2220
33. Steiner E, Fowler PW, Havenith RWA (2002) J Phys Chem A 106:7048
34. Steiner E, Soncini A, Fowler PW (2006) J Phys Chem A 110:12882
35. Steiner E, Fowler PW (2004) Phys Chem Chem Phys 6:261
36. Zanasi R, Lazzeretti P, Malagoli M, Piccinini F (1995) J Chem Phys 102:7150
37. Zanasi R (1996) J Chem Phys 105:1460
38. Lazzeretti P, Zanasi R (1996) Int J Quantum Chem 60:249
39. Lazzeretti P (1989) Chem Phys 134:269
40. Lazzeretti P (1987) Adv Chem Phys 75:507
41. Linderberg J, Öhrn Y (1973) Propagators in quantum chemistry. Academic Press, London
42. Olsen J, Jorgensen P (1985) J Chem Phys 82:3235
43. McWeeny R (1962) Phys Rev 126:1028
44. McWeeny R (1969) Quantum mechanics: methods and basic applications. Pergamon Press, Oxford
45. McWeeny R (1989) Methods of molecular quantum mechanics. Academic Press, London
46. Gell-Mann M (1956) Nuovo Cimento Suppl IV:848
47. Ramsey NF (1953) Phys Rev 91:303
48. Lazzeretti P, Zanasi R (1980) J Chem Phys 72:6768
49. Lazzeretti P, Zanasi R (1977) Int J Quantum Chem 12:93
50. Lazzeretti P, Malagoli M, Zanasi R (1991) Chem Phys 150:173
51. Lazzeretti P (2000) Ring currents. In: Emsley JW, Feeney J, Sutcliffe LH (eds) Progress in nuclear magnetic resonance spectroscopy, vol 36, Elsevier, Amsterdam, pp 1–88
52. Condon EU (1937) Rev Mod Phys 9:432
53. Lazzeretti P, Zanasi R, Cadioli B (1977) J Chem Phys 67:382
54. Sauer SPA, Oddershede J (1993) Correlated and Gauge invariant calculations of nuclear shielding constants, volume 386 of NATO ASI series C. Kluwer, Dordrecht
55. Sauer SPA, Paidarová I, Oddershede J (1994) Mol Phys 81:87
56. Sauer SPA, Oddershede J (1993) Correlated and gauge invariant calculations of nuclear shielding constants. In: Tossell JA (eds) Nuclear magnetic shieldings and molecular structure, volume 386 of NATO ASI series C, Kluwer, Dordrecht, pp 351–365
57. Sauer SPA, Paidarová I, Oddershede J (1994) Theor Chim Acta 88:351
58. Sauer SPA (2011) Molecular electromagnetism: a computational chemistry approach. Oxford University Press, Oxford
59. Faglioni F, Ligabue A, Pelloni S, Soncini A, Lazzeretti P (2004) Chem Phys 304:289
60. Ligabue A, Sauer SPA, Lazzeretti P (2003) J Chem Phys 118:6830
61. Ligabue A, Sauer SPA, Lazzeretti P (2007) J Chem Phys 126:154111
62. Cuesta IG, Marin JS, de Meras AS, Pawlowski F, Lazzeretti P (2010) Phys Chem Chem Phys 12:6163
63. Lazzeretti P (2004) Phys Chem Chem Phys 6:217
64. Ligabue A, Pincelli U, Lazzeretti P, Zanasi R (1999) J Am Chem Soc 121:5513
65. Lazzeretti P, Malagoli M, Zanasi R (1995) J Chem Phys 102:9619
66. Pelloni S, Lazzeretti P (2008) J Chem Phys 128:194305

Theor Chem Acc (2013) 132:1317
DOI 10.1007/s00214-012-1317-5

ERRATUM

Erratum to: Methods of continuous translation of the origin of the current density revisited

P. Lazzeretti

Published online: 30 December 2012

Erratum to: Theor Chem Acc (2012) 131:1222 DOI 10.1007/s00214-012-1222-y

In the original publication of the article, Eqs. (68), (84) and (85) are incorrect. The correct versions of these equations are given below

$$\begin{aligned} \mathbf{J}_p^{\mathbf{d}\times\mathbf{B}}(\mathbf{r}) = & -\frac{ne}{m_e}\int d\mathbf{x}_2,\ldots,d\mathbf{x}_n \\ & \times\Big[\mathbf{d}\times\mathbf{B}\cdot\Psi_a^{\mathbf{d}\times\mathbf{B}*}(\mathbf{r},\mathbf{x}_2,\ldots,\mathbf{x}_n)\hat{\mathbf{p}}\Psi_a^{(0)}(\mathbf{r},\mathbf{x}_2,\ldots,\mathbf{x}_n) \\ & +\Psi_a^{(0)*}(\mathbf{r},\mathbf{x}_2,\ldots,\mathbf{x}_n)\hat{\mathbf{p}}\,\mathbf{d}\times\mathbf{B}\cdot\Psi_a^{\mathbf{d}\times\mathbf{B}}(\mathbf{r},\mathbf{x}_2,\ldots,\mathbf{x}_n)\Big] \end{aligned} \tag{68}$$

$$\begin{aligned} \hat{\xi}^{\Delta}_{\alpha\beta} &= -\frac{ie^2}{8m_e\hbar}\sum_{i=1}^{n}\{\epsilon_{\beta\gamma\delta}[r_\gamma,\hat{u}_{\delta\alpha}]_i+\epsilon_{\alpha\gamma\delta}[r_\gamma,\hat{u}_{\delta\beta}]_i\} \\ &= -\frac{ie^2}{8m_e\hbar}\{\epsilon_{\beta\gamma\delta}[\hat{R}_\gamma,\hat{U}_{\delta\alpha}]+\epsilon_{\alpha\gamma\delta}[\hat{R}_\gamma,\hat{U}_{\delta\beta}]_i\} \end{aligned} \tag{84}$$

$$\begin{aligned} \hat{\sigma}^{\Delta I}_{\alpha\beta} &= \frac{ie^2}{2m_e\hbar}\epsilon_{\beta\gamma\delta}\sum_{i=1}^{n}[r_\gamma,\hat{t}_{I_{\delta\alpha}}]_i \\ &= \frac{ie^2}{2m_e\hbar}\epsilon_{\beta\gamma\delta}[\hat{R}_\gamma,\hat{T}_{I_{\delta\alpha}}] \end{aligned} \tag{85}$$

The online version of the original article can be found under doi:10.1007/s00214-012-1222-y.

P. Lazzeretti (✉)
Dipartimento di Chimica dell'Università degli Studi di Modena e Reggio Emilia, Via Campi 183, 41124 Modena, Italy
e-mail: lazzeret@unimore.it

Theor Chem Acc (2012) 131:1220
DOI 10.1007/s00214-012-1220-0

REGULAR ARTICLE

A simple analysis of the influence of the solvent-induced electronic polarization on the ^{15}N magnetic shielding of pyridine in water

Rodrigo M. Gester · Herbert C. Georg · Tertius L. Fonseca · Patricio F. Provasi · Sylvio Canuto

Received: 13 February 2012 / Accepted: 4 April 2012 / Published online: 4 May 2012

Abstract Electronic polarization induced by the interaction of a reference molecule with a liquid environment is expected to affect the magnetic shielding constants. Understanding this effect using realistic theoretical models is important for proper use of nuclear magnetic resonance in molecular characterization. In this work, we consider the pyridine molecule in water as a model system to briefly investigate this aspect. Thus, Monte Carlo simulations and quantum mechanics calculations based on the B3LYP/6-311++G (d,p) are used to analyze different aspects of the solvent effects on the ^{15}N magnetic shielding constant of pyridine in water. This includes in special the geometry relaxation and the electronic polarization of the solute by the solvent. The polarization effect is found to be very important, but, as expected for pyridine, the geometry relaxation contribution is essentially negligible. Using an average electrostatic model of the solvent, the magnetic shielding constant is calculated as −58.7 ppm, in good agreement with the experimental value of −56.3 ppm. The explicit inclusion of hydrogen-bonded water molecules embedded in the electrostatic field of the remaining solvent molecules gives the value of −61.8 ppm.

Keywords NMR · Chemical shielding · Solvent effects · QM/MM · Electronic polarization effects

Dedicated to Professor Marco Antonio Chaer Nascimento and published as part of the special collection of articles celebrating his 65th birthday.

R. M. Gester · S. Canuto (✉)
Instituto de Física, Universidade de São Paulo,
CP 66318, São Paulo, SP 05315-970, Brazil
e-mail: canuto@if.usp.br

R. M. Gester (✉)
Faculdade de Ciências Exatas e Naturais, Universidade Federal do Pará, Marabá, PA 68505-080, Brazil
e-mail: gester@ufpa.br

H. C. Georg · T. L. Fonseca
Instituto de Física, Universidade Federal de Goiás,
CP 131, Goiânia, GO 74001-970, Brazil

P. F. Provasi
Department of Physics, Northeastern University and I-MIT (CONICET), AV. Libertad 5500,
W 3404 AAS Corrientes, Argentina

1 Introduction

Nuclear magnetic resonance (NMR) is one of the most important experimental techniques for characterizing the structure of organic systems [1]. In more recent years, this status has increased in the area of bio-molecular systems [2, 3]. For this reason, it has attracted considerable theoretical and computational interest. As most experiments are made in solution, a proper treatment of the solvent effect is needed. Continuous theoretical developments made in the recent past are making it possible to include solvent effects [4–15] in the calculation of NMR parameters, such as magnetic chemical shielding. The combined use of molecular mechanics and quantum mechanics (QM/MM) is an important alternative.[1] The QM/MM methodology is becoming a realistic method of choice. One successful possibility is the sequential use of Monte Carlo simulation (MC) to generate the liquid structure and QM calculation on statistically representative configurations [16–18]. For the calculation of NMR chemical shielding, it is important

[1] See the special issue dedicated to QM/MM methods in Advances in Quantum Chemistry, 2010, vol. 59.

to understand the role played by the solvent-induced electronic polarization and geometric relaxation of the reference molecule. The first is the change in the electronic distribution of the reference molecule because of the interaction with the solvent [19–25] and the second is the corresponding change in the molecular geometry that accompanies. In this work, we analyze the influence of these two agents in the calculated ^{15}N magnetic chemical shielding constants σ of pyridine in water. There has been several previous studies on the NMR properties of pyridine, and it is used here as a simple test case. Pyridine is part of several important bio-molecules and is amenable to hydrogen bond with water in one specific site. Geometry relaxation of molecules in solution is important for NMR studies because some molecular properties, like indirect spin–spin coupling constants, show extreme sensitivity to the nuclear arrangement [26]. The geometric relaxation in pyridine is small, but it is caused mainly by the hydrogen bond with water in the N site, which adds interest in the $\sigma(^{15}\text{N})$. Of course, this shielding constant has been studied several times before using different methods. Here we focus simply on the effect of the solute polarization by the solvent. To make it simpler, we assume that the reciprocal solvent polarization by the solute is mild and will not be considered. The solute electronic polarization effect can be included using an iterative method [27, 28]. To include these effects and to analyze them separately, we performed two iterative polarization processes, one relaxing only the charge distribution and another relaxing also the geometry, so that we can compare the rigid and relaxed geometry results. The solvent dependence of the nitrogen shielding constant has been systematically analyzed recently [29]. Combination of different continuum models have been used in four different molecules in several different solvents to assess the reliability of continuum models to predict ^{14}N chemical shifts [29]. Although pyridine was not included in this investigation, some common aspects will be seen related to the role of solute polarization and geometry relaxation. In this work, we use the sequential QM/MM methodology to analyze the role of the electronic polarization of the solute due to the solvent and the geometry relaxation in solution in the calculated $\sigma(^{15}\text{N})$ magnetic chemical shielding constants of pyridine in water.

2 Methodology

A sequential QM/MM methodology was applied to study the magnetic shielding constants of hydrated pyridine. In this approach, the liquid configuration is generated first by classical MC simulations. After that, a subset of uncorrelated configurations is sampled and submitted to QM calculations. The MC simulations were carried out in the NPT ensemble with $T = 25$ °C and $P = 1$ atm with one pyridine molecule and 903 waters. The intermolecular interactions were modeled by the Lennard-Jones (LJ) plus Coulomb potential. For the water molecules, we used the TIP3P parameters [30]. For pyridine, the LJ parameters were extracted from the OPLS force field [31], but the atomic charges were obtained for the pyridine in the solvent environment to consider the solute polarization effects. This is done using an iterative procedure [27, 28]. In such iterative polarization scheme, in the QM step, the solute is permitted to relax both its geometry and charge distribution in the presence of the solvent molecules. The atomic charges are obtained using the MP2/aug-cc-pVTZ calculation with the CHELPG (charges from electrostatic potentials using a grid-based method) [32] fitting of the QM electrostatic potential of pyridine. The solvent molecules surrounding the solute are thus permitted to rearrange according to the new solute charge distribution. During the iterative process, the QM calculations are made using the average solvent electrostatic configuration (ASEC) [28]. For constructing the ASEC, we superimpose 250 uncorrelated Monte Carlo configurations in which the pyridine molecule is surrounded by 300 water molecules represented by point charges. This means that all solute-water electrostatic interactions within a distance of 11 Å are taken into account.

The geometry relaxation in the solvent was performed using the Free Energy Gradient (FEG) method [33–35] in conjunction with the sequential QM/MM process. In practice, at each QM step, after calculating the wave function of the solute including the solvent electrostatic interaction, via the ASEC, we calculate the ensemble average of the first and second derivatives of the energy with respect to the solute nuclear positions. These are then used in a Quasi-Newton scheme (here we used the Broyden–Fletcher–Goldfarb–Shanno (BFGS) algorithm [36–40] implemented in the GAUSSIAN 09 package [41]) to obtain a new molecular conformation in the path to the minimum energy structure. The new solute molecular conformation is used to calculate new atomic charges, again using ASEC, and both geometry and charges are updated for a new MC simulation. The iterative process is repeated until the solute dipole moment and geometric parameters converge.

The details of the FEG approach are well described by Nagaoka et al. [33–35], and we have implemented it in a program called Diceplayer [42], which is an interface between the MC program DICE [43] and QM programs. Using this approach, results for the indirect spin–spin coupling and screening constants of liquid ammonia have been obtained in better agreement with experiment [44]. The FEG method has also been successfully employed by Aguilar et al. [45] to find optimized structures of molecules in solution.

The experimental chemical shift of nitrogen in pyridine can be converted to theoretical shielding scale $\sigma(^{14}N)$ using the nitrogen shielding of nitromethane (−135.8 ppm) as in Ref. [46]. Duthaler and Roberts [47] reported a gas-phase shielding of -84.4 ppm, which is corrected to bulk susceptibility. As NMR measurements are difficult in isolated molecules (vacuum or diluted gas-phase condition), it is common to use the cyclohexane solvent to approximate the vacuum ambient. Duthaler and Roberts also reported a value of −82.9 ppm after considering bulk susceptibility corrections and the extrapolation to infinite dilution. However, comparisons with the theoretical results for the isolated molecule show some discrepancies. Our present results using the B3LYP model with the specially designed aug-pcS-*n* ($n = 1, 2, 3$) basis sets [48] for the isolated pyridine give results for the nitrogen chemical shielding varying between −110.2 and −117.2 ppm. This is far from the gas-phase experiment above [47] with large differences varying between 25.8 and 32.8 ppm. This discrepancy suggests comparison with other quantum chemistry methods and we have also calculated the in-vacuum isolated $\sigma(^{14}N)$ values using the random phase approximation (RPA) [49] and the second-order polarization propagation approximation (SOPPA) [50] as implemented in DALTON program [51]. For instance, using the aug-pcS-2[2] and aug-cc-pVTZ-J [52–56] basis sets, our RPA results give shielding constants of −115.8 and −104.7 ppm, respectively. Our SOPPA/aug-cc-pVTZ-J calculation gives −103.6 ppm for the chemical shielding, what differs appreciably from experiment. Mennucci and collaborators [5] have recently used the B3LYP/6-311+G(d,p) level of theory to obtain a nitrogen nuclear shielding of −102.8 ppm for isolated pyridine. Using the same level of theory, we obtained −103.5 ppm. Thus, there are clear indications that the results for the isolated molecule obtained by theory and experiment show some inconsistencies. Therefore, in this study, we only report the calculated results in aqueous environment.

DFT methods and basis sets have been widely used to calculate magnetic shieldings and spin–spin couplings [15, 20, 57–62]. In this work, we employ the same B3LYP/6-311+G(d,p) model successfully used by Mennucci et al. [5] using the gauge independent atomic orbital (GIAO) [63, 64] approximation to calculate the magnetic constants. In this work, we use the CHELPG scheme for obtaining the atomic charges, and the calculations are performed within the GIAO model both implemented in the GAUSSIAN 09 package [41].

[2] The aug-cc-pVTZ-J basis sets can be downloaded from https://bse.pnl.gov/bse/portal.

3 Results

3.1 Solute polarization

Table 1 shows the calculated and experimental dipole moments of pyridine isolated and in water. The calculated MP2/aug-cc-pVTZ value of 2.33 D is in good agreement with the experimental result of 2.15 ± 0.05 D [65]. The geometry of isolated pyridine obtained at the same MP2/aug-cc-pVTZ level is also in very good agreement with experiment. The N_1–C_2 bond length is calculated as 1.341 Å, whereas the C_2–C_3 and C_3–C_4 bonds are 1.393 and 1.391 Å, respectively, compared with the experimental values of 1.340, 1.395 and 1.394 Å [66] (atomic indices are shown in Fig. 1).

Experimental reports on liquid-phase molecular dipole moments are scarce, because of the natural difficulty of a direct measurement. Theoretical reports can be found only for the pyridine-water clusters [67, 68]. Here, we investigate the solute polarization by the solvent that implies an increase in its dipole moment. This is obtained using continuum and discrete solvent models. The continuum approach uses the polarized continuum model [69] (PCM), while the discrete solvent model uses the solvent molecules treated as point charges only. The iterative polarization is

Table 1 The dipole moment of pyridine calculated at the MP2/aug-cc-pVTZ level of theory

	Isolated		In solution		
	Calc.	Exp. [65]	PCM	Discrete rigid	Discrete relaxed
μ	2.33	2.15 ± 0.05	3.41	3.94	4.38

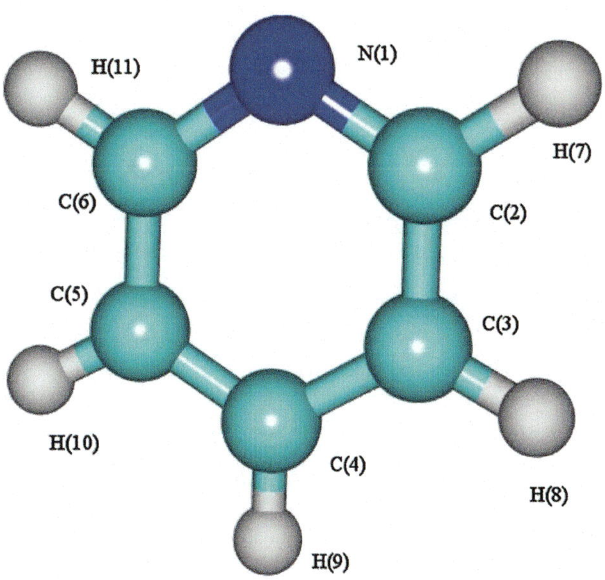

Fig. 1 Pyridine geometry and atomic labels used

used here with and without geometry relaxation, as described in the previous section. The convergence of the calculated value for the rigid case is shown in Fig. 2 for illustration. Table 1 summarizes the results. The PCM approach obtains a dipole moment of pyridine in water of 3.41 D, which represents an increase of 46 % as compared with the gas-phase result. With the iterative scheme representing the solvent as point charges, we obtain dipole moments of 3.94 (increase of 69 %) and 4.38 D (increase of 88 %) for the rigid and relaxed geometries, respectively. The change in geometry is mild, as seen before in similar nitrogen-containing molecules [29], but still affects the calculated dipole moment. We analyze next the influence of this polarization in the solute–solvent hydrogen bonds in the charge on the nitrogen site and hence how these affect the σ (^{15}N) in water.

3.2 Hydrogen bonds

Considering the rigid model of pyridine, we now briefly discuss the effect of the polarization in the coordination of water molecules around the solute. In the unpolarized situation, we obtain the corresponding coordination of 1.62 water molecules. In the polarized case, this coordination changes to 2.02. These coordinated water molecules are not assured to be hydrogen bonded to pyridine. Thus, in addition to the geometric, we have used also an energetic criterion [67, 70, 71] derived from the pairwise interaction energy. This is shown Fig. 3. Thus, for the polarized case, we consider a hydrogen bond when the $R_{\text{N–O}} \leq 3.5$ Å, the angle α(N–OH) $\leq 35°$ and the interaction energy is <-4.0 kcal mol. This geometric criterion is obtained from the radial and angular distribution functions, and the energy criterion is obtained from the pairwise interaction energy distribution (see Fig. 3). Similar analysis is made for the unpolarized case, where the $R_{\text{N–O}} \leq 3.2$ Å, α(N–OH) $\leq 35°$ and the interaction energy is <-3.0 kcal/mol. In the polarized case, we find that 2 % of the configurations make no hydrogen bonds, whereas 36.4 % make one hydrogen bond. The most probable situation corresponds to 60.4 % of the configurations making 2 hydrogen bonds. In the average, as can be seen in Table 2, we find 1.6 hydrogen bonds between the nitrogen atom of pyridine and the hydrogen atom of water. This is considerably larger than the unpolarized situation (0.71 hydrogen bonds) and somewhat larger than using the original point charges of the OPLS model (1.1).

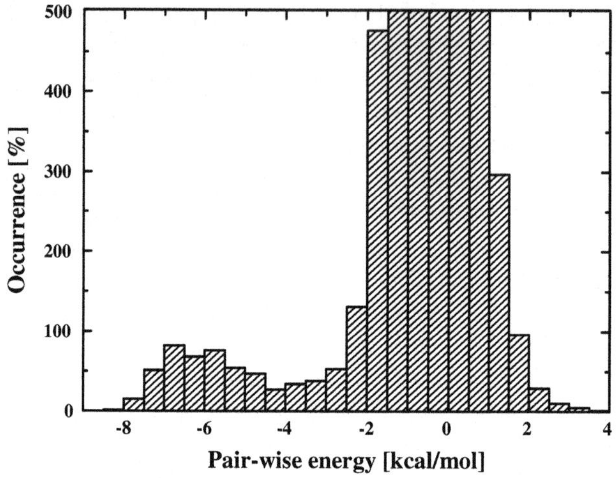

Fig. 3 Histogram of the pairwise interaction energy between rigid pyridine and water in the polarized case

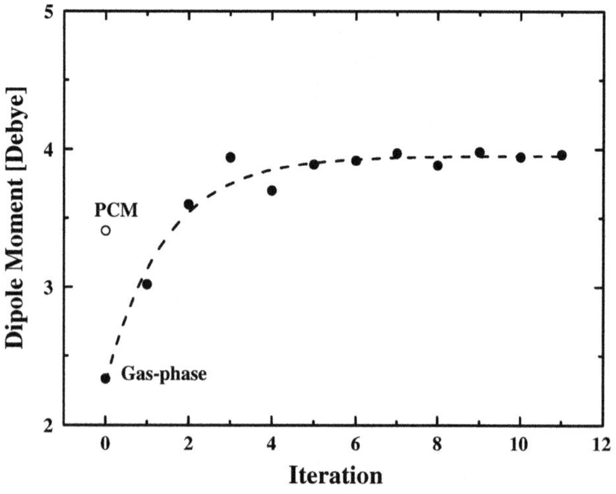

Fig. 2 The calculated dipole moment of isolated and hydrated pyridine. The *empty circle* shows the polarized continuum model prediction. The *black circles* represent the iterative values with rigid geometry, starting from the gas phase

3.3 Chemical shielding

In the following theoretical analysis, we separate the different contributions, and we first analyze the electrostatic part. The PCM and the electrostatic (ASEC) models give the results shown in Table 3. For comparison with experiments, we considered the early measurements of the ^{15}N

Table 2 Statistics of the hydrogen bonds formed between pyridine and water

HBs	Occurrence (%)		
	[Ref. [68]]	Unpolarized[a]	Polarized[a]
0	17	36	2
1	62	56.8	36.4
2	20	7.2	60.4
3	1	0	1.2
Average	1.1	0.71	1.61

[a] Both polarized and unpolarized models use a rigid geometry of pyridine. 250 uncorrelated configurations were analyzed

Table 3 The effects of the solute polarization and geometry relaxation on the calculated nuclear magnetic shielding constant

	σ (^{15}N) in water	
	Rigid	Relaxed
Unpolarized	−82.1	−78.0
Polarized	−59.8	−58.7
PCM[a]	−81.0	
Liquid (Exp.)[b]		−56.3

The magnetic constants are calculated at the B3LYP/6-311+G(d,p) level, while the pyridine (rigid and relaxed) geometries are obtained with the MP2/aug-cc-pVTZ level

[a] The PCM value was obtained using the gas-phase MP2/aug-cc-pVTZ geometry

[b] In cyclohexane solution [59], this value is corrected to susceptibility effects and extrapolated to infinite dilution

shielding constants in solvated pyridine due to Duthaler and Roberts [59]. They report a magnetic shielding in water of −59.9 ppm. After including bulk susceptibility corrections, this value changes to −56.3 ppm. The continuum PCM model estimates a shielding of −81.0 ppm, which differs by 24.3 ppm from the experiment.

We now analyze the polarization effects on the σ (^{15}N). The polarization effect using a fixed geometry is shown in Fig. 4. It can be noted that the theoretical result improve systematically until convergence in the theoretical value of −59.8 ppm, in very good agreement with the experimental value of −59.9 ppm (or −56.3 ppm if correcting for the bulk susceptibility). It is then found here that the polarization effect has a great influence on the calculated value of the magnetic shielding of the nitrogen atom of pyridine in water. Combining the polarization and geometric effects gives the theoretical value of −58.7 ppm. Thus, the separate geometric effect is only −1.1 ppm. This has two components. One is the direct change in the geometry that affects the local magnetic shielding. The other is the indirect contribution that comes from the change in the dipole moment and therefore the coordination of solvent molecules. The combined effect is seen to be very small. The geometric variation is expected to be small, and indeed the largest deviation found is a small increase of the $R(N_1–C_6)$ by 0.006 Å and a decrease in the $A(C_2N_1C_6)$ angle by 0.1°. Contreras and Peralta [72] have shown that for ammonia, the larger contribution to the spin–spin coupling arises mainly from the angle variations. But overall, the present calculations show that geometric effect on the chemical shielding σ (^{15}N) of pyridine in water is very mild, as also noted before [29] for similar systems.

Having analyzed the electrostatic contribution to the magnetic shielding, we now consider the exchange and the van der Waals contributions by considering some explicit solvent water molecules around the reference pyridine molecule. For a systematic consideration, we analyze first the role of the hydrogen-bonded (HB) water molecules. The results are shown in the Table 4. As the results shown above have indicated that the geometry relaxation is unimportant, all calculations are now performed using the Monte Carlo configurations obtained considering the rigid pyridine geometry. The magnetic shielding on the nitrogen atom using only the HB water molecules explicitly is −83.4 ± 0.7 ppm, a value that differs from experiment by ~27 ppm. These calculations are made on the structures previously extracted and use different number of water molecules for each configuration (Table 2). An improvement in the model can be obtained by embedding theses

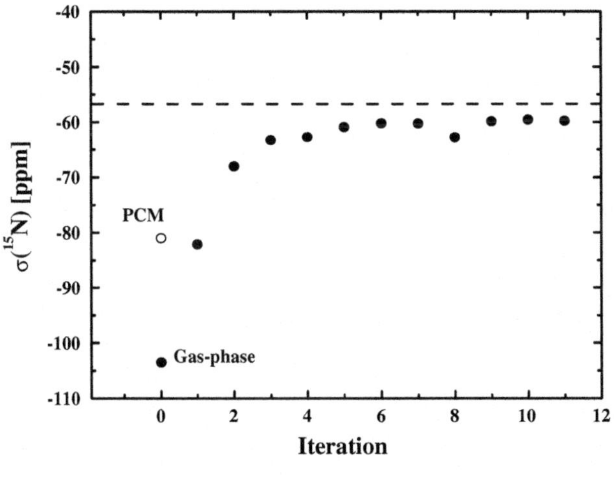

Fig. 4 The evolution of the calculated σ(^{15}N) as a function of the iteration steps obtained for rigid pyridine simulation. The experimental data is shown as the *dotted line*

Table 4 The calculated B3LYP/6-311+G(d,p) and experimental ^{15}N magnetic shielding constants [ppm] in hydrated pyridine

σ (^{15}N) in water					
PCM[a]	ASEC[b]	HB[c]	HB+PC[d]	$6H_2O$+PC[e]	Exp.[f]
−81.0	−59.8 (−58.7)	−83.4 ± 0.7	−62.4 ± 0.7	−61.8 ± 0.9	−56.3 (−59.9)

[a] The PCM values were obtained using the isolated MP2/aug-cc-pVTZ geometry

[b] The values in parenthesis were obtained from in-water relaxed geometry

[c] HB includes only the explicit pyridine-water hydrogen bonds. See text

[d] HB+PC includes the explicit pyridine-water hydrogen bonds and the remaining 320 water molecules treated as point charges. See text

[e] $6H_2O$+SPC includes the pyridine surrounded by the six water nearest the pyridine nitrogen. The remaining 303 water molecules were included as point charges. See text

[f] Value corrected [uncorrected] for bulk susceptibility effects and extrapolated to infinite dilution [59]

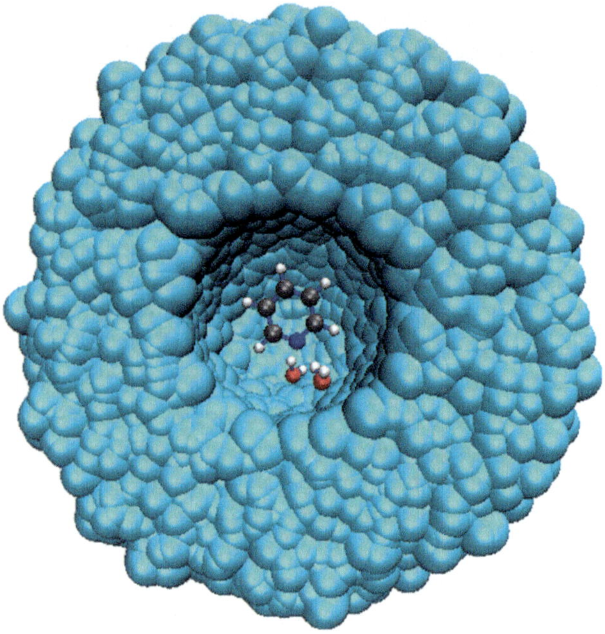

Fig. 5 Pyridine-water hydrogen-bonded complex embedded in the electrostatic field of the remaining water molecules

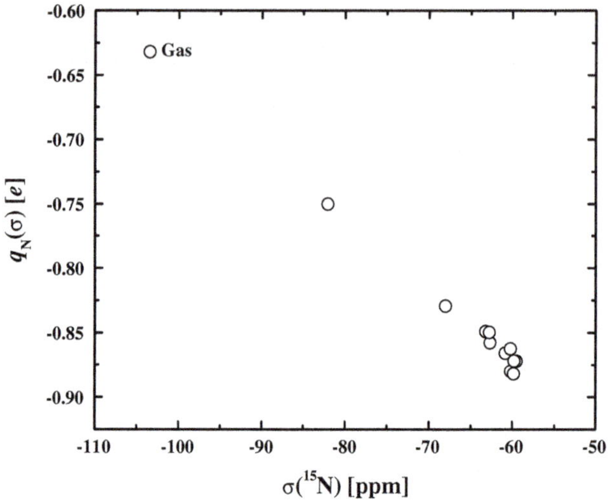

Fig. 6 The electronic charge on the nitrogen atom as a function of the magnetic shielding

hydrogen-bonded structures in the electrostatic field of the remaining water molecules. This model is termed HB+PC, and the situation is illustrated in Fig. 5 that shows one of the configurations used for the calculations of the magnetic shielding. The water molecules in the embedding enter only with their point charges located in the atomic position as described by the TIP3P potential. In HB+PC, a total of 320 water molecules are included as point charges. The calculated chemical shielding is now -62.4 ± 0.7 ppm in fair agreement with the experimental result of -56.3 ppm. This result differs from the simple ASEC electrostatic result by less than 3 ppm. Increasing further the number of explicit water molecules, we see that using six explicit solvent molecules in the electrostatic embedding ($6H_2O$+PC) gives a close result of -61.8 ± 0.7 ppm, only 0.6 ppm of difference with the HB+PC result, showing that the principal contribution comes from the pyridine-water hydrogen bonds plus electrostatic contribution. This is in line with a previous study on diazines [7] where it is concluded that a few explicit solvent molecules would be appropriate, but the most important contribution derives from the long-range electrostatic interaction, as noted before [5].

Finally, we analyze the influence of the polarization in the atomic charge on the nitrogen site. Figure 6 shows that the polarization has a marked influence on the nitrogen charge and this clearly associates with the variations of the magnetic shielding results. One can observe that the correlation is essentially linear with an increase of $q_N(\sigma)$ correlating with the decrease of the magnetic shielding.

4 Conclusions

This work analyses the solute electronic polarization induced by the interaction with water solvent and its implications on the σ (^{15}N) magnetic shielding constant of pyridine. For the isolated pyridine molecule, the dipole moment obtained at the MP2/aug-cc-pVTZ level is 2.33 D, which is close to the experimental value (2.15 ± 0.05 D). In aqueous environment, values varying from 3.41 to 4.38 D were obtained showing some considerable polarization. The influence of this solute electronic polarization on the σ (^{15}N) is calculated. It is seen that the electronic polarization has a great influence in the number of solute–solvent hydrogen bonds and in the atomic charges. Using an iterative procedure to equilibrate the solute charges with the solvent electrostatic field, it is seen that the calculated values of the σ (^{15}N) magnetic shielding constant of pyridine converges to a value that is in close agreement with the experimental value. These variations in the σ (^{15}N) correlate well with the changes in the atomic charge on the nitrogen site. These aspects indicate a redistribution of the solute charges due to the solvent that although not directly observable may be a sensitive probe of the solute polarization due to the solvent and has been of recent and fundamental interest [29, 73–75]. The best result of -61.8 ppm obtained here corresponds to the electronically polarized pyridine surrounded by six water molecules and this entire system embedded in the electrostatic field of the remaining water molecules, and is in good agreement with the experimental results of -56.3 and -59.9 ppm. The solvent shift is, however, overestimated, and we attribute this to uncertainties in the vacuum results.

Acknowledgments This work is partially supported by CNPq, CAPES and FAPESP (Brazil). Additional support from the INCT-FCx (Institute for Complex Fluids) and NBioNet are gratefully acknowledged. RMG thanks CNPq for a doctoral fellowship and P. F. P. thanks UNNE (PICTO-UNNE-227 BID 1728/OC-AR) and CONICET (Argentina).

References

1. Hore PJ (1995) Nuclear magnetic resonance. Oxford University Press, New York
2. Canuto S (ed) (2008) Solvation effects on molecules and biomolecules. Computational methods and applications. Springer, New York
3. Mennucci B, Cammi R (eds) (2007) Continuum solvation models in chemical physics. Wiley, New York
4. Buckingham AD, Schafer T, Schneider WG (1960) J Chem Phys 32:1227
5. Mennucci B, Martinez JM, Tomasi J (2001) J Phys Chem A 105:7287
6. Mennucci B (2002) J Am Chem Soc 124:1506
7. Kongsted J, Mennucci B (2007) J Phys Chem A 111:9890
8. Malkin VG, Malkina OL, Steinebrunner G, Huber H (1996) Chem Eur J 2:452
9. Chesnut DB, Rusiloski BE (1994) J Mol Struct 314:19
10. Cheeseman JR, Trucks GW, Keith TA, Frisch MJ (1996) J Chem Phys 104:5497
11. Fileti EE, Georg HC, Coutinho K, Canuto S (2007) J Braz Chem Soc 18:74
12. Kongsted J, Aidas K, Mikkelsen KV, Sauer SPA (2008) J Chem Theory Comput 4:267
13. Møgelhøj A, Aidas K, Mikkelsen KV, Sauer SPA, Kongsted J (2009) J Chem Phys 130:134508
14. Ligabue A, Sauer SPA, Lazzeretti P (2007) J Chem Phys 126:154111
15. Keal TW, Helgaker T, Salek P, Tozer DJ (2006) Chem Phys Lett 425:163
16. Coutinho K, Canuto S (2000) J Chem Phys 113:9132
17. Coutinho K, Canuto S, Zerner MC (2000) J Chem Phys 112:9874
18. Canuto S, Coutinho K, Trzresniak D (2003) Adv Quantum Chem 41:161
19. Fonseca TL, Coutinho K, Canuto S (2008) J Chem Phys 129:034502
20. Manzoni V, Lyra ML, Gester RM, Coutinho K, Canuto S (2010) Phys Chem Chem Phys 12:14023
21. Aidas K, Møgelhøj A, Kjaer H, Nielsen CB, Milkkelsen KV, Ruud K, Christiansen O, Kongsted J (2007) J Phys Chem A 111:4199
22. Pranata J, Wierschke SG, Jorgensen WL (1991) J Am Chem Soc 113:2810
23. Gao J, Xia X (1992) Science 258:631
24. Jorgensen WL (ed) (2007) J Chem Theor Comput special issue 3:1877–2145
25. Dahlke EE, Truhlar DG (2007) J Chem Theor Comput 3:46
26. Elguero J (2007) Magn Reson Chem 46:356
27. Georg HC, Coutinho K, Canuto S (2006) Chem Phys Lett 429:119
28. Coutinho K, Georg HC, Fonseca TL, Ludwig V, Canuto S (2007) Chem Phys Lett 437:148
29. Ribeiro RF, Marenich AV, Cramer CJ, Truhlar DG (2009) J Chem Theor Comput 5:2284
30. Jorgensen WL, Chandrasekhar J, Madura JD, Impey RW, Klein ML (1983) J Chem Phys 79:926
31. Jorgensen WL, McDonald NA (1998) J Mol Struc 424:145
32. Breneman CM, Wiberg KB (1990) J Comput Chem 11:361
33. Okuyama-Yoshida N, Nagaoka M, Yamabe T (1998) Int J Quantum Chem 70:95
34. Okuyama-Yoshida N, Kataoka K, Nagaoka M, Yamabe T (2000) J Chem Phys 113:3519
35. Hirao H, Nagae Y, Nagaoka M (2001) Chem Phys Lett 348:350
36. Broyden CG (1970) IMA J Appl Math 6:76
37. Fletcher R (1970) Comput J 13:317
38. Goldfarb D (1970) Math Comput 24:23
39. Shanno DF (1970) Math Comput 24:647
40. Shanno DF, Kettler PC (1970) Math Comput 24:657
41. Frisch MJ, Trucks GW, Schlegel HB, Scuseria GE, Robb MA, Cheeseman JR, Scalmani G, Barone V, Mennucci B, Petersson GA, Nakatsuji H, Caricato M, Li X, Hratchian HP, Izmaylov AF, Bloino J, Zheng G, Sonnenberg JL, Hada M, Ehara M, Toyota K, Fukuda R, Hasegawa J, Ishida M, Nakajima T, Honda Y, Kitao O, Nakai H, Vreven T, Montgomery JJA, Peralta JE, Ogliaro F, Bearpark M, Heyd JJ, Brothers E, Kudin KN, Staroverov VN, Kobayashi R, Normand J, Raghavachari K, Rendell A, Burant JC, Iyengar SS, Tomasi J, Cossi M, Rega N, Millam JM, Klene M, Knox JE, Cross JB, Bakken V, Adamo C, Jaramillo J, Gomperts R, Stratmann RE, Yazyev O, Austin AJ, Cammi R, Pomelli C, Ochterski JW, Martin RL, Morokuma K, Zakrzewski VG, Voth GA, Salvador P, Dannenberg JJ, Dapprich S, Daniels AD, Farkas O, Foresman JB, Ortiz JV, Cioslowski J, Fox DJ (2004) Gaussian 03, Revision D 02. Wallingford, Gaussian
42. Georg HC, Canuto S (2009) Diceplayer. University of São Paulo, São Paulo
43. Coutinho K, Canuto S (2009) DICE (version 2.9): a general monte carlo program for liquid simulation. University of São Paulo, São Paulo
44. Gester RM, Georg HC, Canuto S, Caputo MC, Provasi PF (2009) J Phys Chem A 113:14936
45. Galván IF, Sánchez ML, Martin ME, Olivares del Valle FJ, Aguilar MA (2003) J Chem Phys 118:255
46. Facelli JC (2000) Chem Phys Lett 322:91
47. Duthaler RO, Roberts JD (1978) J Am Chem Soc 100:4969
48. Jensen F (2008) J Chem Theory Comput 4:2008
49. Christiansen O, Bak KL, Koch H, Sauer SPA (1998) Chem Phys Lett 284:47
50. Provasi PF, Aucar GA, Sauer SPA (2001) J Chem Phys 115:1324
51. DALTON, a molecular electronic structure program, Release Dalton2011 (2011)
52. Bak KL, Koch H, Oddershede J, Christiansen O, Sauer SPA (2000) J Chem Phys 112:4173
53. Nielsen ES, Jørgensen P, Oddershede J (1980) J Chem Phys 73:6238
54. Packer MJ, Dalskov EK, Enevoldsen T, Jensen HJA, Oddershede J (1996) J Chem Phys 105(14):5886
55. Sauer SPA (1996) Chem Phys Lett 260:271
56. Enevoldsen T, Oddershede J, Sauer SPA (1998) Theor Chem Acc 100:275
57. Keal TW, Tozer DJ (2004) J Chem Phys 121:5654
58. Auer AA (2009) Chem Phys Lett 467:230
59. Becke AD (1993) J Chem Phys 98:5648
60. Lee C, Yang W, Parr RG (1988) Phys Rev B 37:785
61. Adamo C, Barone V (1999) J Chem Phys A 110:6158
62. Alkorta I, Elguero J (2004) Magn Reson Chem 42:955
63. Gauss J (1993) J Chem Phys 99:3629
64. Wolinski K, Hinton JF, Pulay P (1990) J Am Chem Soc 112:8251
65. DeMore BB, Wilcox WS, Goldstein JH (1954) J Chem Phys 22:876
66. Lide DR (1992) Handbook of chemistry and physics, 73rd edn. CRC Press, Boca Raton
67. Fileti EE, Coutinho K, Malaspina T, Canuto S (2003) Phys Rev E 67:061504

68. Malaspina T, Coutinho K, Canuto S (2002) J Chem Phys 117:1692
69. Miertuš S, Scrocco E, Tomasi J (1981) Chem Phys 55:117
70. Stilinger FH, Rahman A (1974) J Chem Phys 60:3336
71. Mezei M, Beveridge DL (1981) J Chem Phys 74:622
72. Contreras RH, Peralta JE (2000) Prog Nucl Magn Reson Spectrosc 37:321
73. Zhan C-G, Chipman DM (1999) J Chem Phys 110:1611
74. Curutchet C, Cramer CJ, Truhlar DG, Ruiz-Lopez MF, Rinaldi D, Orozco M, Luque FJ (2003) J Comput Chem 24:284
75. Marenich AV, Olson RM, Chamberlin AC, Cramer CJ, Truhlat DG (2007) J Chem Theory Comput 3:2055

Theor Chem Acc (2012) 131:1210
DOI 10.1007/s00214-012-1210-2

REGULAR ARTICLE

Theoretical simulations of the vibrational predissociation spectra of H_5^+ and D_5^+ clusters

Alvaro Valdés · Patricia Barragán · Cristina Sanz-Sanz · Rita Prosmiti · Pablo Villarreal · Gerardo Delgado-Barrio

Received: 7 February 2012 / Accepted: 20 March 2012 / Published online: 7 April 2012

Abstract In the present study, the effect of the potential energy surface representation on the infrared spectra features of the H_5^+ and D_5^+ clusters is investigated. For the spectral simulations, we adopted a recently proposed (Sanz-Sanz et al. in Phys Rev A 84:060502-1–4, 2011) two-dimensional adiabatic quantum model to describe the proton-transfer motion between the two H_2 or D_2 units. The reported calculations make use of a reliable "on the fly" DFT-based potential surface and the corresponding new dipole moment surface. The results of the vibrational predissociation dynamics are compared with earlier and recent experimental data available from mass-selected photodissociation spectroscopy, as well as with previous theoretical calculations based on an analytical ab initio parameterized surfaces. The role of the potential topology on the spectral features is studied, and general trends are discussed.

Keywords Potential energy surfaces · Electronic structure calculations · Spectrum simulations

Dedicated to Professor Marco Antonio Chaer Nascimento and published as part of the special collection of articles celebrating his 65th birthday.

A. Valdés · C. Sanz-Sanz · R. Prosmiti (✉) · P. Villarreal · G. Delgado-Barrio
Instituto de Física Fundamental, C.S.I.C.,
Serrano 123, 28006 Madrid, Spain
e-mail: rita@iff.csic.es

P. Barragán
CELIA, Université de Bordeaux-I, UMR CNRS 5107, CEA,
351 Cours de la Libération, 33045 Talence, France

C. Sanz-Sanz · R. Prosmiti · P. Villarreal · G. Delgado-Barrio
Unidad Asociada, UAM-CSIC, Departamento de Química Física, Facultad de Ciencias C-XIV,
Universidad Autónoma de Madrid, 28049 Madrid, Spain

1 Introduction

H_5^+ has been first detected in 1962 [1], and since then several experimental studies have been carried out, although the information available is rather limited to dissociation enthalpies [2–4] and a few vibrational frequencies at energies between 2,500 and 8,000 cm^{-1} [5–7]. Directly related to this work are the experiments reported by Okumura et al. [5] in 1988, later on by Bae [6], and just recently by Duncan et al. [7] on the infrared (IR) photodissociation spectra of the H_5^+. In the earlier investigations, three broad bands have been observed near 4, 000 cm^{-1} and have been assigned to vibrational frequencies of H_5^+, such as fundamental stretching modes of H_3^+ and H_2, as well as combinations or overtones of these modes with the intermolecular $H_3^+ + H_2$ [5]. However, in the latter study the excitation of the shared-proton stretch mode have been found to play a major role in the assignment of the same spectral features [7]. In particular, Duncan et al., in addition to the H_5^+ spectrum in the 2,000–4,500 cm^{-1}, have also recorded for the first time the D_5^+ one in the spectral region of 1,500–3,500 cm^{-1}. They showed that the delocalized and highly anharmonic shared-proton stretch mode carries very large oscillator strength, and its excitations are involved in the spectral transitions at the energy region studied [7]. Just recently, the importance of the proton-transfer stretching in the IR spectra of these clusters has been studied by applying an adiabatic two-dimensional model incorporating the temperature effect [8].

However, despite the apparent simplicity and the effort devoted, very little is still known about the spectroscopic characterization even at low vibrational energies, and finite-temperature properties of H_5^+ and its isotopes [7, 9–12]. For theoretical studies, the main difficulty arises from the very delicate relationship between the interaction potential energy surface (PES) and experimental observations. Comparison of theoretical results with experimental data is subject to the errors coming from the poor, in general, knowledge of the PES, the dynamical approximations adopted, and the uncertainties of the experimental measurements. As the number of atoms increases, the route to accurate interaction PESs is hard, as it depends on the complexity of the system.

Until recently, there was not global potential energy surface available in the literature for the H_5^+, while numerous ab initio calculations at various levels of theory have been reported [13–21], although they were limited to local description of the PES. However, during the last years three surfaces have been reported in the literature claiming a full-dimensional and reliable description of the H_5^+ cluster [22–24]. Two of them have been generated from analytical expressions parameterized to high-level CCSD(T) ab initio data [22, 23], while the third one is obtained from "on the fly" DFT calculations [24].

Simplified models to perform spectra simulations, as the one proposed recently [8], allow to explore the effect of the underlying surface and to evaluate the role of the possible differences between several of them on the dynamics of the system. Therefore, in the present study, by analyzing spectral features, we are able to rationalize the trends within the spectra and relate them to the properties of the topology of the PES of both H_5^+ and D_5^+ complexes. For this purpose, we use the two more recent PESs that were obtained from completely different generation procedures [23, 24].

The plan of this paper is as follows. Section 2 describes the coordinate system and the representation of the Hamiltonian together with computational details for the spectra simulations. In Sect. 3, we present the results obtained and discuss their comparison with previous experimental and theoretical data, while Sect. 4 summarizes our conclusions.

2 Coordinate system, model Hamiltonian operator, and computational details

The coordinate system used for describing the H_5^+ and D_5^+ clusters is shown in Fig. 1. The positions of the centers of mass of the two H_2 monomers, together with the position of the proton, define the two z_1 and z_2 distances with the coordinates used to be defined as $r = z_1 - z_2$ and $R = z_1 + z_2$ with $0 \leq R \leq \infty$ and $-R \leq r \leq R$. The Hamiltonian operator is given as [8]:

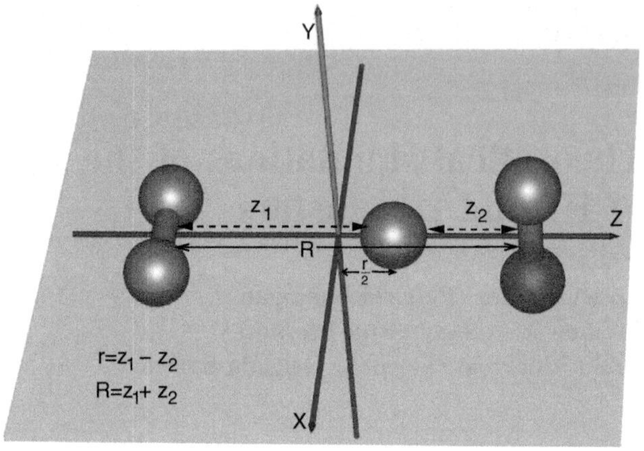

Fig. 1 2D Coordinate system used for the H_5^+ and D_5^+ complexes

$$\begin{aligned}\hat{H} &= -\frac{\hbar^2}{2}\left(\frac{1}{\mu_r}\frac{\partial^2}{\partial r^2} + \frac{1}{\mu_R}\frac{\partial^2}{\partial R^2} + \frac{2}{\mu_{rR}}\frac{\partial^2}{\partial r \partial R}\right) + V(r,R)\\ &= T_r + T_R + T_{rR} + V(r,R), \end{aligned} \tag{1}$$

with $V(r, R)$ the potential energy surface of the system. In our case, $\mu_r = m/5$, $\mu_R = m$ with m the hydrogen or deuterium mass. The reduced masses of the two H_2 or D_2 monomers, for example $\mu_{12} = \mu_{34} = \frac{m}{2}$, are the same, so the $\frac{1}{\mu_{rR}} = \frac{1}{\mu_{12}} - \frac{1}{\mu_{34}}$ is zero. Thus, the crossing kinetic term T_{rR} vanishes, and given the mass difference in the r and R coordinates, we can assume an adiabatic separation by approximating the total wavefunction as $\Phi(r,R) \approx \phi_v(r;R)\psi_{v\epsilon}(R)$. For a given v, the corresponding R-dependent eigenvalues, $W_v(R)$, computed from the

$$[T_r + V(r,R) - W_v(R)]\phi_v(r;R) = 0, \tag{2}$$

act as effective potentials in the Schrödinger equation,

$$[T_R + W_v(R) - \epsilon]\psi_{v\epsilon}(R) = 0. \tag{3}$$

to obtain discrete ($\epsilon < W_v(R \to \infty)$, $\epsilon = E_{vn}$, $n = 0, 1, \ldots$) and continuum solutions ($\epsilon > W_v(R \to \infty)$) accounting for vibrations, n, in the R coordinate. The $\psi_{v'n'}$ vibrational excited states immersed in the continuum of $v'' < v'$ vibrational levels are coupled with the continuum wavefunctions, $\psi_{v''\epsilon}$, as follows:

$$\langle \Phi_{v''\epsilon}|\hat{H}|\Phi_{v'n'}\rangle = \langle \psi_{v''\epsilon}|Q_{v''v'}(R)|\psi_{v'n'}\rangle_R. \tag{4}$$

with the coupling operator $\hat{Q}$ being

$$\begin{aligned}\hat{Q} &= 2\left\langle \phi_{v'}\left|\frac{\partial \phi_v}{\partial R}\right.\right\rangle_r \frac{\partial}{\partial R} + \left\langle \phi_{v'}\left|\frac{\partial^2 \phi_v}{\partial R^2}\right.\right\rangle_r\\ &= 2\frac{\langle \phi_{v'}|\frac{\partial V}{\partial R}|\phi_v\rangle}{W_v(R) - W_{v'}(R)}\frac{\partial}{\partial R}\\ &+ \sum_{v'' \neq v' \neq v} \frac{\langle \phi_{v'}|\frac{\partial V}{\partial R}|\phi_{v''}\rangle\langle \phi_{v''}|\frac{\partial V}{\partial R}|\phi_v\rangle}{[W_{v''}(R) - W_{v'}(R)][W_v(R) - W_{v''}(R)]}\end{aligned} \tag{5}$$

where the Hellmann–Feynman theorem has been used.

The process we pretend to simulate corresponds to the promotion of an initial vibrational state $|\Phi_{vn}\rangle$ to a intermediate or final $|\Phi_{v'n'}\rangle$ one of the H_5^+, as follows:

$$H_5^+(v,n) + \hbar\omega \rightarrow H_5^+(v',n') \rightarrow H_3^+ + H_2 + \epsilon \quad (6)$$

This excitation takes place within the same electronic state by the absorption of a photon with frequency, ω, that matches with the energy difference of these two vibrational states and is in the infrared spectral region:

By considering a Boltzmann distribution over the initial states at temperature T, the line intensity for such transitions is given by

$$I_{v'n',vn} \propto \frac{e^{-(E_{vn}/kT)}}{\sum_{v,n} e^{-(E_{vn}/kT)}} |\mu_{v'n';vn}|^2 \quad (7)$$

and would appear at a the photon frequency $\omega_{v'n',vn} = (E_{v'n'} - E_{vn})/\hbar$. The $\mu_{v'n',vn} = \langle\Phi_{v'n'}|\mu|\Phi_{vn}\rangle$ are the corresponding transition dipole moments, and $\mu(r, R)$ the dipole moment surface of the system. The absorption of a photon that leads to a vibrational excitation of the cluster may be followed by energy transfer to weaker bonds causing the dissociation of the cluster. This process is known as vibrational predissociation and it produces broadening of the corresponding spectral lines. In the Golden rule approximation [25], the half-width associated with the vibrational predissociation of an initial state $\psi_{v'n'}$ into a final continuum state $\psi_{v''\epsilon}$ is given by

$$\Gamma_{v'n'} = \pi \sum_{v''<v'} |\langle\psi_{v''\epsilon}|Q_{v''v'}(R)|\psi_{v'n'}\rangle_R|^2, \quad (8)$$

where the $\psi_{v''\epsilon}$ is calculated for an energy $\epsilon = E_{v'n'} - W_{v''}(R \rightarrow \infty)$, which corresponds, for large R distance, to the relative energy between the H_3^+ and H_2 fragments. Then, by dressing the corresponding lines in the spectrum with Lorentzian functions of $\Gamma_{v'n'}$ widths, and by summing over transitions [26], a continuum spectrum is obtained,

$$I(\omega) = \sum_{v'n'vn} \frac{\Gamma_{v'n'}/2\pi}{\hbar^2(\omega - \omega_{v'n',vn})^2 + \Gamma^2_{v'n'}/4} I_{v'n',vn} \quad (9)$$

which satisfies the condition $\int d\omega I(\omega) = \sum_{v'n'vn} I_{v'n',vn}$.

3 Results and discussion

As we mentioned above, up to date, there are three PESs available in the literature for the H_5^+, [22–24]; however, we consider here the two more recent ones. On the one hand, an analytical representation of the PES based on a TRIM (triatomics-in-molecules) formalism, which has been parameterized to high-level CCSD(T) data [23] and has been previously employed in spectral simulations [8, 27]. On the other hand, an "on the fly" surface based on DFT calculations [24] using a specific hybrid functional, B3(H), which has been specially designed for hydrogen-only systems [28]. As it can be seen, these two surfaces have been generated from completely different procedures, and thus several differences have been found between them. The main ones are the predictions of the D_e well-depth energy: the DFT surface overestimates it compared to the CCSD(T)/CBS results [21], and regarding the barrier height for the internal proton transfer, the TRIM fitted surface overestimates it, while the DFT one underestimates it, taken as reference data the ones from CCSD(T)-R12 calculations [17].

For the present calculations, we employ the DFT-based potential for the H_5^+ [24]. Such representation has been found to correctly describe the overall surface of the cluster [24] and has been used in previous studies [10, 11, 29, 30].

In Fig. 2 (top panel) we show the potential curves as a function of r for the indicated R values. The potential curves are obtained by optimizing at each (r, R) point the bond lengths of the two perpendicular H_2 monomers (see Fig. 1). One can see that the potential curves are symmetric in the r coordinate for small R values, and the proton is moving in a single potential well around the D_{2d} configuration of the H_5^+. As the R distance increases, a double-well potential corresponding to C_{2v} geometries appears. These symmetric wells are separated by a D_{2d} barrier, which gets higher for larger values of R. Also we display in Fig. 2 (middle panel) the $\partial V/\partial R$ derivatives of the potential. We should point out here that the derivatives of the DFT potential show a very smooth behavior; they are symmetric and thus only allow states of the same symmetry (even or odd) to be coupled by the kinetic coupling terms $Q_{v''v'}$ in Eq. 5.

For the spectral calculations, the electric dipole moment surface is also needed, and we compute such surface here by performing DFT/B3(H) calculations, the same ones as for the potential. In Fig. 2 (bottom panel), we present the dipole moment curves as a function of r and R. One can see that for each R the dipole moment does not behave linearly for the whole range of r, although in the region of interest, values of r close to zero, is rather close to it. Also, it is an odd function leading to nonzero transition dipole moment, $\mu_{v'n',vn}$, for states of different symmetry.

In turn, by solving the Eqs. 2 and 3, we obtained the effective $W_v(R)$ potentials together with the corresponding ϕ_v states, as well as the E_{vn} bound or ϵ continuum levels, and their $\psi_{v\epsilon}$ wavefunctions. In Figs. 3 and 4, we present the six lowest $W_v(R)$ potentials, with v = 0–5, and the bound E_{vn}, with n = 0–10, levels up to 8,000 cm^{-1} for the H_5^+ and D_5^+, respectively. The ϕ_v are symmetric and antisymmetric states, and as it can be seen in Figs. 3 and 4, are degenerate for R values larger than 4.5 and 5.0 bohr for the

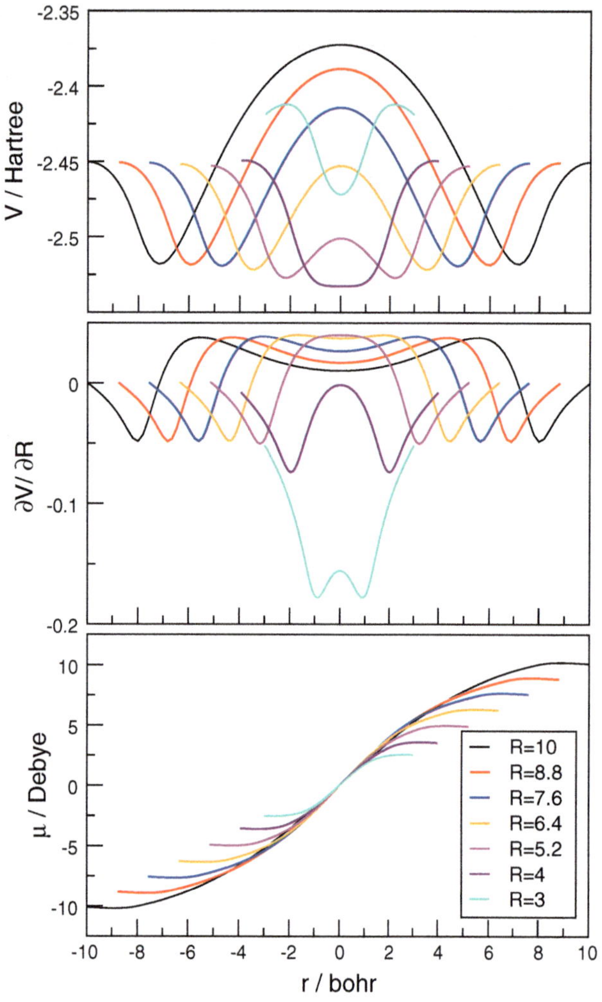

Fig. 2 Potential *V(r, R)* curves (*top panel*), first derivatives with respect to *R* (*middle panel*), and the dipole moment $\mu(r, R)$ (*bottom panel*) for the H_5^+

H_5^+ and D_5^+, respectively. Based on the present 2D model, the dissociation energy, $D_0 = W_0(R \to \infty) - E_{00}$, is estimated to be 3,719 and 3,649 cm^{-1} for each ion. These values are overestimated compared with the full-dimensional ones of 2,455 ± 125 and 2,738 ± 206 cm^{-1} reported previously for the same DFT surface using path integral Monte Carlo (PIMC) [10, 11]. Therefore, we should note that for the H_5^+, the levels with $v < 2$ and the E_{20}, E_{21}, E_{30}, and E_{31} are discrete, while for the D_5^+, the levels show larger anharmonicity and more states, with $v = 0$ and 1 as well as the E_{20}, E_{21}, E_{23}, E_{30}, E_{31}, E_{32}, E_{40}, E_{41}, and E_{50} are actually discrete in this case.

The calculated IR continuum spectra, obtained from Eq. 9 using the DFT surface [24], over a large spectral range (up to 6,000 cm^{-1}) and at temperatures of 1 and 300 K (see top and bottom panels, respectively), are shown in Fig. 5 for the H_5^+ (left panels) and D_5^+ (right panels). For comparison reasons, in the same plot we also present the spectra obtained from the Ref. [8] with the analytical CCSD(T) parameterized surface [23], as well as (see inset plots) the experimental IR photodissociation spectra [7]. One can see that for the H_5^+ both calculated spectra at $T = 1$ K show the same spectral bands, with the energy position of the peaks to be blue-shifted by about 150 cm^{-1} for the one computed with the DFT surface. Also, the first small-intensity band, which has been assigned [8] to a (0,1) → (3,1) transition does not appear now in the spectrum obtained with the DFT surface. This is due to its higher dissociation energy, 3,719 cm^{-1}, for the DFT surface compared to the 3,449 cm^{-1} for the analytical TRIM-based one [8]. As we mentioned above, the (3,1) is now a bound level (see Fig. 3), while for the analytical surface this state is lying above the dissociation limit [8]. For the low temperature ($T = 1$ K) spectra of the D_5^+, we observed a similar behavior, with the main bands to be shifted to the higher frequencies, while again, the transitions to the red do not appear, due to the different dissociation energies of the PESs. In the top panels of Fig. 5, we indicate with arrows the energy positions of the present 2D theoretical threshold dissociation, estimates for both H_5^+ and D_5 using the DFT/B3(H) surface, as well as the reported ones for the analytical PES [8].

By comparing now the spectra obtained at $T = 300$ K, one can see that the features from the initial state at $T = 1$ K are conserved; however, new bands appear for higher energies due to the population of higher states according to the Boltzmann distribution. The transition energies, line intensities (see Eq. 7), and predissociation half-widths (see Eq. 8), for the main lines at $T = 300$ K, are listed in Table 1, together with the corresponding

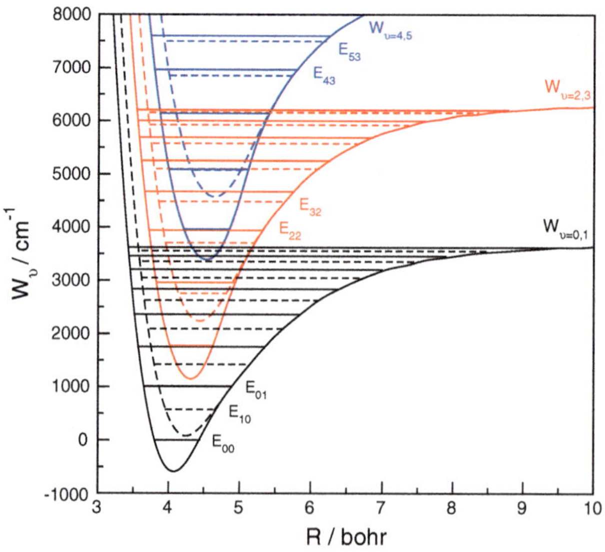

Fig. 3 Adiabatic effective $W_v(R)$ potential curves, and the corresponding E_{vn} energy levels for the H_5^+. *Solid lines* are for even $v = 0$, 2 and 4 values, while *dashed lines* for odd $v = 1$, 3 and 5 ones

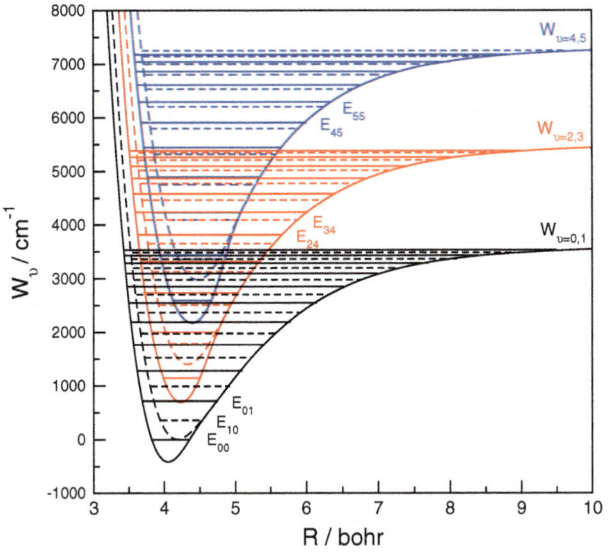

Fig. 4 Same as Fig. 3 for the D_5^+

assignment for each transition, for both H_5^+ and D_5^+ cations. The frequencies of the peaks assigned to the same transition in the spectrum obtained for H_5^+ and D_5^+ with the analytical and DFT surfaces are also shown in the bottom panels of Fig. 5.

In the inset plot of Fig. 5, we present comparison with the experimentally observed spectra. The calculated spectrum for the H_5^+ is in reasonable agreement, regarding the three main bands at 2,603, 3,520, and 4,232 cm^{-1}, assigned recently to excitation modes of the shared-proton mode [7]. The band at 3,904 cm^{-1} is missing from the both calculated spectra. We should note that this peak was previously assigned to the HH stretch [5, 7] and such motions are not taken into account in the present 2D model.

For the D_5^+ case, we can see that the calculated spectra are more congested than the H_5^+ ones; although the spectra from both surfaces present the same features, the one obtained with the DFT PES is shifted to higher frequencies. The comparison with the experiment shows the two main bands shifted to the red and with much lower intensity. Again the assignment to the features around 5,000 cm^{-1} corresponds to progression of the shared-proton mode (see Table 1). We should point out here that the agreement with the experiment for the D_5^+ looks worse than for H_5^+ for both PESs, and this is probably due to the higher anharmonicity of the D_5^+, as we can see in Fig. 4, indicating that the applied 2D model might not be very adequate in this case.

Finally, we should comment that the predissociation half-widths (see Table 1) for both surfaces are almost the same, and we found that are close to the experimental value estimated to 50 cm^{-1} [6]. On the other hand, we should also mention that the present estimates for the half-widths are larger by orders of magnitude than the ones from a previous theoretical calculation [31]. Predissociation half-widths strongly depend of the potential coupling between intra- and intermolecular modes and thus of the PES used. Moreover, in this latter study [31] the employed effective vibrational 3D model Hamiltonian includes only the HH stretching and intermolecular *R* modes, without taking into account the shared-proton mode [31]. Thus, such discrepancies should be attributed to the approaches, both for the potential and for the adiabatic kinetic energy operators,

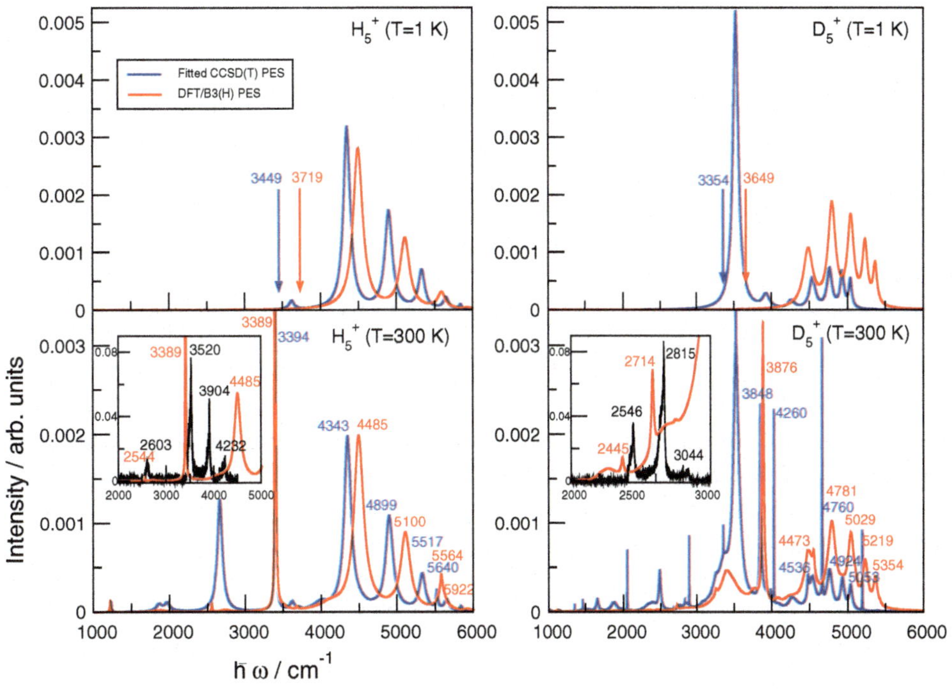

Fig. 5 IR calculated spectra at $T = 1$ K (*top panels*) and 300 K (*bottom panels*) for the H_5^+ (*left panels*) and D_5^+ (*right panels*) using the DFT/B3(H) surface [24] (*red lines*). The comparisons with previous simulations [8] employing the analytical surface [23] (*blue lines*) and with the experimental data [7] are also shown (*black lines* in the *inset plots*). The dissociation energy thresholds for the 2D systems are indicated with the corresponding arrows for each PES

Table 1 Assignments, frequency (in cm^{-1}), intensity (in a.u.), and predissociation half-widths (in cm^{-1}) for the main bands of the calculated IR spectra at $T = 300$ K of the H_5^+ and D_5^+ using the DFT/B3(H) surface

H_5^+				D_5^+			
$(v,n) \rightarrow (v',n')$	$\hbar\omega$	I	$\Gamma_{v'n'}$	$(v,n) \rightarrow (v',n')$	$\hbar\omega$	I	$\Gamma_{v'n'}$
(3,0) → (4,0)	1,209	3.8 (−1)	6	(3,2) → (4,2)	1,103	1.5 (−1)	12
(1,2) → (4,0)	1,870	1.5 (−4)	6	(3,3) → (4,3)	1,246	1.0 (−1)	13
(1,1) → (4,0)	2,544	4.3 (−4)	6	(3,1) → (4,2)	1,722	4.2 (−3)	12
(0,2) → (3,2)	2,727	5.0 (−3)	76	(3,0) → (4,2)	2,445	1.3 (−4)	12
(1,0) → (4,0)	3,389	9.3 (−4)	6	(1,2) → (4,2)	2,714	1.9 (−4)	12
(1,1) → (4,1)	3,673	3.7 (−4)	15	(1,1) → (4,2)	3,246	1.9 (−5)	12
(0,0) → (3,2)	4,485	1.4 (−4)	76	(0,1) → (3,4)	3,382	1.3 (−4)	79
(0,0) → (3,3)	5,100	5.9 (−5)	72	(0,1) → (3,5)	3,755	7.4 (−5)	75
(1,0) → (4,2)	5,564	5.1 (−5)	17	(1,0) → (4,2)	3,876	4.7 (−5)	12
(0,0) → (3,5)	5,922	7.7 (−7)	47	(0,1) → (3,6)	4,063	2.3 (−5)	58
				(1,1) → (4,4)	4,458	2.2 (−5)	13
				(0,0) → (3,5)	4,473	7.7 (−6)	75
				(1,0) → (4,3)	4,541	3.4 (−6)	13
				(0,0) → (3,6)	4,781	1.0 (−5)	58
				(0,0) → (3,7)	5,029	7.8 (−6)	50
				(0,0) → (3,8)	5,219	4.4 (−6)	41
				(0,0) → (3,9)	5,354	2.2 (−6)	31

Numbers in parenthesis are power of 10

adopted in each theoretical treatment. However, similar conclusions on the D_5^+ results have been also reported in this vibrational predissociation dynamics study of the H_5^+, using a completely different adiabatic approximation [31] than the present one.

4 Summary and conclusions

The IR photodissociation spectra of the H_5^+ and D_5^+ clusters are simulated by employing a two-dimensional adiabatic quantum model to describe the internal proton-transfer mode for these systems. Earlier and recent experiments using mass-selected photodissociation spectroscopy have been carried out, and several spectral bands have been observed and associated with excitations of the shared-proton mode. Given the importance of this motion to the assignment of IR photodissociation spectra for both H_5^+ and D_5^+ cations, we investigate here the effect of the underlying PES, employed in the theoretical simulations, on these spectral features. By comparing simulations with two completely different generated potential surfaces, namely TRIM/CCSD(T) fitted PES and DFT/B3(H) one, and analyzing the behavior of the emerged features in the spectra, we could evaluate the quality of the PESs and the influence of the errors in their generation to the dynamics of the systems under study. The analytical surface has been parameterized to high-level CCSD(T) data, and in general, it is expected to be more accurate than a DFT-based one. In one case, the source of errors is associated with the deviations obtained during the fitting procedure and the number and type of configurations of the computed ab initio data, while in the other one it has been shown that a realistic DFT-representation heavily relies on the choice of the functional.

With this in mind, we consider the comparison between the analytical and "on the fly" surfaces. As seen both of them reproduce quite similar theoretical spectra for both H_5^+ and D_5^+ clusters, with the peak positions, intensities and bandwidths of the obtained features being quantitatively comparable. Systematically, the bands calculated with the DFT surface for the H_5^+ appear shifted to higher frequencies by 150 cm^{-1}, and this is mainly attributed to the higher dissociation energy predicted by the DFT surface. We found that both theoretical spectra show a reasonable good agreement with the experimental spectrum of H_5^+ at $T = 300$ K. Both fitted surfaces for D_5^+ predict a more congested spectra than for the H_5^+ cluster. This is consistent with the deeper adiabatic effective, W_v, curves, producing very low intensities for the region between 1,500 and 3,500 cm^{-1} of the experimental bands. This indicates a possible limitation of the 2D adiabatic model for the heavier D_5^+ cluster. Regarding the potential surfaces, we may conclude that both of them could reproduce the main bands of the IR spectrum at energies between 3,000 and 6,000 cm^{-1} and associate them with excitations of the shared-proton mode. The qualitative differences between them are mainly reflected only on the position of the peaks in the H_5^+, while in the spectra of the D_5^+ some variations are also found on the intensities.

Unfortunately, no more experimental results are yet available at the low-energy regime, where we should expect somehow larger quantitative differences between the two surfaces, and thus higher-dimensional models should be developed and employed to count with the highly fluxional nature of these cations. Work in this direction is in progress.

Acknowledgments This paper is dedicated to Prof. M. A. C. Nascimento on the occasion of his 65th birthday, with whom we had the opportunity to have several fruitful discussions. The authors would like to thank Prof. Duncan for providing his experimental results and the Centro de Calculo (IFF), CTI (CSIC), and CESGA for allocation of computer time. We would like also to thank Ricardo Pérez de Tudela for helpful discussions. This work has been supported by the Consolider-Ingenio 2010 Programme CSD2009-00038 (MICINN), the MICINN grant FIS2010-18132, and FIS2011-29596-C02-01, and COST Action CM1002 (CODECS). P. B. acknowledges a postdoctoral fellowship from the Spanish "Fundación Ramón Areces".

References

1. Dawson PH, Tickner AW (1962) J Chem Phys 37:672
2. Beuhler RJ, Ehrenson S, Friedman L (1983) J Chem Phys 79:5982
3. Hiraoka K (1987) J Chem Phys 87:4048
4. Hiraoka K, Mori T (1989) J Chem Phys 91:4821
5. Okumura M, Yeh LI, Lee YT (1998) J Chem Phys 88:79
6. Bae YK (1991) Chem Phys Lett 180:179
7. Cheng TC, Bandyopadyay B, Wang Y, Carter S, Braams BJ, Bowman JM, Duncan MA (2010) J Phys Chem Lett 1:758
8. Sanz-Sanz C, Roncero O, Valdés A, Prosmiti R, Delgado-Barrio G, Villarreal P, Barragán P, Aguado A (2011) Phys Rev A 84:060502-1–4
9. Acioli PH, Xie Z, Braams BJ, Bowman JM (2008) J Chem Phys 128:104318
10. Pérezde Tudela R, Barragán P, Prosmiti R, Villarreal P, Delgado-Barrio G (2011) J Phys Chem A 115:2483
11. Barragán P, Pérezde Tudela R, Prosmiti R, Villarreal P, Delgado-Barrio G (2011) Phys Scr 84:028109
12. Valdés A, Prosmiti R, Delgado-Barrio G (2012) J Chem Phys 136:104302
13. Yamaguchi Y, Gaw JF, Remington RB, Shaefer HF III (1987) J Chem Phys 86:5072
14. Farison M, Chermette H, Farizon-Mazuy B (1992) J Chem Phys 96:1325
15. Ignacio EW, Yamabe S (1998) Chem Phys Lett 287:563
16. Barbatti M, Jalbert G, Nascimento MAC (2000) J Chem Phys 113:4230
17. Müller H, Kutzelnigg W (2000) Phys Chem Chem Phys 2:2061
18. Prosmiti R, Buchachenko AA, Villarreal P, Delgado-Barrio G (2001) Theor Chem Acc 106:426
19. Moyano GE, Collins MA (2003) J Chem Phys 119:5510
20. Barbatti M, Nascimento MAC (2003) J Chem Phys 119:5444
21. Prosmiti R, Villarreal P, Delgado-Barrio G (2003) J Phys Chem A 107:4768
22. Xie Z, Braams BJ, Bowman JM (2005) J Chem Phys 122:224307
23. Aguado A, Barragán P, Prosmiti R, Delgado-Barrio G, Villarreal P, Roncero O (2010) J Chem Phys 133:024306
24. Barragán P, Prosmiti R, Roncero O, Aguado A, Villarreal P, Delgado-Barrio G (2010) J Chem Phys 133:054303
25. Roncero O, Beswick JA, Halberstadt N, Villarreal P, Delgado-Barrio G (1990) J Chem Phys 92:3348
26. López-Durán D, de Lara-Castells MP, Delgado-Barrio G, Villarreal P, di Paola C, Gianturco FA, Jellinek J (2004) Phys Rev Lett 93:053401
27. Aguado A, Sanz-Sanz C, Villarreal P, Roncero O (2012) Phys Rev A 85. Art ID 032514. doi:10.1103/PhysRevA.00.002500
28. Chermette H, Razafinjanahary H, Carrion L (1997) J Chem Phys 107:10643
29. Barragán P, Prosmiti R, Villarreal P, Delgado-Barrio G (2011) Int J Quantum Chem 111:368
30. Barragán P, Prosmiti R (2012) Int J Quantum Chem. doi: 10.1002/qua.24026
31. Spirko V, Amano T, Kraemer WP (2006) J Chem Phys 124:244303

Printed by Printforce, the Netherlands